IET HISTORY OF TECHNOLOGY SERIES 6

Series Editor: Dr B. Bowers
Dr C. Hempstead

Technical History of the Beginnings of Radar

Other volumes in this series:

Volume 4   **The history of electric wires and cables** R.M. Black
Volume 6   **Technical history of the beginnings of radar** S.S. Swords
Volume 7   **British television: the formative years** R.W. Burns
Volume 9   **Vintage telephones of the world** P.J. Povey and R. Earl
Volume 10   **The GEC research laboratories 1919–1984** R.J. Clayton and J. Algar
Volume 11   **Metres to microwaves** E.B. Callick
Volume 12   **A history of the world semiconductor industry** P.R. Morris
Volume 13   **Wireless: the crucial decade 1924–34** G. Bussey
Volume 14   **A scientists war – the diary of Sir Clifford Paterson 1939–45** R.J. Clayton
            and J. Algar (Editors)
Volume 15   **Electrical technology in mining: the dawn of a new age** A.V. Jones and
            R.P. Tarkenter
Volume 16   **Curiosity perfectly satisfied: Faraday's travels in Europe 1813-1815**
            B. Bowers and L. Symonds (Editors)
Volume 17   **Michael Faraday's 'Chemical Notes, Hints, Suggestions and Objects of
            Pursuit' of 1822** R.D. Tweney and D. Gooding (Editors)
Volume 18   **Lord Kelvin: his influence on electrical measurements and units**
            P. Tunbridge
Volume 19   **History of international broadcasting, volume 1** J. Wood
Volume 20   **The early history of radio: from Faraday to Marconi** G.R.M. Garratt
Volume 21   **Exhibiting electricity** K.G. Beauchamp
Volume 22   **Television: an international history of the formative years** R.W. Burns
Volume 23   **History of international broadcasting, volume 2** J. Wood
Volume 24   **Life and times of Alan Dower Blumlein** R.W. Burns
Volume 26   **A history of telegraphy: its technology and application** K.G. Beauchamp
Volume 27   **Restoring Baird's image** D.F. McLean
Volume 28   **John Logie Baird: television pioneer** R.W. Burns
Volume 29   **Sir Charles Wheatstone, 2nd edition** B. Bowers
Volume 30   **Radio man: the remarkable rise and fall of C.O. Stanley** M. Frankland
Volume 31   **Electric railways, 1880–1990** M.C. Duffy
Volume 32   **Communications: an international history of the formative years** R. Burns
Volume 33   **Spacecraft technology: the early years** M. Williamson

Technical History
of the Beginnings
# of Radar

S.S. Swords

The Institution of Engineering and Technology

Published by The Institution of Engineering and Technology, London, United Kingdom

First edition © 1986 Peter Peregrinus Ltd.
Reprint with new cover © 2008 The Institution of Engineering and Technology

First published 1986
Reprinted with new cover 2008

The Institution of Engineering and Technology
Michael Faraday House
Six Hills Way, Stevenage
Herts, SG1 2AY, United Kingdom

www.theiet.org

British Library Cataloguing in Publication Data
Swords, Sean S.
  Technical history of the beginnings of radar
  (History of technology series no. 6)
  1. Radar – History
  I. Title II. Series
  621.3848'09'04   TK6574.2

ISBN (10 digit) 0 86341 043 X
ISBN (13 digit) 978-0-86341-043-7

Printed in the UK by Short Run Press Ltd, Exeter
Reprinted in the UK by Lightning Source UK Ltd, Milton Keynes

To the Memory of Commandant Matt Murphy and Captain Phil Harrington

And moving thro' a mirror clear
That hangs before her all the year,
Shadows of the world appear.
There she sees the highway near,
    Winding down to Camelot:
There the river eddy whirls,
And there the surly village-churls,
And the red cloaks of market girls,
    Pass onward from Shalott.

Alfred Lord Tennyson

Contents

Preface   xi

Acknowledgments   xiii

1   **Radar etymology**   1
     References   3

2   **Radar fundamentals**   5
     2.1   Introduction   5
     2.2   Scattering and scattering cross-section   6
     2.3   Radar range equation   9
         2.3.1   Pulse radar   10
         2.3.2   Bistatic radar range equation   11
     2.4   Multipath propagation   12
         2.4.1   Electromagnetic reflection coefficient   15
         2.4.2   Horizontal polarisation   16
         2.4.3   Vertical polarisation   17
     2.5   Multipath effects and radar coverage   17
         2.5.1   Gap filling   17
     2.6   Choice of polarisation   18
     2.7   Fundamentals of a pulse radar system   19
         2.7.1   Range and azimuth resolution   22
         2.7.2   Pulse radar signals   24
     2.8   Antenna duplexing – TR/ATR switches   26
     2.9   Nomenclature of displays   28
     2.10   Direction finding: split-beam method   30
     2.11   Secondary radar   33
     2.12   Over-the-horizon radar   36
     2.13   Frequency bands   38
     References   39

3   Precursors of radar   42
  3.1   Introduction   42
  3.2   Christian Hülsmeyer   43
  3.3   Hugo Gernsback   45
  3.4   Hans Dominik   46
  3.5   Nikola Tesla   47
  3.6   Detection of aircraft   48
    3.6.1   Ignition detection   48
    3.6.2   Sound locators   48
    3.6.3   Passive and active infra-red detection   49
  3.7   Marconi's speech to the American Institute of Electrical
    Engineers   51
  3.8   Albert Hoyt Taylor and Leo Clifford Young   52
  3.9   Measurement of distance   52
    3.9.1   Ionospheric sounding   53
    3.9.2   Geodetic surveying   59
    3.9.3   Aircraft altimeters   60
    3.9.3.1   Alexanderson altimeter   62
    3.9.3.2   Western Electric radio altimeter   63
    3.9.3.3   Altimeter of Sadahiro Matsuo of Tohoku Imperial
    University, Japan   67
  3.10   Three British events   68
    3.10.1   O. F. Brown's memorandum   69
    3.10.2   L. S. B. Alder   69
    3.10.3   W. A. S. Butement and P. E. Pollard   70
  3.11   Patents   74
  3.12   Mystery rays   75
  References   79

4   The début of radar   82
  4.1   Beginning of radar in Great Britain   82
  4.2   Beginning of radar in Germany   91
  4.3   Beginning of radar in the United States   101
    4.3.1   United States Army Signal Corps   112
    4.3.2   Aircraft Warning Service   118
    4.3.3   Establishment of the NDRC   118
  4.4   Beginning of radar in France   120
  4.5   Beginning of radar in Italy   126
  4.6   Beginning of radar in Japan   130
  4.7   Beginning of radar in Russia   135
  4.8   Beginning of radar in Holland   142
  4.9   Beginning of radar in Hungary   144
  References and Bibliographies   147

5	**The British story**		174
	5.1	Introduction	174
	5.2	Other inventors of radar	176
	5.3	Air defence before the Second World War	176
	5.4	The Daventry experiment	180
	5.5	Development of the first radar system	186
		5.5.1 Navy radar	187
		5.5.2 Army radar	188
	5.6	CH (Chain Home) system	188
		5.6.1 Frequencies	189
		5.6.2 Polarisation	197
		5.6.3 Radiogoniometer	198
		5.6.4 Direction finding	200
		5.6.5 Height finding	201
		5.6.5.1 Measurement of elevation angle	202
	5.7	Transmitters	207
		5.7.1 TF3 transmitter	219
		5.7.2 Chain Home transmitter T3026	221
		5.7.3 Transmitter circuits	222
		5.7.4 Transmitter antenna arrays	225
		5.7.5 Performance diagrams	229
	5.8	Receivers	230
	5.9	RDF beam technique – CHL and GCI	236
		5.9.1 IFF and GCI	242
	5.10	Airborne radar – RDF2	243
	5.11	Organisation	253
	5.12	Radar countermeasures	254
	References		254
6	**Significance of the magnetron**		258
	References		267
7	**Conclusions**		270
	References		272
Appendixes			
Appendix A	**Reciprocity principle**		273
	References		274
Appendix B	**Retarding field generators**		274
	References		276
Appendix C	**Super-regenerative receivers**		276
	References		278

Appendix D   Watson-Watt's two memoranda   278
Appendix E   Watson-Watt's memorandum to the CSSAD on the state of RDF
             research, 9th September 1935   285
Appendix F   Method of deriving height curves for a Chain Home station 292
Appendix G   Two Maps   295
Appendix H   Report issued by Telecommunications Research Establishment,
             1941   296
Appendix I   Table of Japanese Naval Radars   299
Appendix J   Military characteristics – Detector for use against aircraft (heat
             or radio) (United States Army)   305
Appendix K   Mathematical analysis   306
             Scattering and the scattering matrix
             Radar range equation
             Electromagnetic reflection coefficient of ground
             Multipath effects
             Pulses and their spectra
             References
Name Index   319
Subject Index   323

Preface

This work is an analytical record of the emergence of radar in those countries where independent development occurred. It purposely confines itself to before the cavity magnetron era of radar.

After a short discussion on what radar is, the physical principles underlying pulse radar systems are discussed in some detail. The principal forerunners of radar, from Tesla's concept in 1900 of a method for detecting the presence and movement of distant objects, to Chester Rice's experiments with microwaves in the 1930s, are treated in chronological order. The emergence of radar in the United Kingdom, Germany, the United States, France, Italy, Japan, Russia, Holland and Hungary is then examined, and, with the exception of the latter two countries, an annotated bibliography is provided in each case.

The work in the United Kingdom is then chosen for more detailed attention. The political and technical backgrounds to the development of the Chain Home and related systems are treated in detail, as is the emergence of the early metric airborne radar sets. This comprehensive account of British radar permits a certain amount of attention to be given to a description of the functioning of key circuits and of the techniques used; thereby it is hoped to convey some perception of the radio-electronic technology of the time.

The emergence of the resonant-cavity magnetron was a turning point in radar history. The story of its origin, prefaced by an account of earlier types of magnetron, is told.

Some overall conclusions are drawn and a number of appendices of historical and technical interest are provided.

Acknowledgments

This history of radar could not have been undertaken without the help and assistance offered by so many people.

Firstly, I am indebted to Professor B. K. P. Scaife, who encouraged me to undertake the task and who then made it so relatively easy by searching out and procuring the most covert of the source material; his help, guidance and support since the work was begun are greatly appreciated.

I am also deeply grateful to the following who contributed in so many ways to its realisation: Dr Vladimir Adamec; B. M. Adkins; Dr D. K. Allison; Nino Arena; the Austrian Embassy, London; Professor N. Balazs; the late Major R. G. Bartelot; Colonel Martin Bates; Dr Zoltán Bay; L. H. Bedford, C.B.E.; Captain Beeching, RN; Cajus Bekker (Hans-Dieter Berenbrok); Colonel J. P. Berres; Dr H. A. H. Boot; Dr E. G. Bowen, C.B.E., F.R.S.; Dr B. Bowers; Dr G. H. Brown; Lieutenant-Commander D. N. Brunicardi; Rear-Admiral Sir Lindsay Bryson; H. Burkle; Professor N. Carrara; L. C. Castioni; G. C. Chalou; M. le Chef du Service Historique de la Marine, Vincennes; the late Dr R. C. Chirnside; R. W. Clarke; Professor J. F. Coales, C.B.E., F.R.S.; Dr R. Coelho; Dr W. T. Coffey; Cultural Office, Dublin; Embassy of the USSR; Mrs Margrit Cronin; Dr Pierre David; Squadron Leader Davis; Defence Research Information Centre, St. Mary Cray; Colonel H. de Buttet; T. Delimata; M. le Directeur de l'Establissement Technique Central de l'Armement, Paris; C. I. Dolan; Air Marshall Sir Herbert Durkin; the late Sir Eric Eastwood, F.R.S.; Dr Tokuo Ebina; Embassy of Switzerland, Dublin; Lieutenant-Commander F. Eustace, RN; Dr Carolina Evangelista; Major Douglas T. Fairfull, USAF; Professor M. Federici; D. G. Fink; Dr F. Fox; A. J. Francis; Dr Kurt Franz; French Embassy, London; Professor D. G. Frood; K. Geddes; Dr Gießler; Colonel M. G. Goodeve-Ballard; Professor G. Goubau; J. F. Gough; Professor H. E. Guerlac; Professor R. Hanbury-Brown; H. Hart; G. E. Hasselwander, Min. Rat. a.D. Dr Ing. F. Hentschel; Professor K. Higasi; Professor T. J. Higgins; Major-General B. P. Hughes, C. B., C.B.E.; Fleet-Chief Hughes, RN; Judy Inglis; the late Professor L. Janossy; Professor R. V. Jones, C.B., C.B.E., F.R.S.; Dr H. R. L. Lamont; Professor G. Latmiral;

Dr W. B. Lewis; Mrs C. S. Lonergan; Professor Sir Bernard Lovell, O.B.E., F.R.S.; Dr A. C. Lynch; Dr J. L. Maksiejewski; Professor K. Matsumaru; C. A. May; Dr H. H. Meinke; Miss G. Milch; A. M. J. Mitchell; Dr G. Molinari; Professor Akio Morita; Dr S. Nakajima; N. Nerecher; E. M. NicPhaidin; Dr A. Norberg; Dr Minoru Okada; Dr Bolesłow Orlowski; P. O'Rourke; Antonella O'Shea; Dr R. M. Page; Dr H. C. Pauli; Professor A. H. Piekara; Frau Plendl; Alfred Price; Air Commodore Probert; Dr E. H. Putley; Sir John Randall, F.R.S.; Colonel Rinaldo Rinaldi; Brigadier Robertson; Professor C. Rosatelli; Professor E. Roubine; Professor Dr W. T. Runge; Captain Philip Russell; Russian Department, Trinity College Dublin; Professor W. G. S. Scaife; G. E. Schindler; Professor Ch. Schmelzer; H. Sekino; E. C. Slow, O.B.E.; Professor C. P. Smyth; Dr Ir. M. Staal; Colonel W. F. Strobridge; Professor C. Süsskind; T. J. Swords; Professor U. Tiberio; Fritz Trenkle; R. Van der Hulst; A. Verbraech; V. A. Volkov; Lieutenant-Colonel J. M. Walsh; Professor M. V. Whelan; Miss White; the late A. F. Wilkins, O.B.E.; the late Sir Fredrick Williams, F.R.S.; H. L. Wilson; H. S. Young, C.B.E.; Captain H. W. Young, RN.

My thanks to C. G. R. Lyons for producing the line drawings and to Irene O'Neill for her patience in typing the manuscript.

Transcripts of Crown copyright records in the Public Record Office appear by permission of the Controller of Her Majesty's Stationery Office.

The courtesy of the staff of the Public Record Office, Kew, London, was appreciated, as was the assistance of the staff of the Archives, Churchill College, Cambridge, and, in particular, the help of Miss Marian Stewart.

Photographs are reproduced by permission of the following: AEG Aktiengesellschaft (No. 3); Deutsches Museum München (No. 1); Hollandse Signaalapparaten B.V. (No. 12); Imperial War Museum (Nos. 18, 19); Mucchi Editore s.p.a. (Collezione E. Andò) (No. 11); Dr. S. Nakajima (Nos. 13, 14); The Science Museum (Nos. 16, 17, 20, 21); Herr Fritz Trenkle (Nos. 4, 5); United States Naval Research Laboratory (Nos. 2, 6, 7, 8, 9, 10); VAAP – Copyright Agency of the USSR (No. 15).

Finally, I wish to thank my wife, Dympna, and all the family for their forbearance and encouragement; especially Jacinta who helped to compile the index.

Chapter 1

Radar etymology

The term 'Radar', an acronym for radio detection and ranging, was devised by the United States Navy in November 1940 [1]. This code name was suggested by both Lieutenant-Commander S. M. Tucker and Lieutenant-Commander F. R. Furth who took steps to have it put into effect. A letter dated 19th November, 1940, and signed by Admiral H. R. Stark, Chief of Naval Operations, made the word official. It came quickly into use and later, in 1943, was by agreement officially adopted by the Allied Powers. Since the end of the Second World War, it has received general international acceptance.

An earlier terminology in the United States was 'radio echo equipment'. In Britain it was given the code name RDF (which most easily translates to radio direction finding and was then a term coined deliberately to mislead) and for a while, particularly in Army circles, it was also known as 'Cuckoo'.

The German term, now Funkmessgeraet, was originally Dete-Geraet. The French termed it DEM (détection électromagnétique) and the Italians RDT (radio detector telemetro). The Japanese, however, did not adopt one all-embracing name but rather distinguished between detectors and locators.

Generally, no one has difficulty in deciding what constitutes a radar system, new or old. The use of electromagnetic waves from the radio-frequency part of the spectrum in conjunction with their reflection from objects or from targets of interest, coupled with the constant speed of propagation of the radiation, is the essence of all radar systems. The use of timed pulses, a common location and common antenna for transmitter and receiver, narrow sweeping beams of radiation and cathode tube displays constitutes the most common type of radar system, but is nevertheless but one form of a central concept. No one has any difficulty in deciding that a long range surveillance equipment, a police traffic monitor, a missile tracking system, and so forth, are all examples of radar. Where there may be difficulty is in trying now to adjudicate on the significance, relevance, merits or precedence of the early work on radio-wave detection of objects.

Although the concept of detecting objects using electromagnetic waves is probably inherent in the very understanding of their propagation characteristics, and thus might be traced to Hertz, the first formal mention of the idea appears to have been made by Nikola Tesla. (It is difficult to visualise precisely what Tesla had in mind. Detection of objects was to be accomplished by means of 'terrestrial stationary waves', which in turn were part of his 'world system' of wireless transmission [2, 3] which had been conceived from experiments he carried out in Colorado in 1899.) In an article entitled 'The problem of increasing human energy' published in the *Century Illustrated Monthly Magazine*, June 1900, he wrote [2]:

> When we raise the voice and hear an echo in reply, we know that the sound of the voice must have reached a distant wall, or boundary, and must have been reflected from the same. Exactly as the sound, so an electrical wave is reflected, and the same evidence which is afforded by an echo is afforded by an electrical phenomenon known as a 'stationary wave' – that is, a wave with fixed nodal and ventral regions. . . Stationary waves in the earth will mean something more than mere telegraphy without wires to any distance. They will enable us to attain many important specific results, impossible otherwise. For instance, by their use we may produce at will, from a sending station, an electrical effect in any particular region of the globe; we may determine the relative position or course of a moving object, such as a vessel at sea, the distance travelled by same, or its speed. . . .

Tesla's concept [2] 'in any particular region of the globe . . . (to) determine the relative position or course of a moving object . . . the distance travelled by same or its speed' was, like his concept of a world-wide radio system, a comprehensive one, an ambitious one, and, even today, realisable only in part.

Actual radars did not, as will be seen, directly emerge from visionary writing, after-dinner speeches or clever patents, but from people who, with a faith akin to that of Marconi's in developing workable radio, went ahead and discovered experimentally that aircraft and ships had significant scattering cross-sections.

Under the pressures of the Second World War, all manner of radar systems emerged. The diversification was principally in function and frequency.

With the object of distinguishing friend from foe there emerged also secondary radar, where a combined receiver and transmitter unit, called a transponder, was installed in one's own aircraft. When interrogated by a signal from a ground radar, the transponder transmitted a reply back to the ground station. This resulted in an enhanced and possibly an encoded signal being received by the ground set. Further derivatives of radar occurred when two or more well spaced ground stations co-operated with an aircraft or ship to create radio-navigational position lines. These lines of position on the earth's surface may be straight lines, great circle paths or hyperbolae, depending on the principles underlying the particular system being used. A navigator could fly or 'home' along a particular position line, or, by observing at any given time that he was on the intersection of two or more position lines, he could determine his position. By the end of the war there were in existence Gee, Loran, Oboe, POPI, Decca and Consol (or Sonne), to mention the more common aids [4]. There also emerged Doppler radar, a self-contained aircraft aid not dependent on ground stations, which computed the true velocity and track of an aircraft over ground. Indeed there is possibly no radio-navigational aid in existence which had not at least been conceived before the end of the Second World War.

Pulse circuits and pulse techniques predominated, but the common ground in the various radio detection and radio-navigational systems was the use of accurate and precise time measuring circuits coupled to the constant speed of propagation of radio waves.

Another offshoot from radar which emerged immediately after the Second World War were the twin sciences of radar astronomy and radio astronomy. If, for the sake of a balanced overview of the history of radar technology to date, one wished to gauge what developments have taken place since 1945, one would look primarily at improvements in crossfield and linear-beam valves, the increased understanding and implementation of signal-processing techniques, the developments in phased-array antennae and possibly the emergence of over-the-horizon high-frequency radar systems.

Louis N. Ridenour, writing his preface in 1946 to 'Radar system engineering' [5], the first volume of the Radiation Laboratory Series, commented: 'Radar is a very simple subject, and no special mathematical, physical or engineering background is needed to read and understand this book'. Those who helped service the radar equipments of the war years might, in a qualified sense, agree with this, but their counterparts today could in no way accept his statement that radar is a very simple subject. Nevertheless, in a field where reliability and conservatism are key philosophies, in civil air traffic control, the expert of 35 years ago could still feel at home examining a modern airways surveillance radar; and this may be read as a tribute to the tremendous developments during the few years from 1939 to 1945.

Although television had been invented before the war and its circuits had much in common with those of radar, still the emergence of radar can be seen as a distinct jump in the progress of both radio science and electronics. Radar added a completely new and fresh dimension to electronic engineering.

References

1   GEBHARD, Louis A: 'Evolution of naval radio-electronics and contributions of the naval research laboratory', NRL Report 8300, p. 170, Naval Research Laboratory, Washington, D.C.
2   PAPOVIĆ, V., HORVAT, R., and NIKOLIĆ, N: 'Nikola Tesla – Lectures, Patents, Articles' (Nikola Tesla Museum, Beograd, 1956, pp. A109–A152. Reprint available from MRG, Archer's Court, Stonestile Lane, Hastings, England)

3 Reference 2, pp. A153–A161 (reprint of 'The transmission of electric energy without wires', *Electl. Wld.*, March 5, 1904)
4 For an understanding of the role of radio in navigation since the Second World War, the following four texts can be recommended:
(*a*) SMITH, R. A.: 'Radio aids to navigation' (University Press, Cambridge, 1947)
(*b*) SANDRETTO, Peter C: 'Electronic avigation engineering' (International Telephone and Telegraph Corporation, New York, 1958)
(*c*) Report of Electronic Subdivision Advisory Group on Air Navigation, Report Identification Symbol TSELC-SP2, Air Materiel Command Engineering Division, Wright Field, Dayton, Ohio, 1946
(*d*) KAYTON, M., and FRIED, W. R.: 'Avionics navigation systems' (John Wiley, New York, 1969)
5 RIDENOUR, Louis N: 'Radar system engineering' (Radiation Laboratory Series Vol. 1), (McGraw-Hill, New York, 1947)

Chapter 2

Radar fundamentals

2.1 Introduction

The diversity of radar systems in use today, viewed either from the point of operational function or from the point of equipment design, makes it unrealistic even to attempt an overview of modern radar methodology. If one wishes to explore the breadth of modern radar technology, a study of a text such as Skolnik's 'Radar handbook' [1] is helpful.

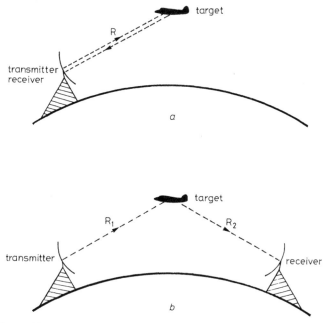

Fig. 2.1 *Ground-based systems*
a Monostatic radar
b Bistatic radar

By far the most common type of radar, however, is a monostatic pulse radar and this is also the type which predominated in the pre-war and war years wherever radar was being developed. The range equation for such a radar will be introduced and used as a vehicle for underlining the fundamental principles and parameters of the system.

The detection of aircraft by ground-based systems constitutes the first use of radar and is probably the easiest situation to visualise. Fig. 2.1*a* depicts a monostatic system and Fig. 2.1*b* a bistatic one. A monostatic radar is one where the energy from the target is received at the transmitting source with generally a common antenna used for transmitting and receiving. A bistatic radar employs separated sites for transmitting and receiving with a separation between the sites usually of the order of the target ranges being dealt with; a transmitting and receiving antenna separated within the one site for purposes of isolation would not constitute a bistatic system. Bistatic, and, indeed, multistatic systems are employable in diverse roles such as missile tracking, planetary surface exploration or the location of natural atmospheric radio-wave sources [1].

2.2 Scattering and scattering cross-section

The scattering cross-section σ, also called the radar cross-section, of a target describes and quantifies the function of a target in back-scattering radio-frequency energy to the radio site. One can define the cross-section σ by the equation

$$\sigma = 4\pi \frac{\text{power directed towards receiver/unit solid angle}}{\text{power per unit area incident at the target}}$$

The hypothetical mechanism is that a plane wave is incident on the target, which captures power from an area σ of the incident field and then re-radiates it isotropically.

The cross-section is measured in units of area and may be formally defined as

$$\sigma = \lim_{R\to\infty} 4\pi R^2 \frac{S_s}{S_i}$$

$$\sigma = \lim_{R\to\infty} 4\pi R^2 \left(\frac{E_s}{E_i}\right)^2$$

(2.1)

where S_s is the scattered power flux density at a distance R from the target, S_i is the power flux density in an incident plane wave at the target, and E_s, E_i are the magnitudes of the corresponding electric field intensities.

The limit as $R\to\infty$ is taken to ensure that a true far-field assessment at a

range R from the target is made where E_s varies directly as $1/R$. This consideration will cancel or counterbalance the R^2 in the formula. As was mentioned above, inherent in the definition of cross-section is the assumption that the target reradiates isotropically. Furthermore, the key to the determination of σ is the evaluation of E_s.

'Scattering' and 'diffraction' refer to the same physical process. When an object scatters an electromagnetic wave the scattered field is defined as the difference between the total field in the presence of the object and the field that would exist if the object were absent, but with all other conditions unchanged. The diffracted field is the total field in the presence of the scattering body.

In physical optics the quantitative study of the scattering of electromagnetic waves preceded the advent of radar by over half a century. Starting with Tyndall's experimental study of the scattering of light from aerosols (1869) one finds the scattering of light waves from various types of particle was studied by many workers but was treated principally by Rayleigh (1871, 1881, 1910, 1914, 1918), Mie (1906), and Debye (1915) [2, 3]. One way in which the concepts of scattering derived in optics are of relevance in the radar case is in the terminology which may be applied to the dimensions of a target relative to wavelength. When the dimensions of the scattering object are much less than a wavelength, one refers to *Rayleigh* scattering. As the target size increases to become comparable with a wavelength (say from 0·1λ to 1λ), one speaks of a *Mie* or *resonance region* of scattering. When the target dimensions are large compared to a wavelength, cross-sections may be computable by the methods of geometrical optics and one may refer to an *optical region* for the scattering behaviour of the target. Fig. 2.2 [4] depicts the radar cross-section of a

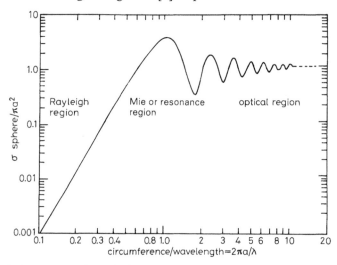

Fig. 2.2   *Radar cross-section of sphere*
a = Radius λ = wavelength

spherical target as a function of its circumference measured in wavelengths [$2\pi a/\lambda$], a being the radius of the sphere (a sphere is, of course, a very idealised target to select). The magnitude of the cross-section varies from virtually nothing (in the Rayleigh region where $\sigma \propto (1/\lambda^4)$) to approximately πa^2 in the Mie region where the cross-section, σ, is oscillating. The exact calculation of the scattering cross-section requires the solution of Maxwell's equations subject to appropriate boundary conditions on the surface of the scatterer. Analytical and experimental methods of determining radar cross-sections are adequately dealt with in the literature [5–7].

The scattering cross-section depends on target size and shape, target material, the aspect from which it is viewed or illuminated, the radar frequency and the polarisations of the transmitting and receiving antennae. In practice, matters may not be so complex, given the use of a single frequency, one antenna of known polarisation and a target with one or two common aspects, such as an aircraft flying broadside or head-on to a ground radar. The cross-sections of some targets whose dimensions are large relative to wavelength are represented by quite simple expressions. For instance, a sphere of radius a viewed from any aspect will have $\sigma = \pi a^2$; a circular cylinder of length L and radius a irradiated at an angle θ to the broadside will have (see Fig. 2.3)

$$\sigma = \frac{a\lambda}{2\pi} \frac{\cos\theta \, \sin^2 (k L \sin\theta)}{\sin^2\theta} \tag{2.2}$$

where $k = 2\pi/\lambda$. As an indication of the order of values found in practice, a large passenger aircraft could present a cross-section of 300 m^2 when broadside on to a radar ($\lambda = 10$ cm).

A rigorous evaluation of cross-section would require a knowledge of the scattering properties of a target for a set of linearly polarised waves incident on the target not only for every aspect of the target, but for incidence at every direction with respect to a set of fixed axes. A brief outline of the procedure adopted is given in Appendix K. In many cases, however, for symmetrical targets and for smooth targets for instance, the task may be greatly simplified.

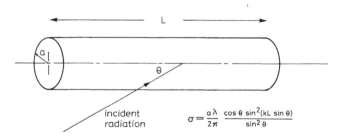

Fig. 2.3   *Radar cross-section of circular cylinder when dimensions are large relative to wavelength*

A good insight into the method of radar scattering research is provided, together with useful historical references, in the August 1965 issue of the *Proceedings of The Institute of Electrical and Electronics Engineers* [8].

2.3 Radar range equation

The basic range equation for a monostatic pulse radar appears in virtually all introductory treatments of radar. It appears in several different but equivalent forms, and, at best, will give quite an accurate prediction of the maximum range to be expected from a system. Range prediction is not an exact exercise. Certain indeterminacies can be present in some of the parameters employed, but, nevertheless, the range equation will at least provide guidelines for outline designs and for making working comparisons between various systems.

The paper of Omberg and Norton [9] on 'The Maximum range of a radar set' is generally considered to be the first comprehensive treatment of the factors determining the maximum range of a pulse radar equipment. It was published originally in February 1943 as a United States Army Signal Corps report, and, when declassified, was presented at a Washington Section of the Institute of Radio Engineers meeting in 1946 and then published with some additions in 1947. Quite a comprehensive literature now exists on the range equation [10]. As might be expected, a statistical aspect must be brought into the determination of maximum range or, rather, into the factors which enable a confirmation of the presence of a target. This arises both because of the molecular processes which generate system noise and because of the inherent indeterminacy in the whole detection mechanism which generally includes an operator watching a cathode-ray tube display.

Ugo Tiberio of Italy was perhaps the first radio scientist to study experimentally and theoretically the reradiation properties of targets [11]. In 1933 Guglielmo Marconi, while testing a microwave installation for the Vatican between the Vatican City and Castel Gandolfo, observed a beat effect from nearby moving objects. Conscious of the importance of the phenomenon, he gave a demonstration to some senior military staff, among whom was General Luigi Sacco. General Sacco requested Professor Ugo Tiberio to make a report on 'The equation of radiotelemetria' [11]. The study was completed in 1935 and resulted in the paper 'The measurement of distance by means of ultra-short waves (wireless range-finding)' [12]. The paper mentions the usefulness of reflection and diffraction phenomena and discusses a frequency-modulated continuous wave system for distance measurement. It also includes calculations of the intensities of echoes from a half-wavelength dipole, and from flat rectangular targets when the latter are placed both at right angles to, and inclined to, the direction of propagation. In each case the resultant field strength was shown to be of the general form:

$$F = \gamma \frac{E}{d^2}$$

where the coefficient γ has the dimensions of length and depends on the nature of the target. F is the electric field strength at the receiver and E is the cymomotive* force at the tramsmitter measured in volts. Undoubtedly the basic range equation was known to radio scientists working on radar in the various countries before the war. One form of the equation, for instance, was used by Butement and his team at Bawdsey Research Station in 1938 in designing a coastal defence (CD) radar. The effects of reflection by the sea were included in their formula [13].

If one postulates a common transmitting and receiving antenna of gain G, and denotes the transmitted signal power and the received signal power at the antenna terminals by P_t and P_r, respectively, then for a target of cross-section σ at a ranged R using a wavelength λ, one can show that (see Appendix K)

$$R = \left(\frac{P_t G^2 \sigma \lambda^2}{(4\pi)^3 P_r} \right)^{1/4} \tag{2.3}$$

Likewise it can be shown that R_{max}, the maximum range obtainable, is given by the expression

$$R_{max} = \left(\frac{P_t G^2 \sigma \lambda^2}{(4\pi)^3 (S/N)_{min} k T_s B_n} \right)^{1/4} \tag{2.4}$$

where $(S/N)_{min}$ is the signal-to-noise power ratio at the antenna terminals corresponding to a minimum detectable signal, k is the Boltzmann's constant $(1\cdot38 \times 10^{-23} \text{ J/K})$, T_s is the system noise temperature of the receiver, and B_n is the receiver noise bandwidth.

2.3.1 Pulse radar

The transmitted signal from a pulse radar can, in its simplest form, be visualised in the time domain as a series of RF pulses, each of duration τ and each containing the transmitted energy in sinusoidal form. The frequency spectrum of such a series of pulses is dealt with in the following. It is seen to be, in practice, a continuous spectrum with most of the energy centred in a bandwidth of the order of $1/\tau$. If the receiver bandwidth is reduced appreciably below this value, both the signal power and the noise power will be reduced, whereas if the bandwidth is much greater than this, the noise power will increase more rapidly than the signal. Thus one can speak of an optimum value of bandwidth, $B_{n(opt)}$.

$$B_{n(opt)} = \frac{\alpha}{\tau}, \quad \text{where } \alpha \sim 1$$

* Cymomotive force is a measure of the radiating ability of an antenna. It is the product of field strength and distance and so is independent of distance from the antenna. See SACCO, L. and TIBERIO U.: 'Sul modo di esporre e di impiegara i dati di irradiazione e propagazione', *Alta Frequenza*. 1935, **14**, pp. 668–687.

(This bandwidth-pulse-duration reciprocal relationship is dealt with extensively in the literature, but in the context of the early radars the paper of Haeff [14] and the experimental results due to Lawson [15] of the M.I.T. Radiation Laboratory are informative.)

Norton and Omberg [9] used a term V which they called the pulse visibility factor and defined it as

$$V = \frac{E_{min}}{F k T} \tag{2.5}$$

where E_{min} is the minimum available pulse energy from antenna, matched to the receiver, required to make the pulse just visible in the presence of noise, T is the ambient temperature of receiver and F is the noise factor of receiver.

The visibility factor was expressed as $V = V_0 C_b$ where V_0 was the lowest possible value of V corresponding to an optimum value of noise bandwidth; $C_b \geqslant 1$. As is shown in Appendix K, the visibility factor may be incorporated into the radar range equation.

At this juncture is is necessary to stress that certain concepts in the range equation will not be developed here further. In particular, the introduction of the visibility factor (which today might be superseded by a more general detectability factor) and the constant C_b open up an avenue into statistical detection theory which is more relevant to post-war developments. Undoubtedly many of the concepts which affected S_{min} were comprehended, at least qualitatively, during the war. The paper of North [16], which originally appeared as a classified RCA Laboratories' Technical Report PTR-6C in June 1943, introduced the concept of the matched filter* and the statistical criteria for detection. For further developments which have taken place in this area, reference may be made to the bibliography given by Barton [10].

2.3.2 Bistatic radar range equation

A bistatic radar is, as indicated in Fig. 2.1*b*, one where the transmitter and receiver are widely separated. It is an interesting fact that all the early experiments on radar in France, Germany, Great Britain, Italy, Japan, the United States and Russia used bistatic systems. The use of a transmitter and receiver (or, indeed, groups of transmitters and receivers) in bistatic configuration allows, in its most basic form, the interference between the direct signal from the transmitter and the scattered signal from the target to be recognised. This provides the ability to detect the entry of a target into a designated zone and the description 'fence coverage' is often associated with this form of bistatic radar. Their greater versatility and handiness has assured the almost exclusive use of monostatic radars from pre Second World War

---

* A matched filter is one which is matched to a particular signal and which filters so as to obtain signal recognition in the presence of noise. Its frequency response maximises the output peak-signal to mean-noise power ratio for a fixed input signal-to-noise energy ratio. The IF amplifier of a receiver is regarded as a filter with gain.

days to the present time, and only in the 1950s was an interest in bistatic systems reawakened and then only for special applications. Skolnik provides a useful comparison of the two types of system [17].

The range equation for a bistatic radar is very similar to the monostatic equation. The target cross-section σ is replaced by a bistatic one σ_b which generally differs in value from the monostatic cross-section. σ_b is a measure of the energy forward scattered in the direction of the receiver and its formal definition is covered by the expression already given in eqn. 2.1:

$$\sigma = \lim_{R \to \infty} 4\pi R^2 \; \frac{S_s}{S_i}$$

The equation for a bistatic pulse radar may, as indicated in Appendix K, be written

$$(R_1 R_2)_{max} = \left(\frac{P_t \, \tau \, G_t \, G_r \, \sigma_b \, \lambda^2 \, F^2_t \, F^2_r}{(4\pi)^3 \, k \, T_s \, V_0 \, C_b \, L} \right)^{\frac{1}{2}} \tag{2.6}$$

and for a CW radar the form would be

$$(R_1 R_2)_{max} = \left(\frac{P_t \, G_t \, G_r \, \sigma_b \, \lambda^2 \, F^2_t \, F^2_r}{(4\pi)^3 \, k \, T_s \, (S/N)_{min} \, B_n \, L} \right)^{\frac{1}{2}} \tag{2.7}$$

F_t and F_r are the pattern propagation factors* for transmitting and receiving antennae, respectively.

The form for a pulse radar may be retained in the case of a CW radar if τ is allowed to signify the finite duration time of an echo signal corresponding to the time when the target is being illuminated by the scanning radar beam. The performance of a CW radar will fall significantly short of that predicted by the range equation, the minimum detectable signal being 20 to 30 dB above the ideal. Leakage from transmitting antenna to receiving antenna is the cause and will occur if the two antennae are any way close together, or if one common antenna with a directional coupler is used.

2.4 Multipath propagation

Terrain and the troposphere (and perhaps the ionosphere) will modify the free space conditions of a land based radar. If one is concerned with a modern tracking radar, then the variations in tropospheric refraction may well be of concern, as would a full consideration of the spherical earth effects on the behaviour of the waves. The level of complexity of the analysis of either maximum radar range or of some other aspect of target detection will depend on operational demands. In the circumstances here, where an understanding

---

* Pattern propagation factor is a scalar quantity which expresses, for the electric field intensity, the difference between the actual situation and an ideal situation, where free-space conditions coupled with the use of the maximum of the main beam of the antenna are assumed.

of the basic functioning of the early radars is sought, the effects of the troposphere can be ignored and a simple flat earth representation of multipath behaviour, as shown in Fig. 2.4c, is also quite adequate. Specular reflection from a smooth plane surface will be assumed to occur.

A few observations can be made, in passing, on the subject of propagation. Most of the concepts and criteria are as relevant to radio as they are to radar. Much of the ground work towards understanding propagation at VHF and higher frequencies had been done before 1935. This included some experimental work. The war demanded from the developers of radar more knowledge and understanding of propagation effects, particularly at centimetric wavelengths, and this meant in-depth experimental studies. Typical of these is

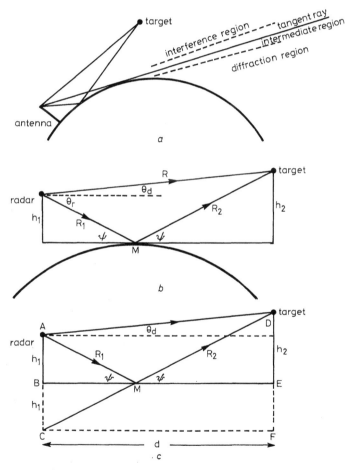

Fig. 2.4 *Earth reflection*
a General ground – air situation
b Modified earth's curvature model
c Flat earth model

Megaw's [18] account of work carried out in Britain from 1941 onwards, while Chapter 4 of Kerr's 'Radiation laboratory' text [19] gives in considerable detail an account of transmission experiments carried out over sea, fresh water and land in the United States. A worthwhile paper based on pre-war experiments carried out by the Bell Telephone Laboratories is that of Schelleng, Burrows and Ferrell [20], while an annotated bibliography in the text of Livingston [21], where over three hundred items are given, is recommended.

The importance of theoretical studies on propagation during the war years was underlined by T. L. Eckersley [22] when he addressed the Radiolocation Convention of the Institution of Electrical Engineers in London in March 1946. He said:

> The development of radar during the war as a major operational device is most usually regarded as a triumph of applied research. I should like to stress the debt that radar owes to pure research, a debt that is very often forgotten in the glare of the war-time practical achievements. The production of apparatus has been remarkably good, but its utility has depended upon the makers knowing along what lines to develop it.
>
> It is the business of the theoretical man to advise the Engineer, other things being equal, what wavelength to use for a given project, and to explain such phenomena as anomalous propagation, whereby on the one hand echoes may be obtained from objects close to the ground far beyond the visual range, while on the other hand no echoes may be obtained from an aeroplane well above the line of sight. It is his function to provide a theoretical description of propagation, and it is his work that forms the basis for practical development.

Reference to Fig. 2.4c could suggest that in considering multipath effects four possible ray paths might be taken into account, namely:

(*a*) transmitter direct to target and then direct from target to receiver;
(*b*) reflected path from transmitter to target and same path back to receiver;
(*c*) transmitter direct to target and reflected path back to receiver;
(*d*) reflected path to target and direct path back to receiver.

However, the same result is obtained by considering only two paths – the direct one and the path reflected from the surface of the earth – with the return signals from the target transmitted back over the same two paths.

Several geometrical constructions for spherical earth radar-target configurations are possible. One could suppose target altitude and range much greater than radar antenna height, or one need not restrict either the height or range of the target relative to the radar. The simple construction in Fig. 2.4b would allow for refraction of the rays so that with the rays shown as straight lines, a modified earth's curvature would be used. In the flat earth model of Fig. 2.4c the following are ignored:

(a) any phase difference at the antenna during transmission between the electric fields radiated along the direct and along the reflected paths;

(b) any difference in antenna vertical radiation pattern for the direct and reflected rays;

(c) any difference in target cross-section for the direct and reflected rays.

In addition, at the point of reflection M, full mirror-like or specular reflection is assumed with no diffuseness, and hence weakening of the reflected signal, owing to roughness. At the point of reflection also, because flat and not convex earth conditions are assumed, a ray divergence factor is not included.

Fig. 2.4a is a schematic of the general ground-to-air radar situation. The interference region is that where the direct path field and the reflected path field can add either constructively or destructively. The diffraction region is where echo signal returns, albeit weak ones, may be obtained from beyond the radio horizon and signal strengths are not predictable by ray theory.

Field strength in the interference region and in the diffraction region are amenable to calculation. As will be seen, ray interference methods indicate that for all terrain conditions and radars (except for a metric radar with vertical polarisation over the sea) the first null occurs in the vertical coverage pattern at zero elevation angle (θ_d of Fig. 2.4b, reflection angle $\psi \sim 0°$). In practice this is not so, since some field remains at zero angle.* The ray tracing method thus breaks down, and between the interference region and the diffraction region is a region termed by Kerr [19] the intermediate region, for which special methods of interpolation are required.

It goes without saying that the multipath situation as depicted in Fig. 2.4, although typical, is idealised in that, with a sloping site, for instance, more than one reflected ray path would be present between radar and target.

2.4.1 Electromagnetic reflection coefficient

It is assumed that the rays striking the ground are composed of uniform plane waves, that the ground is flat and uniform and that the electric field of the waves is either vertically polarised or horizontally polarised (with oblique polarisation the electric field may be resolved into a vertical component and a horizontal component.) The term ground may cover land, sea water or fresh water.

Refer to Fig. 2.4b where a wave is reflected at M. A change of magnitude and a change of phase may take place because of reflection. The terrestrial electric constants ε_r (relative permittivity) and σ (conductivity) at the point M,

---

* The following is usually taken as the minimum limit of angle ψ at which reflection methods can be used [23].

$$\psi = \left(\frac{\lambda}{2\pi Ka} \right)^{1/3}$$

where K = equivalent earth radius coefficient, which is generally taken to be 4/3 in temperate latitudes; a = earth's radius. This corresponds to about 18 minutes of arc at 100 MHz.

in conjunction with the plane of polarisation and the value of ψ, determine the value of the reflection coefficient. The electromagnetic reflection coefficient Γ relates the reflected field E_r to the incident field E_i with $E_r = \Gamma E_i$, such that

$$\Gamma = \rho e^{-j\phi}$$

where

$$\rho = \left| \frac{E_r}{E_i} \right|$$

and $0 \le \rho \le 1$ and $-\pi < \phi \le \pi$ and where positive ϕ means that E_r lags E_i in phase.

The properties of Γ_H and Γ_V, the reflection coefficients for horizontally polarised and vertically polarised waves, respectively, are dealt with in some detail in Appendix K and depicted broadly in Fig. 2.5.

The behaviour of the electric field on reflection may be summarised as in the following.

2.4.2 Horizontal polarisation

Near grazing incidence $\Gamma \cong -1$ for all types of ground. The modulus of Γ decreases with increasing ψ and with increasing frequency; it decreases to,

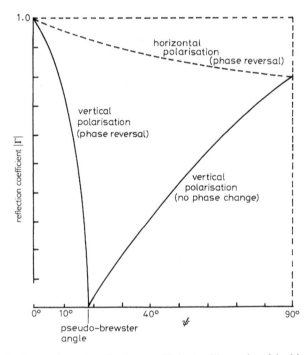

Fig. 2.5   *Typical variation of reflection coefficient with angle of incidence and with polarisation*

typically, 0·8 for sea and 0·5 for 'good' ground. The phase of Γ always remains close to $-\pi$.

2.4.3 Vertical polarization

For a perfect conductor $\Gamma = + 1$. Otherwise the modulus of the reflection coefficient decreases from unity at grazing incidence with increasing ψ, passes through a minimum at the pseudo-Brewster angle [24] and then returns gradually to its value for normal incidence (which in many cases is close to unity). The phase of Γ is $-\pi$ at grazing incidence. This lagging angle decreases rapidly near the pseudo-Brewster angle and maintains its decrease as ψ increases to normal incidence.

2.5 Multipath effects and radar coverage

The influence of multipath propagation is two-fold; a modification of the range equation and a change in the antenna vertical radiation pattern.

It is shown in Appendix K that the ratio of the power received from a target under multipath conditions, as depicted in Fig. 2.4c, to what would be received under free space conditions is

$$16 \sin^4 \frac{2\pi h_1 h_2}{\lambda d}$$

thus modifying the power by a factor which can vary in value from 0 to 16. Because of the fourth power relationship, as depicted in eqn. 2.4, the range will vary correspondingly in value from 0 to 2.

It is shown also in Appendix K that the interaction of the direct and reflected waves will produce a lobed structure in the antenna elevation pattern with maxima occurring (see Fig. 2.4b) at the following angles expressed in radians:

$$\theta_d = \frac{(2n + 1)\lambda}{4h_1} , \; n = 0,1,2,...$$

and minima at

$$\theta_d = \frac{n\lambda}{2h_1} , \; n = 0,1,2,...$$

2.5.1 Gap filling

A metric radar antenna operating at 200 MHz and placed 10 m above the ground would produce lobing at 2·9°, 8·6°, 14·3° etc., whereas a centimetric radar operating at 3 GHz placed at the same height would produce lobing at 0·14°, 0·43°, 0·72° etc. In the metric case in particular, considerable gaps in vertical coverage could occur. With vertical polarisation and with most types of ground a natural form of gap filling would be effected because of imperfect nulls. Again the use of antenna arrays with both vertically polarised elements and horizontally polarised elements could be effective above the pseudo-Brewster angle. The most common method of gap filling was to employ two antenna systems at different heights above the ground with the gaps of one system interlaced with the lobes of the other.

2.6 Choice of Polarisation

One might expect optimum signal returns from a target when the plane of polarisation of the electric field corresponded to the direction of the principal lines of the target. The use by the British of horizontal polarisation for their early air-defence radar was influenced by the fact that in their earliest calculations of field strengths, they had regarded an aircraft's wings as a tuned horizontal dipole [25]. The principal line of ships also lies in the horizontal, while tall buildings or the ground tend to reflect vertically polarised waves best. Therefore, for ground equipments which must detect aircraft or ships, horizontal polarisation gives generally better results. A. B. Wood, who was a noted physicist [26] on the British Admiralty Staff, gave particular thought to the question of polarisation in 1936.

It is interesting to note [27] that the British did experiment at Orfordness in 1936 with 40 MHz vertically polarised waves and subsequently regarded themselves lucky to have discontinued with the work. This was because, at a later date, vertical polarisation was found to give greater returns (unwanted) from objects on the ground and from the sea. The following was written by R. Hanbury-Brown [28] and describes some results obtained in 1938 during British research on 200 MHz ASV (aircraft to surface vessel) radar. The work was part of a programme initiated by E. G. Bowen in 1936. The quotation is given because it underlines the main consideration for the choice of polarisation, ground return, and because the information was obtained at such an early date.

> It was necessary to decide which polarization should be used. The effect of polarization was investigated by fitting an aircraft with both horizontal and vertical aerials, and also by violently altering the flight attitude of an aircraft carrying a fixed aerial system. It was found that the detection ranges of ships were similar with both horizontal and vertical polarization, but the scattered energy which was reflected by the sea was less with horizontal than with vertical. For this reason horizontal polarization was adopted for ASV.

It is worth mentioning also that the same preponderance of sea scatter when using vertical polarisation was obtained with the CHL (Chain Home Low) and the AMES type 11, both land-based sets working on 200 MHz and 600 MHz, respectively [29].

By comparison, it is interesting to note that many of the German shipborne surface watching radars used vertical polarisation. These included the experimental FuMO22 (Seetakt) which was on board the *Admiral Graf Spee* and other GEMA* manufactured shipborne sets such as the FuMO23 and FuMO24 which operated on 368 MHz. Land-based sets such as the GEMA-type FuMO52 operating between 187 and 220 MHz also employed vertically polarised antenna arrays, whereas some sets such as the LORENZ-type FuMO62 ('Hohentwiel') operating at 556 MHZ used horizontal polarisation.

The difficulties experienced by many of the early workers, including the Germans, in the use of vertically polarised radar with sea targets has been referred to by Bowen [30]. Today most marine sets use horizontal polarisation in preference to vertical polarisation, but the optimum choice is found to depend very much on sea conditions. One could, however, generalise and say that sea clutter is less with horizontal polarisation, but that for targets which are low in the water, a better signal-to-clutter ratio is obtainable with vertical polarisation.

Brief reference must be made to circular polarisation. Circularly polarised plane waves find their greatest use in discriminating against weather clutter arising from back-scattering from precipitation particles, particularly rain drops. When a right hand circularly polarised wave, say, is scattered from spherical rain drops, the sense of polarisation is changed and hence rejection will occur when the echo returns to the common transmitter/receiver antenna. Other distributed targets, such as aircraft, will, because of multiple reflections, send back both right handed and left handed circularly polarised echo signals. Hence these targets will be clearly seen in the presence of weather clutter.

2.7 Fundamentals of a pulse radar system

There are some further features of radar, and of a typical pulse radar in particular, which must be alluded to in order to clarify later descriptions of equipment. The concepts dealt with will be treated in a fundamental and qualitative manner. Fig. 2.6 gives a basic schematic of a pulse radar set, and the fundamental timing operations of the system are outlined in Fig. 2.7.

The radio frequency (RF) amplifier stages occurred in early metric equipments, but later, in the first microwave sets, the signal generally went straight from the antenna into a crystal mixer. One reason for this was that available microwave amplifiers were noisy. With the advent of low noise parametric amplifiers, RF amplifier stages have reappeared in microwave receivers. Most radar receivers are, and have been, of the superheterodyne type. The principal aims in receiver design were a low noise 'front end', a

---

* GEMA = Gesellschaft für Elektroakustische und Mechanische Apparate.

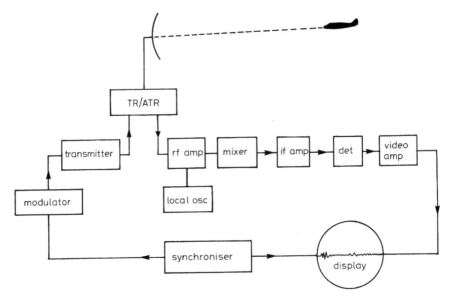

Fig. 2.6   *Schematic of a pulse radar set*

suitable value for the intermediate frequency (IF)* adequate IF bandwidth, and a judicious division of the receiver's amplification between the RF, IF and video amplifier stages. The local oscillator could be a triode oscillator, as in early metric equipments, or a klystron oscillator. The synchroniser is shown so as to indicate that synchronism is necessary between the firing of the transmitter modulator and the initiation of the display time base. In a modern radar, it would also generate clock pulses and shift pulses for a signal processor. The transmitter could be a set of power triodes with Lecher line [31] tuning for metric working, or a magnetron for higher frequencies, or more recently, a high power klystron might be used.

Fig. 2.7 illustrates the basic action of the system. Assume that radar transmissions travel at 3×10^8 m/s or 186, 240 statute miles/s, or 161, 798 nautical miles/s. Using the third identity and taking the time to and from the target into account, one can say that 1 μs is equivalent to a range of 0·0809 nautical miles, or 492 ft. This is equal to 12·4 μs/nautical mile. At instant A the transmitter fires. Suppose the oscillation is a sinusoidal one lasting for 1 μs. The pulse repetition frequency is taken to be 400 so that AC in Fig. 2.7 represent a time interval of 2500 μs. this is equivalent to a range of 202 nautical miles. However, the time base operates for a period AB with BC representing

---

* One would expect that the lower the IF the better from a stability point of view. There is another consideration, however. A pulse width of 1 μs with an IF of 5 MHz, say, would mean only 5 radio-frequency oscillations per pulse and this would not give sufficient 'body' to the pulse. An IF of 20 MHz or more in this case would be adequate.

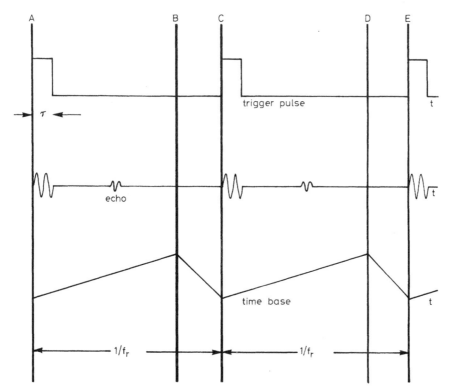

Fig. 2.7   *Basic timing operations in pulse radar*
P_t = Transmitter peak pulse power
P_{av} = Transmitter average power
f_r = Pulse repetition frequency (PRF)
$P_t\tau = P_{av}\dfrac{1}{f_r}$
$P_{av} = \underbrace{P_t\tau f_r}_{\text{duty cycle}}$

Suppose τ = 2 μs, PRF = 1000
Then duty cycle = 0·002 = 0·2%

a resting period of, say, 300 μs. AB (2200 μs) represents the maximum range of the set, which, here, would be just 178 nautical miles.

Suppose the radar is a ground-based airfield surveillance radar with an antenna rotating at 12 rev/min. This means an angular motion of 0·13° between the transmission of a pulse and the return of its echo from a target at 150 nautical miles. This is the minimum antenna azimuthal beamwidth that could be used for this range. The number of pulses striking a point target per rotation of the antenna is given by the relationship

$$N_{sc} = \frac{\theta f_r}{\omega} \tag{2.9}$$

where N_{sc} is the number of pulses per scan on the target; f_r is the pulse repetition frequency (PRF); θ is the azimuthal beamwidth in degrees (beamwidth is half power points beamwidth; these are the two points on the principal radiation lobe of the antenna at which the radiated (or received) power drops to one half of that at the peak of the lobe), and ω is the rotation rate of antenna, or scan rate, in degree/s. (Inspection of equation 2.9 shows that the dimensions of N_{sc} are those of time (θ/ω) multiplied by a repetition rate). N_{sc} influences $P_{r(min)}$ of eqn. K10 and $P_{r(min)}$ is found to be proportional to $1/\sqrt{N_{sc}}$.

There are certain restrictions in the use of eqn. 2.9. If the antenna rotates such that $\omega > f_r\theta$, there will be sectors in any one rotation which will fail to be illuminated. Also, too narrow a beamwidth would mean that echoes returning from targets near maximum range would fail to be received.

Detectability of targets will also be affected by pulse width, and, in general, one can appreciate that it will be very much dependent on a process of summation using the information obtained from groups of pulses over a period of time. In a pulse radar special integration circuits may be used both before detection and after detection. Post-detection integration is the more common. However, even where special circuits are not employed, as in the early radars, a form of integration will take place in any case. This is due to the persistence of the phosphors used in the cathode-ray tube displays and also due to a physiological mechanism of integration in the optical system of the observer (which experiments showed to be of the order of a few seconds).

2.7.1 Range and azimuth resolution

Refer to Fig. 2.8a. P and Q are two targets on the same bearing from the radar and separated by 492 ft. A pulse width of 1 μs is equivalent to a distance of 984 ft. It can be seen that, when P and Q are separated by this particular distance, the trailing edge of a 984 ft long pulse reflected from P will just coalesce with the leading edge of the pulse being reflected from Q. This demonstrates that two targets separated by a distance $\tau c/2$ (half pulse width), or greater, will appear as two distinct echoes. (τ is the pulse duration and c is the electromagnetic velocity constant).

In Fig. 2.8b, L and M represent two targets separated by the azimuth beamwidth θ. As the antenna rotates counter-clockwise and the leading edge of the beam moves from M to L to V, a continuous return will be received and no particular discrimination will be possible between the returns from L and M. If L and M were separated by more than the beamwidth then resolution would be possible. The same criterion pertains, of course, to the beamwidth in elevation. Sometimes in relation to a point target* and to a pulse radar one speaks of a radar resolution cell which is the volume $R^2 \, \delta\theta\delta\phi(c\tau/2)$, where R is the range of the target and $\delta\theta$ and $\delta\phi$ are, respectively, the azimuth and elevation angular beamwidths in radians.

---

* A point target is one which is small compared both to the spatial dimensions of the antenna beam that illuminates it and to the range equivalent of the pulse width being used.

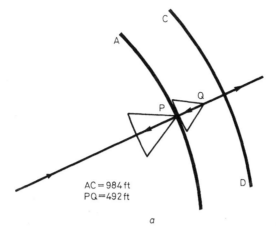

AC = 984 ft
PQ = 492 ft

a

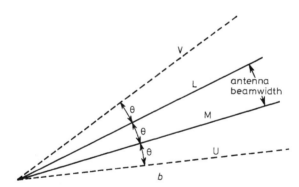

b

Fig. 2.8   *Resolution*
a Range
b Azimuth

The treatment given above on resolution is adequate for an understanding of early radars, and, at this point, it can be noted that an adequate equipment specification, in the context of making an assessment of pre-war and early wartime radars, would comprise transmitter peak power, pulse width, PRF, minimum detectable signal, antenna gain, antenna polarisation and antenna beamwidth and rotation rates.

One could refer to the resolution being discussed here as the inherent resolution of the radar set. With a cathode-ray tube display this inherent resolution may or may not be fully realised. Consider a typical plan position indication (PPI) display (for example, Fig. 2.11) where the spot size is such that 200 spots can be resolved along the radial time base. The ideal is to have the pulse length resolution and the spot size resolution equal. If the radial sweep represents a range of R nautical miles, then the number of pulse lengths resolvable is $(12 \cdot 4R)/\tau$, where τ, is in μs. This pulse length resolution and the spot size resolution are equal when $(12 \cdot 4R)/\tau = 200$.

If $\tau = 1 \mu s$, $R \cong 16$, thus a PPI with a range of 16 miles will have full inherent range resolution. If this same display is used to indicate a 100 mile sweep then the resolution will be reduced by a factor of approximately six. For further elucidation of this aspect of resolution, reference can be made to Ridenour [15] and to Soller, Starr and Valley [32].

A rigorous treatment of resolution examines it in the context of the extraction of information from a received waveform. The detection of the presence of targets from the received signal and then the extraction of information concerning these targets are the fundamental functions of a radar receiver. Range, angular position and velocity, for instance, are three of the basic characteristics of targets and the concepts of accuracy and precision as used in their measurement are well understood (the characteristic whereby a radar can see a small change in the position of a target would be referred to as 'precision'). As in all physical measurements, a statistical element necessarily enters into the estimation of any target parameter. With multiple targets there is the added requirement of distinguishing one target from another, which means the distinguishing of one signal return from another. This leads to the formulation of a 'range ambiguity function' and to an analytical treatment of the subject [33, 34]. The pulse radars of concern here emitted a periodic pulse train, whereas later developments gave attention to the design of special signal waveforms.

2.7.2 Pulse radar signals

The following problem was posed in a well known 1935 textbook on communication engineering [35]:

> Determine the amplitude and phase spectra for the function cos *pt* whose amplitude is modulated by the Morse dot shown in Fig. 195. Show that, as δ is allowed to approach infinity, the amplitude spectrum approaches the single line at the angular frequency *p*.

δ was the duration of the Morse dot. The signal which is transmitted and received by a simple pulse radar may be represented by a succession of such pulses occurring every $1/f_r$ seconds. (No doubt the above exercise and other similar ones involving Morse dots would today be couched in more relevant terms.)

Fig. 2.9a illustrates a succession of rectangular unidirectional impulses, but without an enveloped sinusoidal waveform. This pulse train might today be referred to as the periodic gate function. In Appendix K its frequency transform will be examined, as will the frequency transform of a single pulse,

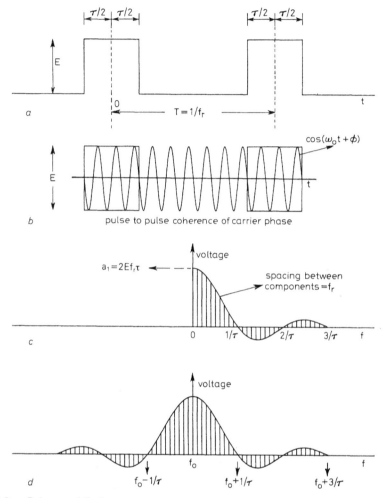

Fig. 2.9   *Pulses and their spectra*

and, finally, the frequency transform of a train of recurrent radio-frequency pulses will be discussed.

The results may be summarised as follows. The pulse train of Fig. 2.9a becomes, in the frequency domain, the one-sided line spectrum of Fig. 2.9c, the envelope of which is the sinc function. An equivalent two-sided spectrum, with positive and negative frequency components, may be constructed: in transforming from the one-sided to the two-sided spectrum the amplitudes of all corresponding spectral components are halved. The components of the line spectrum are separated by a frequency interval of f_r Hz. If then, as depicted in Fig. 2.9b, a train of such pulses modulates the amplitude of a continuous sinusoidal signal, the two-sided line spectrum of Fig. 2.9d results. The centre frequency of this pattern is the carrier frequency of the sinusoidal signal.

A numerical example will demonstrate what takes place in the frequency domain. Consider just the pulse train of Fig. 2.9a. Suppose $f_r = 400\,Hz$, $\tau = 1\,\mu s$ and the peak value of the pulses is E V. The resulting frequency spectrum is as in Fig. 2.9c. As shown in Appendix K, the amplitude of the fundamental frequency is $2Ef_r\tau$ volts. The frequency of this fundamental component is 400 Hz and there are 2500 components between zero frequency and the first zero of the spectrum at 1 MHz.

Fig. 2.9d can be taken as the model for the frequency spectrum which is both transmitted and received by a simple pulse radar. In practice, the ideal line spectrum as portrayed does not occur, but rather a continuous spectrum with more or less the same envelope as determined by the sinc function. This is owing to the following:

(a) there is a lack of phase coherence in the RF sinusoidal waveform between successive pulses;
(b) the transmitter frequency will vary randomly by amounts sufficient to blur the line spectrum;
(c) the pulse width, τ, and the PRF, f_r, are also liable to vary in a random manner.

This situation in no way poses a problem. Provided the pulse width, τ, does not vary appreciably, then the positions of the zeros will remain fixed and inspection of the spectrum shows that a network which transmits all frequencies between the first zero points ($-1/\tau < f < 1/\tau$) will transmit most of the power. The time function (pulse shapes) will also be reproduced with adequate faithfulness.

Even a cursory glance at the literature will make it clear that the broad criteria for the handling of pulsed signals were well understood in the early 1930s [35, 36]. Implementation of principles in the design of satisfactory receivers required some effort, however. One might cite the difficulties of Robert Page at the Naval Research Laboratory in 1934 [37] with the bandwidths of cascaded amplifiers and the inspiration he obtained from the then recently published paper of René Mesny [38].

2.8 Antenna duplexing – TR/ATR switches

The efficient use of a common antenna for transmission and reception is now taken so much for granted that it is worth stressing that the design and implementation of duplexing systems for the various frequency bands was no obvious thing and required some effort. No real assistance was obtainable from radio-communication practice, although on occasion, as in shipboard installations, a common antenna was used. The communication receiver was connected directly to the antenna when the transmitter was not operating and its input was then short-circuited by a relay when the transmitter was in operation.

The purpose of a duplexer is two-fold. It must prevent damaging energy from the transmitter reaching the receiver, and, of lesser importance, it must ensure that the returned echo signals go only to the receiver and are not dissipated in the transmitter input circuits. The circuit which protects the receiver is generally referred to as a TR (transmit–receive) unit. Different techniques are used for open-line, coaxial and waveguide feeders.

Precedence for the development of a duplexer probably goes to the United States Naval Research Laboratory, who, in July 1936, incorporated one in their 200 MHz radar which was then being developed [39]. Duplexing seems to have been used at Bawdsey in 1938 during the development of Coastal Defence (CD) radar and to have been re-invented early in 1940 when the Chain Home Low (CHL) system was being worked on [40]. Also, in Australia in May 1940 a duplexer unit was successfully operated with a low power transmitter [41].

The principle of operation of a typical TR/ATR unit for a metric radar will now be outlined. Refer to Fig. 2.10, where the lengths of transmission line, DF, AD, AB and BC are all $\lambda/4$ in extent. S_1 at B and S_2 at F are spark gaps. In practice, at frequencies where this configuration was employed, many types of spark gap, or equivalent devices, were tried and those used included high emission thermionic diodes, tungsten discharge gaps in air or in argon at high

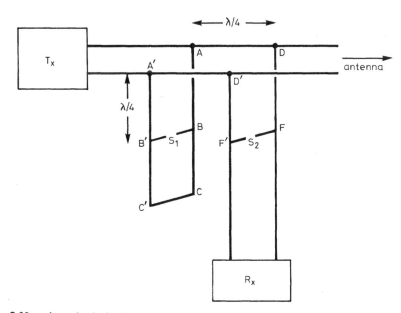

Fig. 2.10 *A method of effecting duplexing*

pressure, and low pressure glow discharge tubes [42]. The primary requirements of the spark-gap device were low striking voltage, short starting time
and rapid deionisation. The basic action of the circuit depended on the
transforming properties of half-wave sections and of quarter-wave sections of
transmission lines and was as follows:

During transmission
S_1 and S_2 conduct
Short circuit of S_1 at B means open circuit at AA'
Short circuit of S_2 at F means open circuit at DD'
Hence transmitted energy will flow straight from transmitter by A, by D and
into the antenna.

During reception
Short circuit at CC' means short circuit at AA'. This means an open circuit
across DD' in the direction of the transmitter. Hence energy will flow
downwards into the receiver.

The connection to the receiver in practice might include another spark-gap
and a balance-to-unbalance (BALUN) [43] (open-line to coaxial line)
transformer.

2.9 Nomenclature of displays

There are many methods of displaying targets on a cathode-ray tube. The most
basic is the type A display, where a horizontal time base indicates range and a
target echo signal is displayed as a vertical 'blip' on the time base. In the
Ground-Controlled Approach (GCA) system, the glide path of an aircraft
coming into land was very realistically portrayed by a type E, or range–height,
display. Fifteen types of display are illustrated in Skolnik [44].

Fig. 2.11 shows four of the more common types of display. Type A, or
range–amplitude, display is the most basic, while possibly the most common
and certainly the most useful is the type P, or plan position indicator (PPI)
display. It shows the radar's position as the centre of the display and the
position of targets are then presented in R,θ (range, bearing) co-ordinates.
The essentials of the system are the rotation of the polar diagram of the
antenna and then, in synchronism with this, the rotation of the display time
base. Echoes appear as an intensity modulation of the cathode-ray tube
screen. The PPI display, whether for a ground-based radar or for an airborne
set, provides a two dimensional map which realistically displays permanent
echoes and moving targets. There are many ways of achieving the rotating time
base. The most obvious method is to actually rotate a coil which is placed about
the cathode-ray tube neck and which provides the range sweep. The coil is
rotated about the tube in synchronism with the rotation of the antenna.

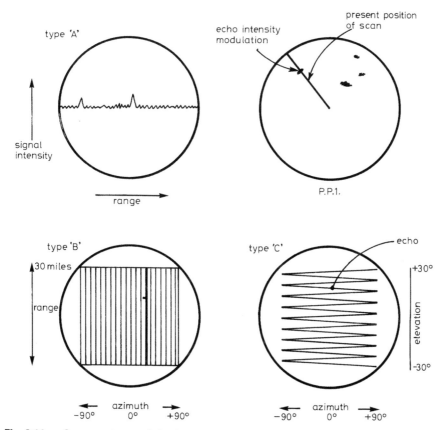

Fig. 2.11   *Common types of display*

Credit for precedence in the development of the PPI goes to E. G. Bowen [45–47]. The suggestion for PPI appears to have been made by him in the autumn of 1935 when the question of direction finding within the proposed British ground-radar system was being discussed. Whereas R. A. Watson-Watt and A. F. Wilkins advocated the floodlighting concept for radar, Bowen then favoured a rotating beam system from which a PPI type of display would naturally follow. When Bowen was in charge of the small group, which since 1936 had been successfully involved in developing airborne radars, he applied himself towards the end of 1938 to implementing the PPI concept for use in aircraft. Again, in November 1939, he proposed the use of PPI in GCI (ground control of interception) equipment.

Robert Page in the United States had the following to say about what, early in 1939, prompted him towards the development of a PPI [48]:

When the XAF, with its boxlike indicator for measuring range, made the famous cruise on the *U.S.S. New York* in January–February 1939, the system for determining the angle of bearing was awkward . . . When I tried to chart the position of every vessel in a formation of a hundred ships, I spent an hour, and by the time I had located the last ship, the first ones had changed position enough so that they did not fit the chart. A s a result, I could never complete a correct chart. It was obvious that a faster method of determining the bearings of many targets was needed.

The concept of the PPI itself, apart altogether from any considerations given to the method of its implementation, may well have preceded the work of any of these radar pioneers. It is interesting to refer to a very small item which appeared in the November 1935 edition of the magazine *Radio Craft* [49]. The article itself (The new 'mystery ray') which contains the item will be mentioned later in Chapter 3. As a footnote to a photograph which shows, presumably, a Dr Spitz in front of a large circular screen which seems to have a map overlay with a rectangular co-ordinate grid, the following is written:

Dr S. Spitz, of Burbank, Calif., developed this amazing machine, which indicates as a moving spot of light, the location of an airplane; the plane's sound automatically actuates the device. The same principle probably could be applied to silent 'planes, the actuation then being obtained by means of centimeter (the 'mystery ray' mentioned in this article) waves reflected from the 'plane!

2.10 Direction finding: split-beam method

The principle of direction finding by means of antenna lobe switching can readily be grasped by reference to Fig. 2.12.

The line of shoot of the receiving antenna array was changed rapidly from OB to OA to OB, and so on. At the same time, the X-plates voltages of the A scope display was increased slightly each time the beam pointed to the right. The received signals from the two beam positions are displayed side by side. When the target is in position ON, and thus perpendicular to the plane of the array, the two 'blips' on the A scope display become equal in amplitude.

The radio-goniometer method of direction finding was used in the British Chain Home stations, which operated at wavelengths of 6 m to 13 m, and which will be described in Chapter 3. A metric radar with a rotating antenna which had a narrow horizontal beamwidth measured the azimuth of a target more accurately than did the goniometer. A higher degree of accuracy still was obtainable with the split-beam method. W. A. S. Butement in Britain developed this technique for CD (Coastal Defence) 200 MHz radars early in 1939 [50]. Notes made about his work in April/May 1939 [51] read:

Results have borne out the theoretical calculations remarkably well and have shown that the position of the approximate centre of ship can be determined to an accuracy of the order of ±14 mins. On aircraft, an even

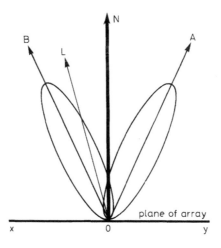

Fig. 2.12   *Split beam principle (lobe switching)*

greater mean accuracy appears to be obtainable, since the echoes generally are clearer and more easily equated for height. The method automatically provides 'continuous following'.

The increase in accuracy in the measurement of azimuth by the use of the split-beam method was so significant that it led Watson-Watt to write [51]:

The inaccuracy of bearing fell from one or two degrees to five or ten minutes of arc. This was indeed a landmark in the history of precise measurement by radio.

The concept employed in the split-beam method was foreshadowed by the runway localiser element of the Lorenz Thick Weather Landing System of the early 1930s, the forerunner of the present day ILS (Instrument Landing System) [52].

The method was also used in Britain in some 200 MHz CHL (Chain Home Low) systems which had separate transmitting and receiving antennae [53, 54].

The beam was swung through an angle of 12°, centred on the perpendicular to the antenna plane, at a rate of between 15 and 20 times a second. The method for achieving this 6° of displacement was the creation of a difference in length, and therefore in phase, in the feeders which came from the right and left halves of the receiving array.

Accuracies in this case appear to have been between ½° and ¼°. (The later CHL sets obtained sufficient accuracy in azimuth by using a common transmitting and receiving array which rotated continuously and by reading the bearing of the centre of the target as displayed on the PPI.)

The split-beam technique was employed in the first United States Army Radar, the SCR-268 [55], and was in use as early as August 1937 in a service test model [56]. In 1937 two separate antenna arrays diverging in direction by 15° were used to produce the overlapping lobes of the type shown in Fig. 2.12. Late in 1939 the phase shift of the signals from two sides of the same array was employed to achieve the same result and with an accuracy of ¹⁄₁₀° [57].

In Germany, Dr Runge and Dr Steppe of Telefunken introduced the same basic principle in 1936, but with a dipole which rotated in a circular path about the focal point of a parabolic reflector [58]. The axis of the antenna lobe thus produced moves on the surface of a cone whose axis coincides with the axis of the reflector (see Fig. 2.13) and the effect, which is an extension of simple lobe switching, is referred to as conical scanning. The Telefunken equipment operated on a frequency of 600 MHz and the experiments preceded the development of the notable 'Würzburg' series of sets.

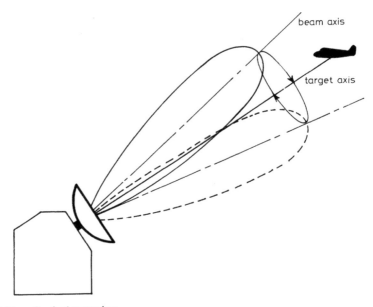

Fig. 2.13   *Conical scanning*

2.11 Secondary radar

Secondary radar, as distinct from the normal or primary radar, is where a target is co-operative, receiving a ground radar's signal and returning what is in effect an enhanced and coded reply to the ground. The system is generally referred to, at least in civilian usage, by the title SSR (secondary surveillance radar). Secondary radar has become an essential element of effective and safe air-traffic control and airborne transponders have become mandatory for most aircraft. The main advantages of secondary radar are that it provides a return signal far stronger than the normal echoes from a primary radar, while not being constrained by ground clutter, and it offers a valuable data link between aircraft and ground. The coded replies of identity can be displayed on a PPI tube as an extra blip beside the primary radar identification, but nowadays a whole range of more sophisticated video presentations are used with targets displayed as numerics or symbols.

The present standard international system of secondary radar, which is used in both civil and military air-traffic control, emerged from the Identification Friend or Foe (IFF) system of the Second World War. Present day secondary air-traffic control radar operates in the L-band of frequencies (see Section 2.13), with a ground interrogator transmitting on a nominal frequency of 1030 MHz and air-to-ground reply from a transponder on a nominal frequency of 1090 MHz. The system is independent of any primary radars being used, although it may be co-located with one and share a common antenna turning gear and mounting. The military side of secondary radar includes the use of transponders on naval vessels, and, indeed, the fitting of interrogators on missile launchers or other low-level air-defence systems. In the latter case, the 'friendly' pilot must surely hope that his transponder is functioning correctly.

A useful reference to present-day secondary radar systems is Honold [59], while the significant design criteria for post-war airborne transponders are well covered in a paper by Harris [60]. An interesting paper from Britain dealing with both IFF and other radar beacons of the war years is that of Wood [61]. Formulae for the interrogation range and the reply range of beacons are found in most general textbooks on radar [62]. As might be expected, secondary radar has its own particular problems and constraints [60].

In Britain, the possibility of distinguishing one's own aircraft from enemy aircraft when both were above cloud, or otherwise out of visual contact, was actively considered long before radar was seriously thought of. In a file [63] which was opened in March 1928, there is a reference to several possible systems. Special 'singing' wires on fighter aircraft, the use of sirens and of devices attached to the aircraft's exhaust and even the special training of listeners to distinguish between various sounds were all advocated. HF/DF (high-frequency direction finding) on the ground and co-operative transmissions from one's own aircraft was another solution that was suggested.

Following on from the deliberations of the late 1920s and early 1930s, the advent of radar in Britain added an extra urgency to the problem of recognising friend from foe. At first, such homely solutions as having their own bombers leave their undercarriages down or give identifying light signals, were considered [63]. In a letter to Air Vice Marshal R. H. M. Saundby, Air Chief Marshal Sir Hugh Dowding could happily write on 12th May 1938: 'I understand that the special RDF attachment is giving very promising results' [64].

The possibility of friendly aircraft identifying themselves was discussed in February 1935 between Watson-Watt and Wilkins before the former composed his memorandum of 27th February 1935, 'Detection and location of aircraft by radio methods' (see Appendix D). Wilkins reasoned that if an aircraft behaved like a resonant dipole then another dipole, keyed and suitably placed, should alter the strength of the received echoes [65]. The relevant portion of Watson-Watt's memorandum reads:

> There will also be, for consideration, the problem whether the interval between detection and engagement may not be best reduced to a minimum by having interceptor craft fitted with a keyed resonating array so that they are readily located by the same methods as those used on the enemy bombers, but discriminated and identified by the intermissions in their 'reflected' field. The interception operation can then be controlled by radio instructions to the interceptors closing them into the positions indicated for the bombers.

This idea was tried out successfully in 1937 when a dipole which could be made or broken at the centre was fitted between the undercarriage struts of an Avro Anson. Keyed dipoles were also fitted to flying boats and to a Heyford bomber. Dowding witnessed some tests, was impressed and was prepared to request that the whole bomber force be equipped with this simple device [66]. Luckily, it was realised that the discrimination afforded by keyed dipoles in the case of single aircraft might not pertain in the case of a formation of aircraft and so 'powered' transponders were next sought. Development started on these transponders in 1937 and production models became available in 1939 [66]. These were directly interrogated by the ground radar station. In what became the IFF MkI and IFF MkII, the frequency of the airborne transmitter and receiver was swept continuously through a wide frequency band, which included the frequencies of all the ground stations concerned and the time for each sweep was a few seconds. Thus, every few seconds a target echo at a ground station might increase in amplitude and indicate the presence of a friendly aircraft. However, by 1941 the number of frequency bands which the airborne set had to accommodate was leading to an impossible situation, and so an indirect interrogation IFF system, with special interrogator and receiver on the ground, was resorted to [67]. (The airborne transponder was the IFF MkIII and operated in the band 157–187 MHz; an American equivalent was the SCR 595.)

In the United States, also, a preoccupation with the identification of friendly aircraft or ships existed before the advent of radar. Recourse was made to special identification procedures, and the possibility of using radio waves was considered [68].

Robert Page recounts [69] that when the XAF radar was taken on sea trials in 1939 a pole with half-wavelength rods was mounted on a destroyer and rotated by a motor; as long as the pole was rotated the XAF could identify the destroyer from among all the other targets. This is mentioned just to indicate that the Americans also resorted to a simple but effective solution to target recognition. However, by 1937 the Naval Research Laboratory were developing their model XAE transponder system which operated at 500 MHz. The co-operating aircraft sent out an omni-directional coded transmission. This was controlled by a contactor operated by a rotating disc, the edge of the disc being notched in accordance with the code. These code discs were easy to change if security required it. On board ship a Yagi antenna, mounted on a rifle stock, was pointed at the aircraft. On receipt of the code the ship sent back an acknowledgment which caused a light to flash on the aircraft and this light could then be seen by the ship.

Following this, the Naval Research Laboratory in 1939 developed a pulse transponder very similar to the British MkI and MkII sets. In 1940 they developed what is claimed to be not only the first transponder to use binary digital techniques to achieve security, but also the first electronic circuitry to employ binary digital methods. This transponder was successfully demonstrated in two-way operation between the Naval Research Laboratory and Fort Washington in May 1940 [70].

These facts, while strictly belonging to the individual histories of just two of the countries which developed radar, have been dwelt upon at some length because they are relevant to what became the main stream of universal IFF development. The sharing of technical information between Britain and the United States affected transponder development in both countries. In 1941, before the attack on Pearl Harbour, there were prolonged discussions in the United States between the Services and with British representatives on the co-ordination and rationalisation of systems [71]. One could summarise the eventual result by saying that the British MkIII set was adopted as the standard United States–British–Canadian IFF during the war and indeed was in use by the United States forces after the war. A MkIV system, developed by the US Naval Research Laboratory, used separate frequencies, 470 MHz and 493·5 MHz, respectively, for interrogation and reply. An argument against its use in Europe was a closeness of its frequencies to that of the German Würzburg set (550 MHz). In September 1942 the Laboratory was directed to produce a new transponder system which would become the MkV/UNB (United Nations Beacon) and operate at approximately 1000 MHz, with separate frequencies for interrogation and reply. Service evaluation was not completed until 1948 and some of the sets did go into service. What is important about this programme is that it led to the present international system of SSR used by both military and commercial aircraft [71, 72].

2.12 Over-the-horizon radar

R. M. Page, in an interview given in 1978 to the Naval Research Laboratory historian [73] and while discussing his last technical contribution to radar, had this to say:

> Oh, yes, I dreamed about assembling megawatts of power and going out after the moon and other heavenly bodies to see if we could get echoes. But it was not military research – it wasn't part of our mission, so I discarded it. It was – call it a 'flash in the pan dream' and discarded right away – that wasn't our business. But, other parts of the Laboratory were interested in getting scientific information from reflections from the moon, and they did so, and they did so with short pulses, so that they were able to resolve lunar features on the radar. They made some very interesting observations and some interesting discoveries as a result. That's outside of our own immediate interest, but that perhaps gave me a little boost towards the idea of increasing the sensitivity of radar, sensitivity to go greater distances and do greater things.
>
> The thing that intrigued me most of all was getting over the horizon. That was the next limitation to radar. The echoes from the moon really had nothing to do with it, but it did spur me on to do things that might not be considered to have immediate military requirement, and we had witnessed with great regret the fact that we could not see below the horizon. Aircraft could fly low over the water and come pretty close to us before we could pick them up, so the idea of getting radar by ionosphere reflection had appealed to me for a long time . . .

This aspiration of Page's was indeed achieved. Development started in 1949 and by 1954 the relatively low powered (50 kW peak, 2 kW average) radar referred to as MUSIC (multiple storage, integration, correlation) was operating successfully [73]. Further on in the narrative quoted above, Page refers to two quite considerable achievements which were necessary before any real development could begin. The first, an electromechanical one, was the design of a magnetic drum (acting as a tapped delay line analogue integrator) which rotated at 180 revolutions per second and recorded, read and erased at frequencies of from 7 MHz to 8 MHz. The second was the design of an extremely narrow notch filter which discriminated between ground clutter and slowly moving targets.

MUSIC operated on 26·6 MHz and 13·5 MHz and was capable of detecting atomic explosions (1957–58) at 1700 miles and missile launchings at 600 miles. A higher powered over-the-horizon radar with acronym MADRE (magnetic drum radar equipment) was initiated in 1958. It delivered 5 MW peak power and 100 kW average power and in 1969 it was detecting aircraft and missile targets at ranges out to 2650 nautical miles over both land and sea [74].

Apart from the obvious military uses of HF radar there have been other post-war developments in this field. Since Crombie's experiments [75] at 13·56 MHz in New Zealand in 1955, the use of high-frequency Doppler radar, operating between approximately 2 MHz and 40 MHz, to investigate the behaviour of the sea surface has become common place among oceanographers [76].

This still rather singular area of radar technology has been alluded to here because its roots are very definitely traceable to the pre-war main stream of radar development.

In long-distance high-frequency transmissions using the ionosphere, the radio waves are considered to be capable of propagating by a series of multiple hops, or a series of reflections between the ionosphere and the ground. A downcoming wave from the ionosphere would be reflected by the earth's surface in a form of specular reflection, but a certain amount of energy could be backscattered and might travel back towards its points of origin, again via the ionosphere. Gebhard [77] refers to the pulsed HF transmission experiments carried out by the US Naval Research Laboratory between 1925 and 1928 and describes how, in 1926, reflections back from prominences on the earth's surface, called 'splashbacks', were observed. He suggests that even at that stage the observers of these 'splashbacks' speculated on the possibility of utilising the phenomenon to detect objects on the earth's surface. At approximately the same period Eckersley in Britain was observing what appeared to be the same type of effect [78].

Wilkins also had some interest in high-frequency backscatter before he became involved in radar development in 1935 and after the war, back at Slough, he returned actively to this interest [79].

Bowen [80], in a letter dated 25th February 1942 to Professor J. D. Cockroft, referred to the German vessels Scharnhorst and Gneisenau and then wrote:

> What about taking a look at them when they work along the coast of Denmark and Norway. About the only way of doing this is by reflection from the E layer on a longish wavelength. It is possible that we did see something of this kind on 50 m at Orfordness in 1935, but missed the significance of it at the time.

He then outlined a specification for a super-CD (coastal defence) which might do the job. This included:

wavelength: 50 metres or longer depending on the time of day;
transmitter: several hundred kW of power 25 μs; 25 Hz repetition rate;
aerial system: dipole a half-wavelength above the ground or arrays of the Bell Laboratory MUSA* type.

* MUSA = multiple-unit steerable antenna. For details see Reference 36, p. 680.

He suggested that an experiment might be made by modifying an MB (mobile base) transmitter for a longer wavelength (they normally operated between 20 and 50 MHz) and seeing what came back from 300 or 400 miles.

Skolnik [81] provides an excellent summary of HF over-the-horizon radar and includes equations for range, and also includes a 'figure of merit' factor used in the comparison of radars.

2.13 Frequency bands

During the war years a letter code, such as P,S,X, was used by the Allies for security reasons to describe particular radar frequency bands. These band designations, with some variations and additions, continued after the war. The nomenclature is useful in that it groups together equipments which have many broad characteristics in common. For further information, reference may be made to Skolnik [1] and Friedman [72].

Band boundaries vary slightly from one designating authority to another and in Table 2.1 are reproduced [82] three present-day systems of designation: a United Kingdom IEE recommended system, a United States system and a NATO (North Atlantic Treaty Organisation) system.

Table 2.1   Systems of designation

Band	Frequency (GHz)	Band	Frequency (GHz)	Band	Frequency (GHz)
United Kingdom IEE recommended system		United States system		NATO system	
L	1–2	P	0·225–0·39	A	0–0·25
S	2–4	L	0·39–1·55	B	0·25–0·50
C	4–8	S	1·55–5·2	C	0·50–1·00
X	8–12	X	5·2–10·9	D	1·0–2
J	12–18	K	10·9–36	E	2–3
K	18–26	Q	36–46	F	3–4
Q	26–40	V	46–56	G	4–6
V	40–60	W	56–100	H	6–8
O	60–90			I	8–10
				J	10–20
				K	20–40
				L	40–60
				M	60–100

References

1 SKOLNIK, Merrill I: 'Radar handbook' (McGraw-Hill, New York, 1970)
2 BERNE, Bruce J., and PECORA, Robert: 'Dynamic light scattering' (John Wiley, New York, 1976)
3 BORN, Max and WOLF, Emil: 'Principles of optics' (Pergamon Press, Oxford, 1975), 5th edn
4 SKOLNIK, Merrill I: 'Introduction to radar systems' (McGraw-Hill, New York, 1962), p. 41
5 MENTZER, J. R.: 'Scattering and diffraction of radio waves' (Pergamon Press, London, 1955)
6 Reference 1, Chap. 27
7 BLAKE, Lamont V.: 'Radar range-performance analysis' (Gower Publishing Company Ltd, Lexington, 1980)
8 *Proc. IEEE*, 1965, **53**, pp. 769–1137
9 NORTON, K. A., and OMBERG, A. C.: 'The maximum range of a radar set', *Proc. Inst. Radio Engrs*, 1947, **35**, pp. 4–24
10 BARTON, David K.: 'Radars. Volume 2. The radar equation' (Artech House, Inc., Dedham, Massachusetts, 1977), contains annotated bibliography
11 CASTIONI, Luigi Carilio: 'Storia dei radiotelemetri Italiani' (Istituto Storico E Di Cultura Dell'Arma Del Genio, Roma, 1974)
12 TIBERIO, U.: 'Misura di distanze per mezzo di onde ultracorte (radiotelemetria)', *Alta Freq.*, 1939, **8**, pp. 305–323
13 Record of scientific work undertaken by BUTEMENT, W. A. S., C.B.E., and submitted to Adelaide University for the degree of Doctor of Science, March 1960
14 HAEFF, A. V.: 'Minimum detectable radar signal and its dependence upon parameters of radar systems', *Proc. Inst. Radio Engrs*, 1946, **34**, pp. 857–861
15 RIDENOUR, Louis N.: 'Radar system engineering' (Radiation Laboratory Series Vol. 1) (McGraw-Hill, New York, 1947), pp. 33, 34
16 NORTH, D. O.: 'An analysis of the factors which determine signal/noise discrimination in pulsed carrier systems', *Proc. IEEE*, 1963, **51**, 1016–27
17 Reference 4, pp. 585–594
18 MEGAW, E. C. S.: 'Experimental studies of the propagation of very short radio waves', *J. Instn. Elect. Engrs*, 1946, **93**, Part IIIA, pp. 79–97
19 KERR, Donald E.: 'Propagation of short radiowaves' (Radiation Laboratory Series Vol. 13) (McGraw-Hill, New York, 1951)
20 SCHELLENG, J. C., BURROWS, C. R., and FERRELL, E. B.: 'Ultra short-wave propagation', *Proc. Inst. Radio Engrs*, 1935, **21**, pp. 427–463
21 LIVINGSTON, Donald C.:'The physics of microwave propagation' (Prentice-Hall, New Jersey, 1970)
22 ECKERSLEY, T. L.: 'The importance of theory in the development and understanding of radar propagation', *J. Instn Elect. Engrs*, 1946, **93**, Part IIIA, pp. 103–104
23 PICQUENARD, Armal: 'Radio wave propagation' (Macmillan, London, 1974), p. 66
24 JONES, D. S.: 'The theory of electromagnetism' (Pergamon Press, Oxford, 1964), p. 372
25 WILKINS, A. F.: 'The early days of radar in Great Britain', Archives, Churchill College, Cambridge, UK
26 ADM 218/259 and ADM 218/260: Public Record Office, Kew, London, UK
27 Interview with A. F. Wilkins, 7th January 1981
28 SMITH, R. A. (Chief Writer): 'A.S.V. (the detection of surface vessels by airborne radar)', unpublished technical monograph on wartime research and development in M.A.P. Bowen Papers, EGBN 2/4, Archives, Churchill College, Cambridge, UK
29 Air Ministry: 'Introductory survey of radar principles and equipment' (AP1093C, London, 1946)

30 An interview with Dr Edward G. Bowen. Conducted by Dr David K. Allison, Historian, Naval Research Laboratory. Part 6 of tape, available at Naval Research Laboratory, Washington DC, USA

31 LECHER, Erns: 'Eine Studie über electrische Resonanzercheinungen', *Annln Phys.*, 1890, **41**, pp. 850–870

32 SOLLER, Theodore, STARR, Merle A., and VALLEY, George E.: 'Cathode ray tube displays' (Radiation Laboratory Series Vol. 22) (McGraw-Hill, New York, 1948)

33 Reference 4, p. 484 *et seq.*

34 BIRD, G. J. A.: 'Radar precision and resolution' (Pentech Press, London, 1974)

35 GUILLEMIN, Ernest A.: 'Communication networks Vol. II' (John Wiley, New York, 1935), p. 507

36 HENNEY, K.: 'The radio engineering handbook' (McGraw-Hill, New York, 1941), p. 723

37 An interview with Dr Robert Morris Page, 26th and 27th October, 1978. Conducted by Dr David K. Allison, Historian, Naval Research Laboratory. Transcript available at Naval Research Laboratory, Washington DC, USA

38 MESNY, René: 'Constantes de temps, durées d'éstablissment, décrément', *Onde Élect.*, 1934, **13**, pp. 237–243

39 GEBHARD, L.: 'Evolution of naval radio-electronics and contribution of the naval research laboratory', NRL Report 8300, Naval Research Laboratory, Washington DC, 1979

40 An interview with Dr Edward G. Bowen. Conducted by Dr David K. Allison, Historian, Naval Research Laboratory. Part 7 of tape, available at Naval Research Laboratory, Washington DC, USA

41 BOWEN, E. G.: 'A textbook of radar' (University Press, Cambridge, 1954), 2nd edn, p. 294

42 COOKE, A. H., FERTEL, G., and HARRIS, N. L.: 'Electronic switches for single-aerial working', *J. Instn Elect. Engrs*, 1946, **93**, Part IIIA, pp. 1575–1584

43 WEEKS, W. L.: 'Antenna engineering' (McGraw-Hill, New York, 1968), pp. 167–180

44 Reference 1, pp. 6–3

45 'Proceedings of Royal Commission on Awards to Inventors', fifth day of Hearing (24th May 1951), pp. 1–4 of transcript. Copy available in Science Museum Library, London, UK

46 Bowen Papers, EGBN 1/5 (draft of proposals for range, and range direction-finding from a single site; signed 9th October 1935), Archives, Churchill College, Cambridge, UK

47 Bowen Papers, EGBN 1/6, Archives, Churchill College, Cambridge, UK

48 PAGE, Robert Maurice: 'The origin of radar' (Doubleday and Company, New York, 1962), p. 152 *et seq*

49 'The new mystery ray', 1935, *Radio Craft*, **7**, pp. 267–269

50 Reference 13, p. 11 *et seq.*

51 Reference 13, p. 30. Quotation is from WATSON WATT, Robert: 'Radar in war and in peace', *Nature*, 1945, **156**, pp. 319–324

52 KEEN, R.: 'Wireless direction finding' (Iliffe and Sons Ltd, London, 1938), 3rd edn, p. 630, references are given to original papers on the Lorenz system

53 AVIA 26/32 (Transmitter and Receiver Aerial Systems for CHL System): Public Record Office, Kew, London, UK

54 TAYLOR, D., and WESTCOTT, C. H.: 'Divided broadside aerials with applications to 200Mc/s ground radio-location systems', *J. Instn Elect. Engrs*, 1946, **93**, Part IIIA, pp. 588–597

55 'The SCR-268 radar', *Electronics*, 1945, **18**, pp. 100–109

56 DAVIS, Harry M.: 'History of the signal corps development of US Army radar equipment, Part II' (Signal Corps Historical Section, Office of the Chief Signal Officer, Washington, DC), p. 34. Microfilm copy available from Photoduplication Service, Library of Congress, Washington DC, USA

57 Reference 56, p. 72 *et seq.*

58 BRANDT, Leo: 'Zur Geschichte Der Radartechnik In Deutschland Und Grosbritannien', p. 30 *et seq.*, (XV Convegno Internazionale Delle Communicazioni Genova 12–15 Ottobre 1967), Istituto Internazionale delle Communicazioni, Genova, 1967

59 HONOLD, Peter: 'Secondary radar' (Hayden and Sons Ltd, London, 1976) Translation of 'Sekundär-Radar' (Siemens, Berlin, 1971)

60 HARRIS, K. E.: 'Some problems of secondary surveillance radar systems', *J. Br. Instn. Radio Engrs*, 1956, **16**, pp. 355–382

61 WOOD, K. A.: '200 Mc/s radar interrogator-beacon system', *J. Instn Elect. Engrs*, 1946, **93**, Part IIIA, pp. 481–495

62 See, for instance, Reference 1, Chap. 38

63 AIR 16/312: Public Record Office, Kew, London, UK

64 AIR 2/2615: Public Record Office, Kew, London, UK

65 WILKINS, A. F.: 'The early days of radar in Great Britain', Archives, Churchill College, Cambridge, UK

66 AIR 2/3049: Public Record Office, Kew, London, UK

67 Air Ministry: *AP1093D, Vol. 1*, Chap. 6, London, 1946. Available under AIR 10/2288, Public Record Office, Kew, London, UK

68 Reference 39, Chap. 6

69 Reference 48, p. 165

70 TERRETT, Dulany: 'The signal corps: the emergency' (Office of the Chief of Military History, United States Army, Washington DC, 1956), Chapter 10

71 Air Ministry: 'Introductory survey of radar principles and equipment', AP1093C, Vol. 1, Chap. 4, London, 1948

72 FRIEDMAN, Norman: 'Naval radar' (Conway Maritime Press Ltd, Greenwich, 1981)

73 Reference 37, p. 181 *et seq.*

74 Reference 39, p. 212 *et seq.*

75 CROMBIE, D. D.: 'Doppler spectrum of sea echo at 13.56Mc/s', *Nature*, 1955, **175**, pp. 681–682

76 A good survey of work carried out between 1955 and 1980, together with a useful bibliography, is given in the paper: SHEARMAN, E. D. R.: 'Remote sensing of the sea-surface by Dekametric radar', *The Radio and Electronic Engineer*, 1980, **50**, pp. 611–623

77 Reference 39, p. 45

78 ECKERSLEY, T. L.: 'Short wave wireless telegraphy', *J. Instn Elect. Engrs*, 1927, **65**, pp. 600–644

79 Interview with A. F. Wilkins, 7th January, 1981

80 Bowen Papers, EGBN 1/14, Archives, Churchill College, Cambridge, UK

81 SKOLNIK, M. I.: 'Introduction to radar systems' (McGraw-Hill, New York, 1980), 2nd edn, p. 529 *et seq.*

82 'Newnes radio and electronic engineers' pocket book' (Butterworth and Company Ltd, London, 1978), 15th edn

Chapter 3

Precursors of radar

3.1 Introduction

In a paper [1] 'A terrain clearance indicator' read before a Chicago meeting of the Institute of Aeronautical Sciences on 19th November, 1938, and published in the *Bell System Technical Journal*, the authors Lloyd Espenschied and R. C. Newhouse made what is considered a key observation. Discussing the FM type of radio altimeter and referring to early efforts in this area, they said:

> The evolution of this method is interesting because it illustrates how one art is built upon another, and also the familiar story of separate inventors arriving at the same answer almost simultaneously, actually somewhat in advance of the existence of instrumentalities having the characteristics required to make the invention practically serviceable.

No more apt comment could be made today of the proposals, the patents and the experiments which preceded the actual development of radar in the various countries in the 1930s. In this regard, one thinks of the proposal of Christian Hülsmeyer in Germany in 1904, of the observations of A. Hoyt Taylor and Leo C. Young in the United States in 1922 and of the specifications of L.S.B. Alder at the Royal Navy Signal School in 1928, to mention but a few. A reader of radar history might well regret the apparent lack of interest shown by their superiors to the suggestions of A. Hoyt Taylor to the United States Navy in 1922 or to the reports of W. A. S. Butement and P. E. Pollard to the British Army in 1931. On reflection, however, he might consider it just as well that proper development programmes were not undertaken with the technology of the time. Each reader must eventually form his or her own judgement and one can do no better than record events as accurately and as objectively as possible. In appreciating technological events of the past, hindsight from the present has indeed its value, but it is essential to regress to the past where one's horizon becomes that of the time in question.

3.2 Christian Hülsmeyer

Mention has been made already of Nikola Tesla and of his conception in 1900 of the possibility of employing radio waves not only to detect, but also to measure the movement of distant objects. The man and the date are at least worth noting. Whether his words reflect only the wishful thinking of a visionary or whether they are in some measure based on knowledge obtained in his experiments at Colorado Springs is difficult now to assess. The history of radio technology might benefit from a critical study of the Colorado Spring's epoch of Tesla's life.

As we move along the years, and, of course, with our present hindsight, the 30th April 1904 is the next significant date. On that day, Christian Hülsmeyer in Düsseldorf, Germany, applied for a patent for his 'telemobiloscope' which was a transmitter–receiver system for detecting distant metallic objects by means of electrical waves [2]. Two significant features about the telemobilo-scope are that the apparatus was designed principally as an anti-collision device for ships and that it did work quite successfully [3–7].

Christian Hülsmeyer was the son of a farmer. His interest in collision prevention arose after observing the grief of a mother whose son was killed when two ships collided. After a period teaching in Bremen, where he had the opportunity of repeating Hertz's experiments, he joined Siemens. In 1902 he moved to Düsseldorf to concentrate on his invention. He became acquainted with a merchant from Cologne, was given 5000 marks and founded the company 'Telemobiloscop-Gesellschaft Hülsmeyer und Mannheim'. The first public demonstration of his apparatus took place on the 18th May 1904 at the Hohenzollern Bridge, Cologne. As a ship on the river approached, one could hear a bell ringing. The ringing ceased only when the ship changed direction and left the beam of the apparatus. All tests carried out gave positive results. The reaction of the press and public opinion were favourable. Nevertheless, neither the naval authorities nor public companies showed any interest.

In June 1904 he was given the opportunity by the director of a Dutch shipping company to display his equipment at various shipping congresses at Rotterdam. At this stage his equipment was detecting ships at ranges up to 3000 m, and he was planning new apparatus which would function up to 10 000 m. He took out a fourth and final patent on the 11th November 1904 [8], but after this no further publicity was given to his work and no further experimentation was carried out. He seems to have become somewhat embittered by the apathy of the experts, particularly during the First World War, when he felt that his invention could have been put to so much use.

Later he was successful as an engineer in Düsseldorf, where he died at the age of 75 in 1957. In 1955, although too ill to attend, he was honoured at a congress in Munich on Weather and Astro-Navigation (Flug-Wetter-und Astro Funkortungs-Tagung).

A brief description of his apparatus would say that it probably operated on a

Christian Hülsmeyer's equipment of 1904 on display in the Deutsches Museum, Munich.

French (SFR) 16 cm obstacle detector CW radar on the deck of the S.S. NORMANDIE, New York harbour, 1935.

wavelength of 40–50 cm. The transmitter used a Righi-type spark gap (part of which was immersed in oil) fed from an induction coil. The radiated pulses were beamed by a funnel-shaped reflector and tube which could be pointed in any desired direction. The receiver used a coherer detector and a separate vertical antenna, which, because of a semi-cylindrical movable screen, was also directional. Basically, the apparatus was designed to detect the presence of an object in a particular direction. The question of determining distance was later solved, in principle anyway, by a modification which aimed at beaming the radiation at any desired angle of elevation or, rather, of dip. Knowledge of the height of one's own transmitting antenna above the surface of the water and of the angle of vertical dip at which an object was detected would, by simple calculation, give the range of the object. Perhaps the most ingenious aspect of the inventor's later apparatus was his awareness that the equipment might respond to other than its own transmissions and his safeguarding against it by a time limiting electromechanical mechanism. The receiver responded to a first transmission's signal only if, after a predetermined interval, it received the signal from a second transmission.

Most of his apparatus was destroyed in 1919 but some, principally a receiver, is on exhibition at the Deutsches Museum, Munich.

3.3 Hugo Gernsback

The next person of note must surely be Hugo Gernsback, who published in 1911 his romantic story RALPH 124C 41+. Gernsback was born in Luxemburg in 1884, went to the United States in 1904 and died in 1967. Among his achievements he launched a monthly magazine *Modern Electrics* which later became *The Electrical Experimenter*; he designed batteries and he marketed home radio sets. However, his fame lies in the world of science fiction and today just as there are Oscar awards in the world of the cinema, so are there Science Fiction Achievement Awards, which are named Hugos in his honour [9].

RALPH 124C 41+ was subtitled 'A romance of the year 2660' and was published in *Modern Electrics* in issues from April 1911 to March 1912, inclusive. The serial was later published in book form [10]. In the story there are several quite accurate predictions other than that of radar, and, unlike those of other writers of fantasy such as Jules Verne, they are all physically plausible.

Gernsback breaks the narrative of his tale to describe the detection apparatus used by the hero, Ralph. Here, but without the diagram in the text, is the description:

A pulsating polarized ether wave, if directed on a metal object can be reflected from a bright surface or from a mirror. The reflection factor, however, varies with different metals. Thus the reflection factor from silver is 1000 units, the reflection from iron 645, alomagnesium 460, etc. If, therefore, a polarized wave generator were directed toward space, the waves would take a direction as shown in the diagram, provided the parabolic wave reflector was used as shown. By manipulating the entire apparatus like a searchlight, waves would be sent over a large area. Sooner or later these waves would strike the metal body of the flyer, and these waves would be reflected back to the sending apparatus. Here they would fall on the actinoscope (see diagram 1), which records only reflected waves, not direct ones.

From the actinoscope the reflection factor is then determined, which shows the kind of metal from which the reflection comes. From the intensity and the elapsed time of the reflected impulses, the distance between the earth and the flyer can then be accurately and quickly calculated.

3.4 Hans Dominik

Mention has been made of the experimental work of a German engineer, Christian Hülsmeyer, and of the writings of a man who was primarily an author of fiction. A brief tribute is now due to another German who was both an engineer and a writer of fiction. The man was Hans Dominik [11, 6] and the critical year of his work on a radar-type device was 1916. Linked with the work of Hans Dominik was another German, Richard Scherl, who had emigrated to South America in 1910, who had worked there on aircraft and who had then returned to Germany in the autumn of 1915. Richard Scherl provided the ideas and Hans Dominik did the experimental work.

Their object was to build a machine that would ensure the detection of an enemy target in darkness. They gave the name 'strahlenzieler' (ray aiming device) to the project. Their reasoning was that every metal surface reflected electromagnetic waves, that the wavelength of the rays and the unevenness of the surface were important, and that, to obtain mirror-like reflections from a surface with minute unevenness, wavelengths of the order of 10 cm should be used. The task then was to generate 10 cm waves and to construct apparatus that could utilise the waves that had been reflected from a target. Dominik approached some friends in Siemens and Halske. He met Professor Raps who gave him a job there. Professor Raps, however, did not wish to get involved with the project and a director, Fiedler, inventor of the flame thrower, was really the first person to listen to him. Dominik received financial support and built a 10 cm wavelength apparatus which employed a spark-discharger type of transmitter. He was able to use the range, which Fiedler had for his flame thrower, to perform successful experiments and with the notes from these experiments he approached the navy in February 1916. Some interest was shown but nothing of a concrete nature resulted. In May 1916, having been asked how soon he could have equipment ready and having replied that it would be 6 months, he was told that this would be too late for the First World War. As in the case of Hülsmeyer, Dominik decided at this point that this was the end of the matter and handed everything over to Director Fiedler. Fiedler kept an interest in the project for some time but his main aim was the development of the flame thrower, so that nothing really ever came of Scherl's work and Dominik's ambition.

3.5 Nikola Tesla

In 1917 Tesla earned mention once more when, in the August edition of *The Electrical Experimenter* edited by Hugo Gernsback, an account was given of an interview with him on methods of subjugating enemy submarines [12]. The caption under a sketch of two ships and a submerged submarine in the first page of the article reads:

> Nikola Tesla, the famous Electric inventor, has proposed three different electrical schemes for locating submarines. The reflected electric ray method is illustrated above; the high-frequency invisible electric ray, when reflected by a submarine hull, causes phosphorescent screens on another, or even the same, ship to glow, giving warning that the U-boats are near.

The text of the interview is interesting. A magnetic detector for locating submerged craft was discussed and here his confidence in its feasibility was based on observations he had made on ships on the river Seine many years earlier. When Tesla described his radar-type idea, an 'electric ray' method for locating metal objects, he appears not to have taken into consideration the attenuation suffered by radio waves when propagating through water. Nevertheless, his ideas are revealing, and two quotations from the paper are deemed of interest. The first contains the concept of pulse modulation:

> Suppose, for example, that a vessel is fitted with such an electric ray projector. The average ship has available from say 10 000 to 15 000 H.P. The exploring ray could be flashed out intermittently and thus it would be possible to hurl forth a very formidable beam of pulsating electric energy, involving a discharge of hundreds of thousands of horse-power. The electric energy would be taken from the ship's plant for a fraction of a minute only, being absorbed at a tremendous rate by suitable condensers and other apparatus, from which it could be liberated at any rate desired.

The second quote shows an appreciation of both the principles of scattering and of bistatic working:

> To make this clearer, consider that a concentrated ray from a searchlight is thrown on a balloon at night. When the spot of light strikes the balloon,

the latter at once becomes visible from many different angles. The same effect would be created with the electric ray if properly applied. When the ray struck the rough hull of a submarine it would be reflected, but not in a concentrated beam – it would spread out; which is just what we want. Suppose several vessels are steaming along in company; it thus becomes evident that several of them will intercept the reflected ray and accordingly be warned of the presence of the submarine or submarines.

3.6 Detection of aircraft

Apart from actual visual sighting of an aircraft (which even today may on occasion be done by an air-traffic controller equipped with binoculars) there are four known methods of determining the approach of an aircraft. The first and obvious one is by using a sensitive sound locator. Next, one might try to detect infra-red radiations emanating from the aircraft's engines. Thirdly, one might try to detect the electromagnetic waves radiated by the ignition system of the aircraft, provided the aircraft had a petrol engine. Finally, one could have recourse to radar. All four methods were tried and, at one time or another during the Second World War, all methods, except the detection of ignition radiation, were used.

3.6.1 Ignition detection
This method was experimented with during the First World War and the French certainly considered it between 1925 and 1928 [13] until they eventually realised that good screening of the ignition system made the method impracticable.

In Marconi's address of 1922 he said [14]:

> Incidentally I might mention that one of these short wave receivers will act as an excellent device for testing, even from a distance, whether or not one's ignition is working all right. Some motorists would have a shock if they realised how often their magnetos and sparking plugs are working in a deplorably irregular manner.

Perhaps the most interesting feature about this possible method of detection is that it led the then Major E. H. Armstrong of the United States Signal Corps, while seeking to design a very sensitive receiver for this purpose, to his particular development of the superheterodyne receiver in 1918 [15, 16].

3.6.2 Sound locators
Sound locators [17, 18] were an integral part of land-based anti-aircraft defence systems until the outbreak of the Second World War. Indeed, they were used by the Japanese during the Second World War [19].

Sound locating instruments were developed in many countries during the

First World War. Their principle is quite simple. A person's ears are quite an effective sound direction finder, the person generally turning his head in a reflex action until the sound from a source reaches both ears simultaneously or in phase. The use of two trumpets separated a few feet apart extends the discriminatory ability of the ears while also providing an opportunity to catch more sound energy.

The 'Claude Orthophone' [17], a binaural device, was used by the French in 1917. The anti-aircraft experimental section of the British Munitions Inventions Department under A. V. Hill improved the French design and made an instrument with two pairs of trumpets, one pair to measure the azimuthal direction of the sound and the other to measure its angle of elevation. Before the outbreak of the Second World War locators had been perfected which could give an accuracy of ¼° in azimuth for a fixed source of sound and typically 2° for an aircraft. In some cases cathode-ray tube displays were used.

Even for slow moving aircraft, with sound waves in air travelling at about 1100 ft/s, a 'sound lag' occurred resulting in an appreciable difference between the true and apparent positions of a target. Correctors, based on estimates of speed and angular velocities, were fitted to locators to compensate for this time lag. The air, however, is not a clear, still medium and sound waves are affected by such factors as variations in temperature in the atmosphere and winds. In brief, sound locators for the detection of aircraft had little value or potential after the First World War. Admittedly they were still kept under consideration, as evidenced by the acoustical mirror system proposed for the Thames Estuary in 1933 and referred to in Chapter 5.

3.6.3 Passive and active infra-red detection

In contrast with ignition detection and sound location, whose complete histories may be readily appraised, the subject of thermal or infra-red detection and imaging of objects has a breadth and depth that could justify a complete study in itself. Its military uses were explored during the First World War, and, at the present time, continue to be the subject of a variety of areas of research. It could never, because of its inability to operate through cloud, have become a direct alternative to radar.

Early in 1918, Master Signal Electrician Samuel O. Hoffman, and other enlisted men of the United States Signal Corps and Air Service, were assigned to work in the laboratory of Professor George Pegram at Columbia University [20]. In a paper which he read to the American Physical Society in 1919, Hoffman gave the results obtained using a Hilger thermopile, which was a combination of thermocouples, and which was mounted at the focus of various parabolic mirrors and whose output deflected a d'Arsonval galvanometer:

> Early in 1918 work was started on the problem of detecting men and other objects at a higher temperature than their background, in the dark, by

their thermal radiation. Men were detected at ease at 600 ft. A man lying in a depression in the ground at a distance of 400 ft was detected unfailingly as soon as he showed the upper part of his face above the ground. These observations were repeated on different nights and with different backgrounds during April 1918. . . . Secret signaling was accomplished by simply covering and uncovering the face.

The Germans are credited with picking up British torpedo boats off Ostend during the First World War at about 10 km range [21].

Between the wars work was carried out on military uses of infra-red detection in Britain, Germany and the United States. The United States Signal Corps, operating from Fort Monmouth with a passive thermolocator, tracked liners passing in and out of nearby New York harbour. Vantage points near Fort Monmouth, which is at the New York harbour entrance, provided a ready made laboratory for testing out detection equipment on ocean-going vessels. The liner *Mauretania* was tracked to a distance of 23 000 yards in 1934 and in 1935 the *Normandie* was tracked to a distance of 30 000 yards while the Aquitania was tracked through fog to a distance of 18 000 yards. It was also possible to distinguish the *Mauretania*'s dummy smokestack from her three real ones [22].

The following quotation from Edgar W. Kutzscher [23], who had been a Director of Infrared Research and Development of the Electroacoustic Co., Kiel, Germany, is of interest in that it gives quite succinctly the factors that had to be considered in designing an infra-red detection or homing system.

The development of these infrared homing devices was based on comprehensive efforts to use infrared methods and devices for various military applications. Basic laboratory investigations necessary for feasibility studies and for the technical development of military systems started early in the 1930's. This work on basic problems was concerned with:

calculations and measurements on the emission of special infrared transmitters and with the total energy and spectral content of the infrared radiation emitted by various targets of military interest; studies of the attenuation of infrared radiation by the atmosphere;
optical problems;
the design of special optical systems including the development of new optical materials suitable for lenses and windows, the creation of new infrared detectors, investigations of the influence of background radiation;
theoretical and experimental work on methods for scanning optical fields and for modulating infrared radiation;
the research carried out in this field resulted in the successful design and use of practical infrared devices for various military purposes . . .

Furthermore, in addition to the tactical advantage of the passive method, the efficiency of the passive system is far superior to that of the active system. For example, the infrared radiation emitted by a three-engined bomber in the wavelength region of approximately 0·8 to 3·5 μ was found to be from 1000 to 10 000 times greater than that reflected in the active method when a searchlight with a special infrared filter and a mirror of 150 cm diameter was used. This comparison is based on a distance of approximately 10 km. The advantage of the passive over the active method grows with increasing range.

In a paper [21] which gives a concise overview of infra-red techniques since the time of William Herschel, Professor R. V. Jones had this interesting comment to make:

> On 27th April 1937 I made what appears to have been the first flight in which an aircraft was detected by another in flight by infra-red. The range was modest – only 500 yards – but the limitation was not due to the detector but to the effect of bumps on the input transformer that I used to couple the thermopile into the electronic amplifier.

3.7 Marconi's speech to the American Institute of Electrical Engineers

On the 20th June 1922, in New York, Guglielmo Marconi addressed the American Institute of Electrical Engineers and The Institute of Radio Engineers. In his address [24] he stated:

> I propose now to bring to your notice some of the recent results attained in Europe and elsewhere and to call your attention particularly to what I consider a somewhat neglected branch of the art; and which is the study of the characteristics and properties of very short electrical waves.

Towards the end of his address he said:

> Before I conclude I should like to refer to another possible application of these waves which, if successful, would be of great value to navigators.
>
> As was first shown by Hertz, electric waves can be completely reflected by conducting bodies. In some of my tests I have noticed the effects of reflection and deflection of these waves by metallic objects miles away.
>
> It seems to me that it should be possible to design apparatus by means of which a ship could radiate or project a divergent beam of these rays in any desired direction, which rays, if coming across a metallic object, such as another steamer or ship, would be reflected back to a receiver screened from the local transmitter on the sending ship, and thereby immediately reveal the presence and bearing of the other ship in fog or thick weather.
>
> One further great advantage of such an arrangement would be that it would be able to give warning of the presence and bearing of ships, even though these ships be unprovided with any kind of radio.

3.8 Albert Hoyt Taylor and Leo Clifford Young

In September 1922, Dr Albert Hoyt Taylor and his assistant Leo Clifford Young [25, 26], who were stationed at the United States Naval Aircraft laboratory, Anacostia, DC, were carrying out VHF propagation experiments at 60 MHz. They employed a superheterodyne receiver and a 50 W transmitter amplitude-modulated at 500 Hz. Initially tests were carried out in the grounds of the Naval Air Station and audible maxima and minima caused by reflections from steel buildings were observed. The receiver was placed in a car which was driven a few miles from the Station to Haines Point across the Potomac River. These reflection phenomena were again observed from trees and other objects and in particular from the wooden steamer *Dorchester* which passed down the river.

On the 27th September a memorandum drawn up by Taylor was sent to the Bureau of Engineering. It gave details of the experiments and, among several suggestions, proposed that radio beams at about this frequency could be used for the radio detection of objects such as, for instance, enemy vessels passing between two destroyers.

The letter states [27]:

> If it is possible to detect, with stations one half mile apart, the passage of a wooden vessel, it is believed that with suitable parabolic reflectors at transmitter and receiver, using a concentrated instead of a diffused beam, the passage of vessels, particularly of steel vessels (warships) could be noted at much greater distances. Possibly an arrangement could be worked out whereby destroyers located on a line a number of miles apart could be immediately aware of the passage of an enemy vessel between any two destroyers in the line, irrespective of fog, darkness or smoke screen. It is impossible to say whether this idea is a practical one at the present stage of the work, but it seems worthy of investigation.

There is no recorded official reaction to this proposal, and, in so far as the United States Navy were concerned, the next significant event in the radar story occurred on 24th June 1930 when Leo Young and Lawrence Hyland were testing the directional properties of an aircraft antenna system as part of a HF project being undertaken by the Radio Division of the Naval Research Laboratory. This event, which laid the foundation for serious research into radar in the United States, will be recounted in Chapter 4.

3.9 Measurement of distance

In three distinct spheres of activity during the 1920s, electromagnetic distance measuring techniques were being developed quite independently of any intention of detecting distant targets. Each of these areas of ionospheric sounding, geodetic surveying, and radio-altimetry did, however, in some way, assist the emergence of radar.

3.9.1 *Ionospheric sounding*

The experiments of Gregory Breit and Merle A. Tuve in 1925 under the auspices of the Carnegie Institution's Terrestrial Magnetism Laboratory, where they demonstrated the existence of a reflecting ionospheric region in the upper atmosphere, are invariably cited whenever the history of pulse radar is recounted. Indeed, today in the context of radar astronomy one could justifiably say that the system they used constituted a radar. The high frequency transmissions from station NKF of the Naval Research Laboratory were used in the tests. Furthermore, Dr A. H. Taylor, L. A. Gebhard and L. C. Young, who became key people in the Laboratory's radar development, were closely involved in adjusting the transmitter for the experiments. Previous to this work in the United States, Professor E. V. Appleton and M. A. F. Barnett of Britain had, on the nights of 11th December 1924 and 17th February 1925, by utilising the British Broadcasting Company's transmitter at Bournemouth, demonstrated the existence of a sky reflected wave; their's was a less direct, but nonetheless, equally valid method.

We will take a look at these two key ionospheric experiments for two reasons. First, the people associated with ionospheric research in Britain and the United States had close connections with the early creators of radar in their respective countries. Secondly, the methods in ionospheric measurement used at the time are considered to be a very accurate indicator of the then state of the art in circuit components, circuit techniques and radio propagation concepts.

Shortly after the successful early transatlantic radio transmissions of December 1901, an explanation was sought for the attainment of such long ranges over a spherical earth. Both Professor A. E. Kennelly [28] in the United States and Heaviside [29] in England suggested in 1902 that an ionised layer, which would act as a reflecting surface, might exist in the upper atmosphere. Eccles in 1912 and then, more adequately, Larmor in 1924, formulated a mechanism of refraction. Larmor's theory did not take account of the earth's magnetic field, but in 1925 Appleton in Britain and Nichols and Schelleng in the United States remedied this inadequacy. It is of interest that in the early 1920s the short-wave band was freely available to amateurs and was considered worthless for professional communications. The experiments of amateurs coupled to developments in valves were the principal factors in the breakthrough into the use of the higher frequencies.

A common phenomenon of high-frequency propagation is that of 'skip distance'. There is a no-signal zone between the end of the useful ground wave and the beginning of the useful sky wave, as indicated in Fig. 3.1. This skip distance zone had been well documented by 1924. Skip distance was found to be a function of wavelength in a way which corroborated the Appleton–

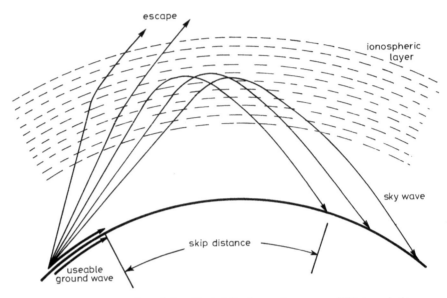

escape

ionospheric layer

sky wave

skip distance

useable ground wave

Fig. 3.1   *Simple illustration of the effect of the ionosphere on an HF transmission*

Larmor concepts. To the user of high-frequency radio the phenomenon is a striking demonstration of the presence of the ionosphere. Nevertheless, by 1925 a more direct proof of the existence of a reflecting layer in the upper atmosphere was deemed desirable. The direct proof would be centred about a measured time lag between a direct and a reflected signal with evidence of a downward path of propagation for the reflected signal.

We will first look at the experiments of Appleton and Barnett [30]. Essentially the experiment was concerned with the measurement of electric field intensity produced at a point by a distant transmitter and an assessment of the variations in intensity produced by interference between a direct ground ray and a reflected atmospheric ray. The receiver consisted of two amplifying RF stages, followed by a square law detector and a galvanometer. The voltage imposed on the detector was assumed to be proportional to the amplitude of the received wave. The current in the detector was proportional to the square of the voltage impressed on it, and, therefore, to the square of the amplitude of the received electric field. Hence the galvanometer reading was proportional to the square of the intensity of the electric field. The reasoning in the experiment was as follows.

The interference between the direct ray and the reflected ray was considered analagous to that pertaining in a Lloyd's mirror fringe system. The ground ray path was denoted as a and the skywave path as a'. If the skywave was N wavelengths shorter than the ground wave, then

$$N = \frac{a' - a}{\lambda} \tag{3.1}$$

where λ is the wavelength. N equal to an integer corresponds to a reinforcement of the received field. If the wavelength is gradually increased by δλ to λ', the number of maxima observed at the receiver should be

$$(a' - a) \left(\frac{1}{\lambda} - \frac{1}{\lambda'} \right) \tag{3.2}$$

or

$$\frac{\delta\lambda}{\lambda^2} (a' - a), \text{ if } \delta\lambda \text{ is small} \tag{3.3}$$

If δλ and λ are known, the value of $(a' - a)$ may be estimated.

δλ was of the order of 5 to 10 metres, the change in wavelength being uniformly produced in times of the order of 10 to 30 seconds.

It was reckoned that, at a distance of 100 miles from the transmitter, the magnitudes of the direct and reflected rays would be equal and that interference phenomena would be most pronounced. Hence the choice of Oxford for the receiving station and the use of the BBC transmitter at Bournemouth in the principal series of tests. A typical result was: change of λ from 385 to 395 metres gave the average number of maxima on the galvanometer as 7·0 and thus a value of $(a' - a)$ of the order of 100 km with the height of the layer estimated at 80–90 km.

To demonstrate that reflected waves came downwards from the upper atmosphere, two receivers at the receiving site, one with a T-shaped antenna (see Fig. 3.2) and the other with a large single-turn loop antenna in the plane of

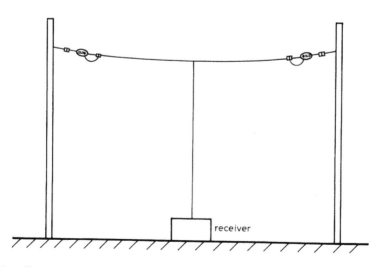

Fig. 3.2   *T-antenna*

propagation, were used during the tests. The spacing between the receivers was varied during the series of experiments from about 90 ft to 300 ft. The arrangement was as in Fig. 3.3 with the plane of the T-antenna perpendicular to the plane of the loop, thus ensuring minimum interaction. It can be shown that at night, when a sky wave exists, the signal current in both antennae will depart from the daylight values and that, furthermore, the variation of the loop signal will exceed that of the T-antenna if the reflected waves arrive at the receiving point in a downward direction. If the angle which these downward-coming waves makes with the horizontal is θ, then the ratio of these signal variations may be shown to be sec θ.

The RF amplifiers of the receivers were adjusted to give equal galvanometer readings at the receiver outputs during the daytime. At night, when fading began, owing to ionospheric reflection, the ratio of loop signal variation to T-antenna signal variation was consistently greater than unity. In tests carried out at Cambridge on the London (2LO) transmission (wavelength = 360 m), the most frequently obtained ratio was 2·85 corresponding to a slope angle, θ, of approximately 69°.

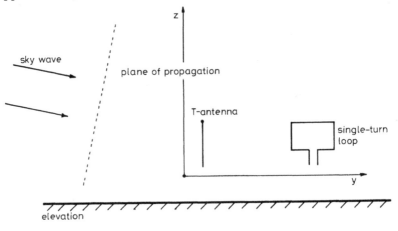

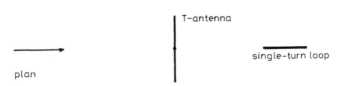

Fig. 3.3   *Antennae layout for Appleton and Barnett experiment*

The experiments of Breit and Tuve [31] are of particular interest in that they utilised a pulse technique and undoubtedly their work was known internationally to all engaged in radio science. We have here selected the work of Appleton and Barnett in Britain and that of Breit and Tuve in the United States because, chronologically, they are considered significant in the evolution of a radar technology. Indeed it is worth recounting that Taylor, who was prominent in the development of radar in the United States, had assisted with the Breit and Tuve experiments [32, 33]. It is likewise interesting to note that Arnold F. Wilkins, who was a vital figure in the development of British radar with Sir Robert Watson-Watt, was for a while associated with Professor E. V. Appleton. It must, however, be stressed that, in the context of ionospheric science, these two important series of tests are but a minute portion of the work that went on at that time and that a very abundant literature exists on ionospheric height measurements [34].

The idea of studying the height of what was then called the Heaviside–Kennelly Layer was due to Breit. Tuve proposed the use of a pulse technique and in this seems to have been influenced by the work of Professor W. F. G. Swann and Dr J. G. Frayne of the University of Minnesota. Their most successful tests were done using the crystal controlled 5 kW transmitter of the Naval Research Laboratory (Station NKF, Bellevue, Anacostia DC). This station was 8 miles south-east of the laboratory of the Department of Terrestrial Magnetism. The transmitter's oscillator was fed by DC but the amplifier valves were fed by an AC supply whose frequency was nominally 500 Hz. Pulses lasting about 1 ms were obtained by adjusting the biasing grid voltages.

This adjustment allowed an active time for the valves of between 1/3 and 1/5 of a cycle. The tests, which were carried out at wavelengths of 71·3 m and 41·7 m, commenced on 28th July 1925 and continued on 6th August, 21st, 23rd and 25th September 1925.

The receiver was a superheterodyne with an intermediate frequency of 50 kHz. It comprised a grid-leak first detector stage, three intermediate-frequency amplifiers and a grid-leak second detector. The receiver fed a power amplifier consisting of four or five 5 W valves in parallel and this amplifier in turn activated an oil-immersed* General Electric oscillograph (of the Duddell type). The received pulses of the form indicated in Fig. 3.4*a* were made visible by means of a rotating mirror and were also photographed using Eastman high-speed NC-type film.

A purely qualitative estimate of the angle between the downward wave and the vertical was achieved by setting up three aerials with horizontal portions at 6, 30 and 50 ft above the ground, as shown in Fig. 3.4*b*. The aerial at 30 ft gave the strongest signal, and, if an interference pattern formed by the downward

---

* Castor oil at room temperature was used for damping.

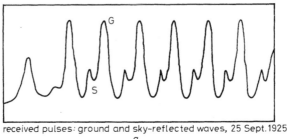

received pulses: ground and sky-reflected waves, 25 Sept.1925

a

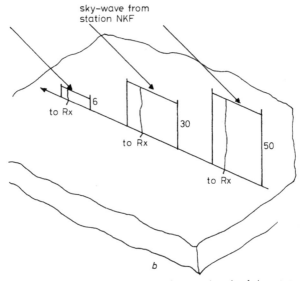

antenna system used to estimate downward angle of sky-wave

Fig. 3.4   *Breit and Tuve Experiment*
     a Received pulses: ground and sky reflected waves, 25th September 1925
     b Antenna system used to estimate downward angle of sky wave

and reflected waves, with a node at the ground is assumed, then for a wavelength of 71·3 m one has:

One-quarter wavelength $= 58$ ft

Angle wave makes with vertical $= \cos^{-1} \dfrac{30}{58} = 60°$

There is an interesting detail regarding the accurate calculation of the pulsing rate of the transmitter. The machine producing the 500 Hz supply had 32 rotor inductors and 2 poles. The speed of the machine, which was 26·4 rev/s, was measured by means of a revolution counter. The counter was checked by means of a whistle which, in turn, had been calibrated against standard tuning forks.

An obvious disadvantage of the modulation method used in the Breit and Tuve experiments was that the quiescent period of the transmitter was only slightly greater than the transmission period and this led to poor range resolution and ambiguity. Breit suggested the use of an asymmetrical arrangement of the Abraham and Bloch type multivibrator. This was used in a later series of tests [35]. The multivibrator supplied about 300 pulse/s with a pulse width of some 200 μs. The pulses were amplified and keyed the intermediate power amplifier which consisted of two 250 W valves in parallel. The power amplifier stage of the Anacostia transmitter was a 20 kW valve. The transmitter was crystal controlled and operated on 4·015 MHz.

There is an account of simple experiments which were carried out successfully at Allahabad, India, between the 3rd December 1934 and the 15th February 1935 [36]. The frequency used was 4 MHz and the aerial was just a half-wave horizontal dipole. The interesting aspect of the experiments which achieved a vertical sounding of the ionosphere was the use of the same aerial for transmission and reception. The output from a small coil which was coupled to the tank circuit of the transmitter was fed into the receiver. The transmitter was based on a Hartley circuit and pulse modulation was achieved by injecting a 50 Hz voltage into the grid-filament circuit. A variable air capacitance of about 0·001 μF shunted by a 10 MΩ resistance was used in the grid circuit. The receiver consisted of two RF amplifiers followed by an anode bend triode rectifier, and one stage of low frequency amplification which fed the vertical plates of a cathode-ray oscillograph employing a thryatron time base.

3.9.2 Geodetic surveying

Geodesy today still uses the classical triangulation methods of a century ago. However, it is true to say that what could be broadly classified as electromagnetic distance measurement (EDM) techniques have not only refined the instrumentation but have also expanded and modified the methodology of the science [37]. EDM uses radar methods, laser beams and optical methods and special variants of commercial radio-navigation systems, and makes use of ground, airborne and artificial satellite platforms.

This is a convenient point to mention, if not indeed to stress, the fact that both the sources of inspiration and the methods of applied science transcend the convenient man-made departments of technology. Heinrich Löwy's United States Patent [38], which was filed in 1923, is mentioned in most historical accounts of radar. His apparatus could be cited as either the first proposed radio-wave geodetic instrument or the first proposed radio altimeter. It is interesting also because it used a pulse principle.

Löwy, an Austrian, regarded his proposed technique as an electric counterpart of H. L. Fizeau's toothed wheel experiments to measure the velocity of light, carried out between Montmartre and Suresnes, Paris, in 1849.

The patent proposes an oscillator feeding into a transmitting antenna. A separate antenna is used for reception and a coupling link between the two antenna is available for neutralisation*. The receiver consists of a detector, an indicator and a gaseous discharge tube which flashes on receipt of a signal.

The modulator is a sinusoidal wave oscillator with an LC tuned circuit. During one half of the oscillation cycle the supply circuit of the transmitter valve is closed by a positive potential from the LC circuit applied to an electronic switch, and likewise on the other half-cycle the supply circuit of the receiver valve is closed. The capacitance of the tuned circuit could be continuously varied between limits by a motor; the same motor also rotates a gaseous discharge indicator tube in synchronism, thus allowing maximum received signal to be related to the modulating signal wavelength and hence to distance, as explained below.

The time required for a 'pulse' of radio-frequency energy to travel to an object, distance H away, and back is

$$t = \frac{2H}{c}$$

where c is the velocity of propagation of radio waves. If one recollects that, while energy is being returned from the reflecting body, the wave train is continuing onwards then one can perceive that a modulation wavelength equal to $4H$ will give maximum reflected signal because then the reception of reflected energy begins immediately at the end of the transmission cycle.

An interesting example [39] of the use of radio waves for purely geodetic purposes was an extensive series of tests carried out in Russia between 1934 and 1936. These included, in 1934, some 900 measurements which were carried out in the North Caucasus between mountain peak and mountain peak and also between mountain peak and valley. In 1935 tests of a similar type were carried out over salt water at the Black Sea near Odessa, while in 1936 distances between islands some 44 km apart were measured.

The method used was to position a transmitting and receiving station at each terminal. One station transmitted on a frequency f_1 and this transmission was received by the other station which duly responded on a frequency, f_2, which was a convenient multiple of f_1. The two frequencies were then compared at the first station and phase difference was calculated. Ambiguity was resolved by changing the value of f_1 between known limits and noting the number of phase reversals that then took place.

3.9.3 Aircraft altimeters
The standard altimeter in aircraft is of the aneroid type and so it measures atmospheric pressure directly. Although the vagaries of the atmosphere might prompt one to infer that such altimeters might be totally unsuitable in aviation,

---

* Neutralisation refers to the technique of compensating for some unwanted coupling reactance in a circuit by introducing an equivalent opposite or neutralising reactance.

this, in fact, is not so. Adjustment for local air conditions in addition to proven flying procedures results in their being a very necessary cockpit instrument. However, since the very early days of flying, and certainly since the First World War, the search has gone on for an instrument capable of measuring, directly, the height of an aircraft above the ground. Today radio altimeters are used in the navigation of aircraft and they find their use principally as a terrain clearance indicator in *en route* flying and as a precise and accurate indicator of height in the final approach to landing.

At this point, in case of confusion arising from the use of the term 'radio altimeter', it should be noted that there is a difference between radio altimeters of the frequency-modulated continuous-wave type, used in terrain clearance devices, and the high-altitude pulse-type altimeters developed during the Second World War. For an introductory understanding of their role in air navigation, one might consult Smith [40] and Sandretto [41].

On the road of development which led to the FM–CW radio altimeters of the Second World War many devices were tried [42]. These may be categorised as sonic altimeters, capacity altimeters and radio altimeters.

Sound, or sonic, altimeters suffered from two basic disadvantages: the low speed of sound in relation to the speed of aircraft, and the amount of sonic

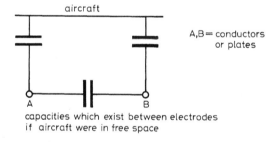

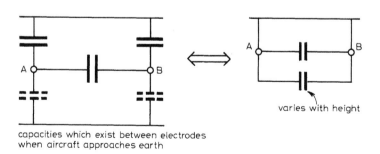

Fig. 3.5   *Capacitance altimeter*

power, and hence apparatus weight, required. In the United States the only device which came near to commercial realisation was one designed by Rice of the General Electric Company. Work started on it in 1929 and United Air Lines flight tested it in 1933 and 1934 but then discontinued its development. Compressed gas from the engine was stored under pressure and, every 2 s, actuated a whistle at 3000 Hz for a duration of 10 ms.

In the capacity altimeter two conductors mounted outside the aircraft fuselage each had an electrical capacity to the aircraft and a mutual capacitance, which was modified by proximity to the ground (see Fig. 3.5). A calculation by Sandretto [42] for a particular configuration gave a change of capacitance from 20·7 pF to 20·2 pF when the aircraft's altitude increased from 10 ft to 100 ft. No commercially available systems seem to have existed before the Second World War (Dr. Ross Gunn of the United States Navy carried out extensive development work on one from 1924 onwards).

Some early radio altimeters attempted to measure altitude by sending out a radio frequency signal and then attempting to measure the intensity of the signal returned from the ground. Because intensity is a function not only of altitude but also of terrain, and because of the existence of standing waves, the method is basically unsound.

We will consider in some detail three examples of pre-war altimeters. Two of them are of American and one is of Japanese origin.

3.9.3.1 Alexanderson altimeter: The altimeter was developed about 1928 by Alexanderson of the General Electric Company. The basic principle involves the use of a regenerative receiver which also acts as a transmitter. The reflected energy from the ground is received and its phase will alter the frequency of the device. The variation of frequency with altitude is illustrated in Fig. 3.6. The altimeter was usable at altitudes in excess of 3000 ft. A

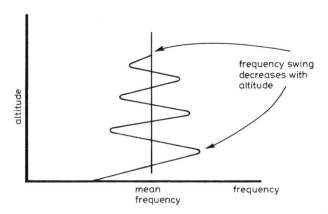

Fig. 3.6   *Behaviour of Alexanderson altimeter*

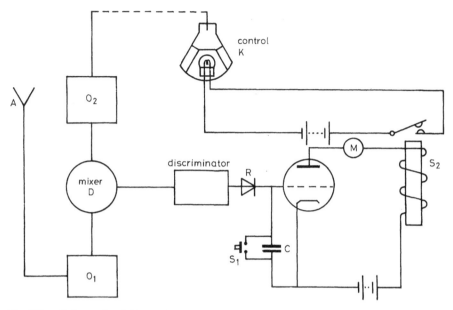

Fig. 3.7   *Schematic of Alexanderson altimeter*

schematic of the altimeter is shown in Fig. 3.7. Two oscillators, O_2 and O_1 (which served both as transmitter and as regenerative receiver), were used and their outputs were mixed in a detector D. The frequency of O_1 was varied by the reflected waves which reached the antenna A. O_2 was an oscillator providing a fixed frequency. It had three fixed frequencies selectable by control K. A slope detector type of frequency discriminator followed D and produced a current proportional to the beat frequency. This current charged C via rectifier R, the combination acting as a peak detector so that each time a high-frequency peak occurred C was charged and retained its charge; each succeeding peak added its charge to C. As charge on C increased, the plate current of triode increased and was recorded on M. There was normally no plate current because of a bias voltage applied to the grid. Meter M indicated altitude. When the current through M reached a predetermined value, the relay S_2 operated and a lamp lit. The pilot then closed switch S_1 to discharge the condenser and he also moved switch K to change the frequency of O_2. The switch also brought a glass of a different colour over the lamp. This obligation on the pilot to closely monitor the altimeter action and to operate switches at possibly a critical phase of flight must have weighed heavily against the use and acceptance of the instrument.

3.9.3.2 Western Electric radio altimeter: This altimeter [43] originated in work carried out by Professor W. L. Everitt of Ohio State University in 1928–1929 under the grant made by the Daniel Guggenheim Fund for the promotion of aeronautics. The principal obstacle towards a successful achievement of a working instrument at that time appears to have been the lack of a valve capable of operating at sufficiently high frequencies.

Some theoretical work was done on the project around 1930 both by Lloyd Espenschied of the American Telephone and Telegraph Company and by R. C. Newhouse who had been a student of Professor Everitt. United Airlines in 1937 negotiated with the Western Electric Company (and hence with Bell Telephone Laboratories) for the development of the altimeter. At that time, Bell Telephone Laboratories had succeeded in producing triodes which could give a stable output of between 5 W and 10 W at a frequency of approximately 500 MHz. Work culminated in the successful 'Weco' altimeter which was demonstrated by United Air Lines in the autumn of 1938. During the

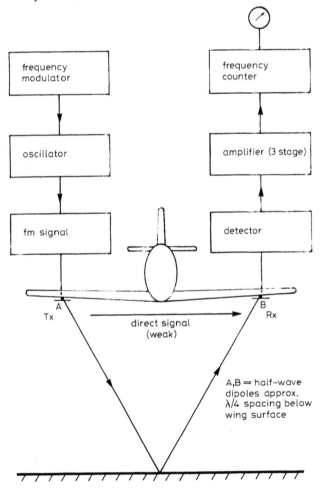

Fig. 3.8   *Western Electric FM radio altimeter*

development stage, a Bell Telephone Laboratories' aircraft flew about 100 flight tests over a period of 7 months.

The low-altitude radio altimeter, radio set AN/APN-1, which was produced during the war years by the Radio Corporation of America, was similar in principle to the Western Electric System.

Fig. 3.8 gives a rough schematic of the latter system. The frequency was varied at 60 Hz between 410 and 445 MHz. This frequency modulation was achieved by a synchronous motor driving the rotor of a small variable condenser which was located at the centre of one pair of Lecher lines. A schematic of the transmitter is shown in Fig. 3.9a. The total output was 10 W.

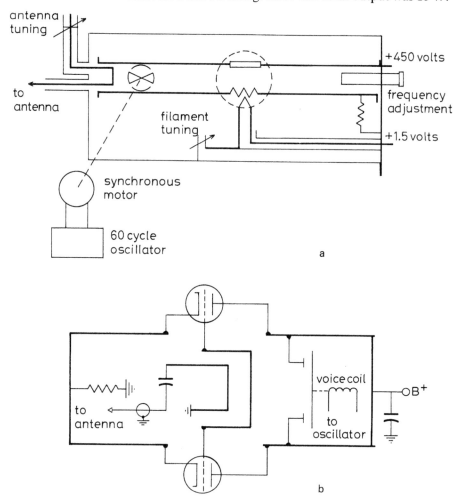

Fig. 3.9   Transmitters
a Western Electric radio altimeter
b RCA AN/APN-1 radio altimeter

The triode had interelectrode capacitances not exceeding 1·5 pF. In Fig. 3.9b is shown the schematic of the Radio Corporation AN/APN-1 transmitter.

We shall now examine the underlying principle of operation of the altimeter. Let Δf be the frequency difference between the highest and lowest frequency values and let this change in frequency occur linearly at F_m Hz. Then

$$\frac{df}{dt} = 2F_m \Delta f \tag{3.1}$$

A beat frequency, F_d, between the direct wave and the reflected wave occurs in the detector of the receiver. For a given height above the ground, H, the length of time required for the signal to travel to the ground and back is

$$t = \frac{2H}{c}$$

The corresponding frequency difference is

$$F_d = 4F_m \, \Delta f \frac{H}{c} \tag{3.2}$$

A glance at this relationship will show why high frequencies of the order of several hundred megahertz and large frequency deviations are required if realistic minimum heights are to be measured. The designer must also take into consideration that, for the complete system including the antenna, there will be limits on the total bandwidth allowable as a percentage of the centre frequency; for the order of deviation used this would have prohibited frequencies below about 100 MHz.

The resolution of the system is given by

$$\left|\frac{F_d}{H}\right| = \frac{4F_m \, \Delta f}{C} \tag{3.3}$$

If $F_m = 60$ Hz and $\Delta F = 25$ MHz, then

$$\left|\frac{F_d}{H}\right| = \frac{240 \times 25 \times 10^6}{3 \times 10^8} = 20 \text{ Hz/m}$$

Hence, at a height of 5000 ft, a frequency swing of some 30 kHz would take place in the signal entering the receiver.

The quotation that follows is from the Espenschied and Newhouse paper of January 1939 [43]. It describes the flight trials of the Western Electric altimeter and is of interest because it highlights the author's awareness of the role or possiblity or radar mapping.

An indication of the character of the surface over which the airplane is flying is given by the variations in the meter reading. A city usually causes rapid fluctuations of the order of fifty feet, depending of course upon the height and the spacing of the buildings. Cultivated farmland causes fluctuations of lower frequencies and amplitude. An isolated high object such as a skyscraper or a chimney is indicated only by a slight meter kick as the airplane passes over it, which may be noticed by the observer. If the airplane passes over only a few feet above the object and the top is large enough to contribute momentarily most of the echo signal received by the airplane, the indication is unmistakable and the correct distance to the object is indicated by the meter. For instance, the gas storge tank near the Chicago airport is an excellent object upon which to demonstrate the altimeter performance. The instrument is useful as a position indicator when approaching an airport on a course which crosses an obstruction of appreciable height and size since the moment of passage over the obstruction is clearly indicated. In fact, use as a position indicator may be one of the altimeter's most valuable applications.

3.9.3.3 Altimeter of Sadahiro Matsuo of Tohoku Imperial University, Japan:
Sadahiro Matsuo describes an apparatus [44] very similar to the Western Electric equipment. Flight trials were not carried out, but, instead, ground-based experiments were undertaken on the football field of Tohoku Imperial University. These tests were sufficient to prove that the altimeter was suitable for low altitude terrain clearance or, as the author put it, 'for blind-landing use in aeronautics'.*

Matsuo's paper, which had been published in Japanese in 1936, warrants consideration because it gives a concise summary of the then available options in absolute altimetry and the reasoning behind the choice of a frequency-modulated principle; and also because there is an interesting account of the apparatus used which included a Barkhausen–Kürz type of transmitter.

A schematic of the transmitting oscillator is shown in Fig. 3.10. Positive voltage was applied to the grid and negative voltage to the anode of the triode. The frequency variation was found to be approximately proportional to the total grid voltage. Frequency modulation was achieved by applying an alternating voltage, E_m, to the grid circuit. The ground was not used as 'earth' but, instead, a vertical metallic net of 7 cm square mesh was employed. The dimensions of this mesh were 10 m by 30 m. Altitude was simulated by moving the altimeter away from the artificial earth. No constructional details are offered, but calculated figures for its divergence angle indicate that the aerial was reasonably directive.

The receiver comprised a triode detector and a two stage RC coupled pentode amplifier, followed by a triode limiter stage and a two valve altitude

---

* It is not relevant to discuss here the maturing in thinking since that time on the whole concept of 'blind landing'. It must be pointed out, for correctness sake, that misconceptions did exist as to the ultimate potential of radio aids, and of that known as the Instrument Landing System in particular, in providing sufficient guidance to a pilot to effect a true 'blind landing'. It is sufficient to say that neither the availability of an accurate altimeter nor the provision of a very accurate and precise set of radiation patterns or beams will alone suffice.

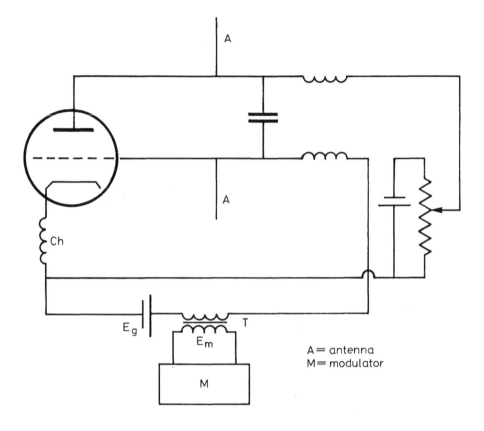

Fig. 3.10   *Schematic: transmitter of Matsuo's altimeter*

indicator circuit. The latter circuit included a milliameter which was the actual indicator of height.

A frequency of about 600 MHz was used with a total deviation of 38 MHz at a modulation frequency of 24 Hz. Altitudes of less than 4 m could be indicated, and for a transmitter power consumption of 3·9 W altitudes of 160 m or more were measurable.

3.10 Three British events

Three events which occurred in Britain in the late 1920s and early 1930s are now referred to. None of the events seems to have influenced in any way the ultimate development there of radar, but all of them are interesting nevertheless and are considered worthy of note.

3.10.1 O. F. Brown's memorandum

On the 4th August 1926, Mr O. F. Brown* submitted a memorandum [45] to the Anti-Aircraft Subcommittee of the Committee of Imperial Defence. It dealt with the application of the cathode-ray oscillograph to anti-aircraft research. (The term 'cathode-ray oscillograph' referred to a mounted tube with power supplies, but without time-base or amplifier circuits.) Brown appears to have been concerned with two things. He was conscious of the problems of anti-aircraft defence and he was also aware of the potential of the new reliable commercially available cathode-ray oscillograph, whose cathode life was over 100 hours as opposed to an average of about 10 hours for other commercial tubes [46, 47].

Brown suggested several applications of the oscillograph including a possible anti-aircraft one. It is worth giving an extended quotation from his memorandum. The key words in the quotation have been italicised.

> . . . the oscillograph may be of service for the examination of acoustic radiation from aircraft in respect of amplitude and wave form, and possibly could be applied in sound-locating devices, dependent on its use as indicator of phase difference. In this connection it may be stated that sensibly distortionless amplification in respect of amplitude and phase is obtainable with the Western Electric cathode-ray oscillograph using one stage of amplification, giving twentyfold magnification at all frequencies up to 10 000, and that phase distortion with two stages of amplification, giving a four-hundred-fold magnification, is usually negligible.
>
> It is believed that a cathode-ray direction-finder could be used at any radio-telegraphic frequency by the adoption of a supersonic heterodyne method of amplification. It is therefore possible that a method of location in azimuth could be based on the use of the cathode-ray direction-finder or the short wave radiation excited in the metal work of aircraft by magnetos or by *secondary excitation in a strong field emitted from a ground transmitter*. The location by observation on normal signalling has already been dealt with.
>
> If low frequency fields from the aircraft can be detected at any useful distance, then doubtless a certain amount of detection and identification data would be obtainable from examination of its wave form, while location by direction-finding would be exceptionally simple [45].

3.10.2 L. S. B. Alder

In 1928 L. S. B. Alder of the Signal School, Royal Navy, saw the possibility of using radio waves for the detection and ranging of ships. He filed provisional

* Department of Scientific and Industrial Research.

specification No. 6433/28 on the 1st March 1928 with Captain J. S. C. Salmond [48]. The specification relates to

> ... methods and means for the employment of reflection, scattering, or re-radiation of wireless waves by objects as a means of detecting the presence of such objects ... The invention may also be used as an aid to navigation or for detecting the approach of enemy craft by observation at known positions of the reflection or re-rdiation of wireless waves by an approaching ship or aircraft. The waves emitted are interrupted periodically or modulated in one of a number of ways [48].

The method proposed was FM–CW with the frequency modulation taking place in an approximately linear manner.

3.10.3 W. A. S. Butement and P. E. Pollard

In 1931 W. A. S. Butement and P. E. Pollard, two members of SEE (Signals Experimental Establishment), Woolwich Common (see Fig. 3.11), proposed a method of using radio waves for the location of ships [49, 50]. They carried out successful experiments using a Barkhausen–Kürz oscillator. They wrote a brief note to the Royal Engineer and Signals Board, and they also informed the War Office of their experiments. Pollard was in fact on secondment to Woolwich

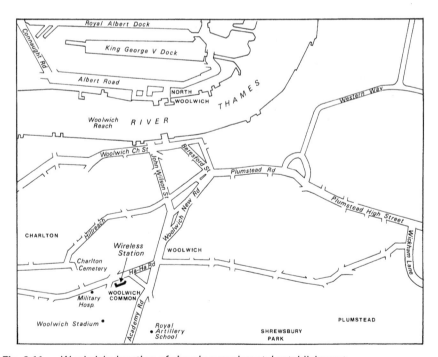

Fig. 3.11   *Woolwich: location of signals experimental establishment*

from the Air Defence Experimental Establishment at Biggin Hill. The technique used was again, as in Löwy's proposal, an electrical analogue of Fizeau's method for measuring the velocity of light.

Butement and Pollard used the shortest wavelength they could generate, which was 50 cm, and, as already mentioned, they used a Barkhausen–Kürz oscillator. This oscillator was in turn switched on and off by another oscillator operating at about 50 kHz, which energised a tuned circuit that was inserted either in the anode or in the grid circuits of the Barkhausen–Kürz oscillator. The whole arrangement acted as a sensitive super-regenerative receiver (see Appendix C). It was discovered that a separate valve for this medium frequency oscillator was not, in practice, required and that quenching* at 50 kHz occurred when only the tuned circuit was inserted. The 50 kHz oscillator was itself modulated at 1000 Hz so that an audible note could be available in the receiver. The quenched oscillator provided sufficient radiation to act as a transmitter of 50 cm waves. The Barkhausen–Kürz oscillator was placed at the focus of a cylindrical parabolic mirror 4 ft wide and 1½ ft deep and with this a sheet of metal 5 ft by 2 ft at a distance of about 100 yd was clearly picked up.

Set out below is a copy [51] of their proposal submitted on the 26th January 1931. This is followed by a copy of a memorandum compiled by both of them, after the war, which discusses their experiments of 1930.

Proposal – for a coastal defence apparatus – submitted in January, 1931, by Messrs W. A. S. Butement and P. E. Pollard, of the Signals Experimental Establishment, for record in the Inventions Book of the Royal Engineer Board

COASTAL DEFENCE APPARATUS

Apparatus to locate ships from the coast or other ship, under any conditions of visibility, or weather.

This apparatus depends on the reflection of Ultra Short Radio waves by conducting objects, e.g. a ship.

It will determine the range and traverse (i.e. the exact position) of the object practically instantaneously.

One station only is required and in one position.

Alternatively several stations may be employed and locations obtained by cross-bearings, as is usual in direction-finding practice.

The apparatus is reasonably certain to operate over a definite range, the magnitude of which is yet to be determined experimentally.

An ultra short-wave transmitter (of wavelength say 50 cms.) and a corresponding receiver, either or both having beam aerial systems, which may be rotated, are supplied with high tension derived from a medium

---

* The term used to describe the suppression of oscillations.

frequency oscillator (frequency of say 50,000 cycles) in such a manner that when the transmitter is radiating, the receiver is inoperative and vice-versa. Any other means of ensuring that the receiver is unaffected by its local transmitter may be employed.

The medium frequency oscillator may be itself modulated at a low frequency (say 1,000 cycles) so that an acoustic note may be obtained in the phones of the receiver if required.

On rotating the moving system no signals will be received until the beam is reflected by some conducting object. Thus the bearing of the object is directly determined.

One method of measuring this is as follows. If the time taken by the signal to perform this journey be a multiple of the period of the medium frequency oscillator, then the reflected signal will arrive when the receiver is inoperative; and no signal will be observed. If the frequency of the medium frequency oscillator is varied the signal in the receiver will pass through a series of maxima and minima. By measuring the frequencies at which any two successive maxima or minima occur a measure of the distance can be obtained directly.

It will be observed that as the high tension supply to the receiver is periodically discontinuous it will operate on the super-regenerative principle, which is desirable, in fact almost essential, for sensitive reception on ultra-short wavelengths.

In addition, low frequency amplification may be provided, sharply tuned to the low frequency modulation of the medium frequency oscillator. In this way low frequency amplification may be employed on a super-regenerative receiver resulting in increased sensitivity and freedom from interference. It must be pointed out that the low decrement of such a low frequency amplifier is in no way detrimental to the success of the apparatus.

A vibration galvanometer tuned to the low frequency or other means may be used to give visual or other indication of the maxima or minima.

It may be found desirable to provide for variation of the frequency of the ultra-short wave transmitter and receiver so that better reflection may be obtained. For example, if on the ship whose position is to be found there is a metal object of some definite length, it may be desirable that a multiple of half the wavelength of the ultra-short wave transmitter shall be equal to this length.

Preliminary work is necessary to measure the magnitude of the reflected wave under these conditions and hence the useful range of the apparatus.

Signed W. A. S. Butement
P. E. Pollard

After the war ended, Butement and Pollard wrote a short memorandum which reads:

1 *Historical:* It is not possible after so long a lapse of time to apportion credit for various points specifically to one or other of us but as the work was done by both of us in close collaboration it should be considered a joint effort.

As a result of discussions between ourselves we wrote a memorandum dated 26th January, 1931, proposing the use of radio for locating ships.

This memorandum was the first as far as we know to propose the use of a method of range measurement which may be described as the electrical analogue of Fizeau's method for measuring the velocity of light.

We were allowed to carry out experiments on a small scale in our own time.

We proposed to use short waves, about 50 cms., this being the shortest wavelength we could generate at that time.

2 *Circuit Details:* We tried various kinds of valves ranging from 'V.24', and its dull emitter equivalent, up to 250 watt valves. We found that nearly as much power could be obtained from a 'V.24' as from larger valves, one of the limiting factors being the temperature of the grid, which we ran at a bright red heat. After considerable tests, an experimental type of 'A.T.40' with 'top connected' grid and anode was found to be the most satisfactory. It was also found that only bright emitter valves were suitable for use in the Barkhausen-Kürz circuit.

The transmitter was a simple B.K. oscillator with the aerial situated at the focus of a cylindrical parabolic mirror about 4 ft. wide and 18 in. deep.

This transmitter was then used as the nucleus of a receiver. We proposed to quench this B.K. oscillator using a separate valve oscillating on a medium frequency, but found that quenching could be effected merely by adding in the anode or grid circuits of the B.K. oscillator, a circuit tuned to the medium frequency.

We arranged to modulate the quenched receiver at 1,000c/s approximately and used a low frequency amplifier to enhance the received signal. The quenched B.K. receiver-oscillator gave sufficient radiation to act as transmitter for preliminary experiments in detecting echoes from various targets.

3 *Experimental results:* We used a sheet of metal about 5 ft. × 2 ft. as a target and observed that the reading of the milliameter in the output circuit varied as the sheet was moved. We then put long leads with chokes on the headphones, so that the man carrying the sheet could listen to the variation of signal strength as he moved. Definite maxima were observed every half wavelength. The limit of space available to us was about 100 yds., at which distance a marked signal was obtained. Signals were also

obtained from masts, etc., about the Establishment and we recollect receiving a good signal from a mast about 100 yards away.

Looking back on this work in the light of our present knowledge of radar, we feel confident that our experiments, if we had been allowed to continue them and had used a bigger aerial would have shown that we could detect a destroyer at a distance of from ½ to 1 mile.

Signed W. A. S. Butement

P. E. Pollard

3.11 Patents

The experiments of Butement and Pollard towards the close of 1930 could have served as an arbitrary but fitting close to this chapter. The 1930s saw the emergence of radar in many nations. In Britain it was invented virtually by Government decree, while elsewhere the same clouds of war that had spurred on the Committee for the Scientific Survey of Air Defence in Britain provided the necessary stimulus for Government agencies and interested commercial concerns. However, there is an exception to this trend, and so, continuing this narrative, we see below that some significant or realisable radar-type propositions were put forward by individual researchers with no thought for any military application of their systems. Looking at some of these leads one to think that in the mid-thirties the development of radar may have been imminent anyway. An article 'Pictorial radio' which was written by C. D. Tuska and published in 1952 is one which covers quite comprehensively a variety of equipments and which lends weight to this view [52]. Tuska was Director of the Patent Department of RCA Laboratories. He divides his study into three periods: the Pioneer Period, starting with Hertz in 1887, the Hopeful Period and the Practical Period.

The items of particular relevance that are mentioned are:

(a) The demonstrations in the United States in 1934, before meetings of the Institute of Radio Engineers, by Irving Wolff and Ernest Linder. They used a split-anode magnetron transmitter, a parabolic reflector and wavelengths of about 9 cm. In 1937 Wolff extended the system and used pulses of less than 1 μs.

(b) William A. Tolson's United States Patent (Nos. 2,130,913) filed in 1934 and Howard I. Becker's United States Patent (Nos. 2,151,549) filed in 1936 for an aircraft landing system. Both patents required the co-operation of fixed low-power transmitters on the ground.

(c) Chester W. Rice's United States Patent (Nos. 2,412,631) of 1936 and Gabriel Moulineer's French Patent (Nos. 813,404), also of 1936, both of which aimed at determining not only the distance but also the azimuth, elevation and speed of objects. Further reference will be made to Rice's work in Section 3.12.

(*d*) A patent filed in 1937 by a Frenchman, Paul Gloess (Australian patent Nos. 108,556), for an airborne device which would indicate on a cathode-ray tube display the contours of the land below and in front of the aircraft. An interesting point concerning the proposed device was the short duty-cycle of 5 μs (PRF of 200 kHz) and a pulse width of only 20 ns.

Also, it is of interest that, in 1933, development work was going on in the United States by both the Navigational Instrument Section of the Air Corps and the Signal Corps Aircraft Laboratory into the question of an aircraft microwave anti-collision device [53]. This work was being done purely from an aircraft navigational or air safety point of view with no thought to general radar development. No great priority was given to the project and difficulties over Air Corps and Signal Corps responsibilities in aircraft electronics did not help at the time. Like the 'blind landing' goal of the Lorenz Beam development, true collision avoidance radar was, in retrospect, a rather ambitious project; collision prediction necessitates not only accurate information on positions and velocities, but also acccurate information on accelerations.

Finally, mention must be made of the patents of Gustav Guanella of Brown, Boveri and Company in Switzerland. Among some one hundred patents attributed to him [54] is one titled 'Distance determining system' filed in Switzerland on 26th September, 1938 and in the United States on 27th May, 1939 (US Patent Nos. 2,253,975). This describes a CW type radar but one that, in concept, was at least a decade ahead of its time. It incorporated all the characteristics of a stored-reference speed-spectrum (SR–SS) system [54].

The above have been concerned with developments and events of the 1930s, knowledge of which was not readily available to the general public. The next section, which closes this chapter, will include some occurrences of the same period which were openly recorded in the literature.

3.12 Mystery rays

Available evidence indicates that the development of radar in each of the relevant countries was undertaken in isolation and under quite strict security. The veil of security appears to have been unintentionally lifted on a few occasions.

The 'Oslo Report' [55, 56] is in a category of its own and seems to have been a very valuable piece of intelligence, which made known to the British in November, 1939, just after the outbreak of the war, much of Germany's intended weaponry including its development of radar.

In the days of pre-war tension during the 1930s, there appears to have been a certain openness among the public, and, indeed, among military authorities, to the possibility of 'death rays' and 'mystery rays'. In Britain we find [57] a letter of the 2nd March 1933 from the President of the Royal Engineers and Signals Board to a gentleman at Maidenhead which begins as follows:

> With reference to an interview which Colonel Evans and Captain Gibson had with you at Maidenhead on 17th Feb. 1933, to discuss your electrical ray, we have to inform you . . .

At this stage of the proceedings the authorities were unavailingly seeking some concrete information and evidence and, understandably, running out of patience with the inventor who claimed to have produced an effective 'death ray'. Some ten days later Colonel C. H. Silvester Evans wrote:

> He claims to have destroyed a beech tree with it from a distance of about 150 yards.

A felled tree was indeed produced as evidence, as was a closed box, together with some witnesses, but the inspection team were unimpressed. After a lapse of some three years, on the 23rd January 1940, the same gentleman resubmitted his invention for consideration, but this time he received but scant attention.

A more promising contact in the area of detection was made [58] when, on the 1st March 1935, Colonel J. P. G. Worledge wrote back to London from Vienna as follows:

> Herr Kolbl, from Eggenburg, South Austria, has made an appointment with the military attache, Major Beaufield and the latter asked Mr. Rowe and I to attend the interview.

The subject under discussion was a microwave detector. Herr Kolbl claimed that his equipment had detected the Vienna–Munich aeroplane at a height of 500 m as soon as it had come in sight over a hill at a distance of about 5 km. The wavelength of operation of the apparatus was given as 0·05 mm which would place it in the far infra-red part of the spectrum. An outline description of the transmitting equipment was given:

> The apparatus so far developed consists of a mirror oscillator of the shock-excited spark type, whose wavelength is determined by the physical dimensions of a dipole and spark gap*.

The inventor claimed that he had discovered the following rather qualitative properties of some common substances.

Aluminium	good
Copper	good
Iron	bad
Lead paints	good

---

* By using a small Hertzian oscillator made of two tungsten cylinders varying in length from a few millimetres to one-fifth of a millimetre and separated by a gap of the order of 0·01 millimetre, E. F. Nichols and J. D. Tear [59] in 1923 succeeded in producing electromagnetic waves of 1·8 millimetres in length.

No further information was forthcoming from the file which closed on the 3rd May 1935.

In 1935 an intriguing article complete with photographs appeared in the magazine *Radio Craft* [60]. The title of the article was 'The new "mystery ray"' and beneath this title was the caption:

> This new ray – developed simultaneously by three World Powers – depends upon the beam-effect and reflection of 'centimeter' length or ultra-ultra-short waves to 'spot' enemy planes, ships, etc. Fortunately, there are valuable commercial applications for this new tool of warfare and destruction.

The three World Powers referred to were Germany, Italy and the United States. Brief but quite accurate reference is made to the United States Signal Corps tests in New Jersey and to experiments being carried out in Germany at the Telefunken Laboratories. Correct mention is made of the use of the split-plate magnetron. A diagram of one of these magnetrons is shown together with a photograph of one beside a match box to illustrate its size. One intriguing reference to a demonstration is given:

> These circumstances make the demonstration of the German mystery rays recently conducted in a suburb of Berlin especially interesting. These demonstrations disclosed to the invited newspapermen, and foreign military attaches, contrary to all expectations – were not units of tremendous dimensions, but tiny devices about as large as a normal match box!.

In June 1936 the *Wireless World* published an article [61] describing the obstacle detection equipment installed on the French liner *Normandie*.* The account beings as follows:

> When the wireless equipment of *S.S. Normandie* was first described, mention was made at the time of a special safety device which had been installed, which enabled the vessel to discover the presence of an obstacle in its path, such as an iceberg or another vessel which, by reason of fog or other obstruction to vision, might otherwise not be detected.
>
> No particulars were disclosed as to the nature of this device, and only a bare reference to it could be made in The *Wireless World* of November 8th 1935. We are now able, through the courtesy of the Société Française Radio-Electrique, who were responsible for the wireless equipment of the *Normandie,* to give an account of this 'feeler' equipment which makes use of the properties of microwaves in a novel manner, and is known as the S.F.R. obstacle detector.

---

* The description includes a photograph, taken in New York harbour, of the transmitter and receiver antenna mountings. (See p. 44).

The account gives a reasonably detailed and an accurate description of the apparatus: 16 centimeter wavelength; continuously emitted signal modulated by 800 Hz [62]; retarded field valve* (which was the UC16) placed at the focal point of a parabolic reflector of 0·75 m aperture and used for both the transmitter and the receiver. The signal, when received back from the target, was detected and amplified and could be received by earphones or displayed on a cathode-ray tube.

The above brief description of the *Normandie* installation is, strictly speaking, an integral part of the story of French radar. It is included here because it is an explicit description, made available to the public in 1936, of a working radar equipment. Likewise, the next item, which concludes this section, could, strictly speaking, be regarded as belonging to the early history of United States radar, but it too was released to the general public in 1936.

Chester W. Rice, of the General Electric Company, introduced his paper 'Transmission and reception of centimeter radio waves' [63] as follows:

> This article will outline a new system of transmitting and receiving centimeter radio waves – that is, wavelengths from 1 to 10 centimeters. It should be considered merely as a progress report in the very complex and, as yet, little understood field of electronic oscillations. In all pioneer fields, we have to get along as best we can without a complete theory or accurate quantitative measuring devices.

The paper describes tests carried out at the General Electric Research Laboratory at Schenectady (New York). A wavelength of approximately 4·8 cm was used. The transmitting valve, which was modulated in turn by tones, voice and music, was a single anode magnetron which effected an actual radiated power of about 3 W. The principal receiver valve was a specially designed triode which was made to oscillate in the Barkhausen–Kürz mode (see Appendix B) and the receiver was judged to be functioning as a self-quenching and super-regenerative detector. Detection of stationary and of moving objects was achieved as part of the test and the usual beat method was used. A small biplane was detected at a distance of one mile. An interesting part of the article is a list of suggested applications of what is referred to as 'radio optics'. These were listed as follows:

1 Point-to-point beam communication.
2 Wide side-band communication over a chain of automatically repeating beam stations spaced 15 to 20 miles apart. This might be termed a *radiation transmission line*. This type of line would appear to be a practical way of distributing television.
3 *Foglight* for navigational purposes.
4 Airplane landing beams and direction markers.
5 All-weather airway beacons.

* See Appendix B.

6 Radio-beam protection of drawbridges, harbours, etc.
7 *Doppler detection* of moving objects including relative line-of-sight velocity and distance.
8 Airplane ground speed indicator using Doppler detection.
9 Radio searchlight using modulation.
10 Radio-echo altimeter.
11 Radio-echo locator for navigation.

References

1 ESPENSCHIED, L., and NEWHOUSE, R. C.: 'A terrain clearance indicator', *Bell Syst. Tech. J.*, 1939, **18**, pp. 222–234

2 Bundespatentamt, München: DRP Nr. 165, 546. Klasse 21/g Gruppe 30/10, 30.4.1904. Hülsmeyer's British patents were: 13 170 (1904), 25 608 (1904), 8 511 (1905)

3 DAHL, A.: 'Radartechnik seit mehr als 60 Jahren', *Ortung und Navigation*, 1964, **2**, pp. 29–34

4 VOLLMAR, Fritz: 'Das "Telemobiloscop" von Christian Hülsmeyer', *Funkschau*, 1974, **12**, pp. 1402–1404

5 V. WEIHER, Sigfrid: 'Christian Hülsmeyers "Telemobiloskop" ein Vorlaufer der RADAR-Ortung', *Nachrtech. Z.*, 1979, **32**, pp. 242–244

6 REUTER, Frank: 'Funkmess' (Westdeutscher Verlag, Opladen, 1971)

7 PHILIPS, V. J.: 'The telemobiloscope; an Edwardian radar', *Wireless World.*, 1978, **84**, pp. 68–70

8 Bundespatentamt, München: DRP Nr. 169, 154. Klasse 21/g Gruppe 30/10, 11.11.1904

9 NICHOLLS, Peter (General ed.): 'The encyclopedia of science fiction' (Granada, London, 1979)

10 GERNSBACK, Hugo: 'Ralph 124C 41+' (Frederick Fell, New York, 1950), 2nd edn

11 LIMANN, Otto: 'Wer war de Zweite? Zur Geschichte der Funkmeßtechnik', *Funkschau*, 1955, **4**, pp. 61–62

12 SECOR, H. Wingfield: 'Tesla's views on electricity and the war', *Electl Expr*, 1917, **5**, No. 4, pp. 229, 230, 270

13 Private notes of Pierre David. Copy given to Professor B. K. P. Scaife of Trinity College, Dublin

14 MARCONI, Guglielmo: 'Radio telegraphy', *Proc. Inst. Radio Engrs.*, 1922, **10**, pp. 215–238. See p. 228

15 TERRETT, Dulany: 'The Signal corps: the emergency' (Office of the Chief of Military History, United States Army, Washington, DC, 1956), p. 20

16 WITTS, Arthur T: 'The superheterodyne receiver' (Sir Isaac Pitman and Sons Ltd., London, 1946), p. 7 *et seq.*

17 JONES, H. A.: 'The war in the air, vol. V' (The Clarendon Press, Oxford, 1935), p. 73 *et seq.*

18 GUERLAC, H. E.: 'Radar in World War II' (unpublished history of Division 14 of the National Defense Research Committee, 1947). Microfilm copy available from Photoduplication Service, Library of Congress, Washington, DC

19 Memorandum: 'Japanese in Burma Air Warning System' (14th August 1944), pp. 26, 27 of File. Reel No. 8125 (File 820.635–1, 12th March 1945). Available: Chief of Circulation, The Albert F. Simpson Historical Research Center, USAF, HOA, Maxwell AFB, AL36112, USA

20 DAVIS, Harry H: 'History of the Signal corps development of US Army Radar Equipment Part I', p. 17 *et seq.*, Historical Section Field Office, Office of the Chief Signal Officer, New York, (1944). Microfilm copy available from Photoduplication Service, Library of Congress, Washington DC

21   JONES, R. V.: 'Some turning-points in infra-red history', *The Radio and Electronic Engineer*, 1972, **42**, pp. 117–126

22   Reference 15, p. 39

23   BENECKE Th., and QUICK, A. W.: 'History of German guided missiles development' (Verlag E. Applehans and Company, Brunswick, Germany, 1957), pp. 201–203

24   MARCONI, Guglielmo: 'Radio telegraphy', *Proc. Inst. Radio Engrs.*, 1922, **10**, pp. 215–238

25   ALLISON, David K: 'The origin of radar at the Naval Research Laboratory: a case study of mission-oriented research and development', a doctoral dissertation, University Microfilm International, Ann Arbor, Michigan, 1980. This dissertation has been revised and published as follows:

ALLISON, David K: 'New eye for the navy: the origin of radar at the Naval Research Laboratory' (NRL Report 8466), Naval Research Laboratory, Washington DC, 1981

26   GUERLAC, H. E.; 'Radar in World War II' (unpublished history of Division 14 of the National Defense Research Committee, 1947), Chap. A-III, Microfilm copy available from Photoduplication Service, Library of Congress, Washington DC, USA

27   This extract from Taylor's memorandum is taken from Reference 25 which also gives details of the primary source and other relevant information. Reference 26 is also informative on this matter.

28   KENNELLY, A. E.: 'On the elevation of the electrically-conducting strata of the earth's atmosphere', *Elec. World and Engr*, 1902, **39**, p. 473

29   HEAVISIDE, O.: 'Telegraphy I, Theory', *Encyclopaedia Britannica*, 1902, **33**, p. 215

30   APPLETON, E. V., and BARNETT, M. A. F.: 'On some direct evidence for downward atmospheric reflection of electric rays', *Proc. R. Soc.*, 1925, **109**, pp. 621–641

31   BREIT, G., and TUVE, M. A.: 'A test of the existence of the conducting layer', *Phys. Rev.*, 1926, **28**, pp. 554–575

32   TAYLOR, A. Hoyt, and HULBURT, E. O.: 'The propagation of radio waves over the earth', *Phys. Rev.*, 1926, **27**, pp. 189–215

33   GUERLAC, Henry: 'The radio background of radar', *J. Franklin Inst.*, 1950, **250**, pp. 285–308

34   TUSKA, C. D.: 'Historical notes on the determination of distance by timed radio waves', *J. Franklin Inst.*, 1944, **237**, pp. 83–102, has useful references

35   TUVE, M. A., and DAHL, O.: 'A transmitter modulating device for the study of the Kennelly-Heaviside layer by the echo method', *Proc. IEE*, 1928, **16**, pp. 794–798

36   BAJPAI, R. R.: 'Recording the ionospheric echoes at the transmitter', *Proc. Natn. Acad. Sci. India*, 1936, **6**, pp. 40–48

37   BOMFORD, G.: 'Geodesy' (Clarendon Press, Oxford, 1971), 3rd edn

38   LÖWY, Heinrich: 'Means for electric proof and measuring of the distance of electrically-conductive bodies', US Patent issued 18th May 1926; Filed 17th July 1923.

39   GUERLAC, Henry: 'The radio background to radar', *J. Franklin Inst.*, 1950, **250**, pp. 285–308; see p. 297, see also Reference 26, p. 34

40   SMITH, R. A.: 'Radio aids to navigation' (the University Press, Cambridge, 1947), Chapter 11

41   SANDRETTO, Peter C.: 'Electronic avigation engineering' (International Telephone and Telegraph Corporation, New York, 1958), Chapter 8

42   SANDRETTO, Peter C.: 'Principles of aeronautical radio engineering' (McGraw-Hill Book Company, New York, 1942). In 1933 in France a system by Dubois-Labourer of CEMA (Constructions Électro-méchanique d-Asnières) was available as was also a system by Florrison of SCAM (Société de Condensation et d'Applications Méchanique, of Paris). In Germany, the 'Behnolot' system, developed as early as 1924, was available in 1935. Its transmitter consisted of a pistol fired at intervals

43   ESPENSHIED, Lloyd, and NEWHOUSE, R. C.: 'A terrain clearance indicator', *Bell Syst. Tech. J.*, 1939, **18**, pp. 222–234

44 MATSUO, Sadahiro: 'A direct-reading radio wave-reflection-type absolute altimeter for aeronautics', *Proc. Inst. Radio Engrs*, 1938, **26**, pp. 848–858

45 AVIA 10/349: Public Record Office, Kew, London, UK

46 WATSON-WATT, R. A., HERD, J. F., and BAINBRIDGE-BELL, L. H.: 'Applications of the cathode ray oscillograph in radio research' (HMSO, London, 1933)

47 JOHNSON, J. B.: 'The cathode ray oscillograph', *J. Franklin Inst.*, 1931, **212**, pp. 687–717. (Some characteristics of the 224 tube were: Argon gas at 0·01 mm pressure was used. This assisted focusing, with focusing occurring at a beam current of about 20 µA. Deflection sensitivity: 1 mm/V. Anode potential: 300 V. Screen material: Mixture of zinc orthosilicate and calcium tungstate; the former produced a green light of high visibility and the latter a blue light of high photographic activity).

48 RATSEY, O. L.: 'As we were: fifty years of ASWE history 1896–1946' (Admiralty Surface Weapons Establishment, Portsdown, 1974), p. 135. Available Naval Historical Library, Empress State Building, London, UK

49 SAYER, A. P.: 'Army radar' (The War Office, London, 1950)

50 Record of Scientific Work undertaken by BUTEMENT, W. A. S., C.B.E., and submitted to Adelaide University for the degree of Doctor of Science. March 1960

51 Extracted with some editing from Reference 50

52 TUSKA, C. D.: 'Historical notes on the determination of distance by timed radio waves', *J. Franklin Inst.*, 1944, **237**, pp. 1–20 and 83–102

53 TERRETT, Dulany: 'The Signal corps: the emergency' (Office of the Chief of Military History, United States Army, Washington, DC, 1956), pp. 91–92

54 SCHOLTZ, Robert A.: 'The origins of spread-spectrum communications', *IEEE Trans.*, 1982, **COM. 30**, pp. 822–854

55 JONES, R. V.: 'Most secret war' (Hamish Hamilton, London, 1978), Chapter 8

56 HINSLEY, F. H. (with THOMAS, E. E., RANSOM, C. F. G., and KNIGHT, R. C.): 'British Intelligence in the Second World War, Volume 1' (Her Majesty's Stationery Office, London, 1979), Appendix 5

57 AVIA 7/2818: Public Record Office, Kew, London, UK

58 AVIA 7/2825: Public Record Office, Kew, London, UK

59 NICHOLS, E. F., and TEAR, J. D.: 'Joining the infra-red and electric wave spectra', *Proc. Natn Acad. Sci.*, 1935, **9**, pp. 267, 299

60 'The new "mystery ray"', *Radio Craft*, 1935, **7**, pp. 267, 299

61 '"Feelers" for ships', *Wireless World.*, 1936, **39**, pp. 623, 624

62 The actual article gives a modulating frequency of 7500 Hz. This may indeed have been the case, but all other accounts give the more likely frequency of 800 Hz

63 RICE, Chester W.: 'Transmission and reception of centimeter radio waves', *Gen. Elect. Rev.*, 1936, **39**, pp. 363–369

Chapter 4

The début of radar

4.1 Beginning of radar in Great Britain

The story of the emergence of radar in Great Britain is told in some detail in Chapter 5. The account confines itself to the pre-cavity magnetron era and concentrates on what were until mid-1940 the main streams of development, namely the decametric Chain Home system and the principal 200 MHz sets, including the early airborne AI (aircraft interception) and ASV (aircraft to surface vessel) equipments. The emergence of the resonant-cavity magnetron in Britain in 1940 and the visit of the Tizard Mission to the United States opened the way for the development of effective microwave radar in the two countries. It is befitting to record the background to the Tizard Mission and also to bring into focus some important occurrences in radar in Britain and the Commonwealth which are not treated in Chapter 5. The circumstances relating to the evolution of the high-power cavity magnetron are outlined in Chapter 7.

Radar emerged in Great Britain in the mid-1930s because of the imminence of war in which the country, and particularly London, would be vulnerable to air attack. The summer Air Exercises of 1934 underlined the fact that Britain was virtually defenceless against air assault. A realisation of this by Air Force officers, by Air Ministry Civil Servants and by politicians, including Winston Churchill, generated mounting pressures which quickly led to positive action. This was what Ronald Clark, in his biography of Sir Henry Tizard, referred to as 'the political birth of radar' [1].

The sequence of events was briefly as follows. H. E. Wimperis had, in 1925, been made the first Director of Scientific Research at the Air Ministry and, in the same year, A. P. Rowe had become his assistant. Sometime in 1934, Rowe became acutely aware of the possibility of war and took it upon himself to study all the files on air defence, 53 in all, that he could find. In no way reassured by what he saw in the files, he wrote to his Chief, H. E. Wimperis, expressing his fears of what would happen in the event of war. In early October 1934, H. E. Wimperis wrote to Professor A. V. Hill, who was Foulerton

Research Professor of the Royal Society at University College, London, requesting a meeting with him. Wimperis wished to discuss the question of radiant energy or 'death rays' as a means of anti-aircraft defence. They lunched at the Athenæum on the 15th October, and, on the 12th November, Wimperis drafted a note which was sent to Lord Londonderry, the Secretary of State for Air, to Air Marshal Sir Hugh Dowding, the Air Member for Research and Development, to Air Marshal Sir Edward Ellington, Chief of Air Staff and to Sir Christopher Bullock, the Secretary of the Air Ministry. In this note he proposed the setting up of a committee which would carry out a scientific survey of what was possible in methods of air defence. He proposed in addition that Henry Tizard, the then chairman of the Aeronautical Research Committee, be considered as chairman of the proposed committee, and that Professor A. V. Hill and Professor P. M. S. Blackett, physicist, be considered as members.

The proposal of Wimperis was sanctioned, and Tizard, Hill and Blackett accepted an invitation from the Air Ministry to serve on the new committee. On the 28th January 1935, the first meeting of the 'Committee for the Scientific Survey of Air Defence' took place at the Air Ministry [2]. Present were the Chairman, H. T. Tizard, Professor P. M. S. Blackett, Professor A. V. Hill, H. E. Wimperis and the Secretary, A. P. Rowe.

It is fitting to give some details concerning each member of the committee as the particular composition of the comittee, which made it equally at home in the world of the scientist, the world of the civil servant and the world of the Services, was a major factor in its effectiveness.

Henry Thomas Tizard was born in Gillingham, Kent, in 1885. He attended Magdalen College, Oxford, where he majored in chemistry and, having graduated with a First in 1908, he did one year's postgraduate work under Nernst in Berlin University from 1908 to 1909. He obtained a Fellowship and Lectureship in Oriel College and, when the First World War broke out, he joined the Army and received a commission in the Royal Garrison Artillery. He transferred in 1915 to the Royal Flying Corps as an Assistant Equipment Officer. He learned to fly and was appointed Scientific Officer in charge of a new airfield at Martlesham Heath, where he developed scientific procedures for flight testing of aircraft. He was promoted to an appointment as Assistant Controller of Equipments and Research in the Air Ministry and in 1919 he was demobilised with the rank of Lieutenant-Colonel. Immediately after demobilisation, he returned to Oxford and was appointed University Reader in Thermodynamics. He became involved with the Department of Scientific and Industrial Research and was its permanent Secretary from 1927 to 1929. He was made a Fellow of the Royal Society in 1926 and in 1927 he became Rector of the Imperial College of Science and Technology, London, which post he held until 1942. In 1933 he became Chairman of the Aeronautical Research Committee. He died in 1959.

Archibald Vivian Hill, the physiologist, was born in Bristol in 1886. He was a member of Trinity College, Cambridge, and obtained the degrees of M.A. and Sc.D. He joined the Army in the First World War and was Director of the Anti-Aircraft Experimental Section, Munitions Inventions Department. He investigated several urgent problems in anti-aircraft gunnery and had available a distinguished group of mathematicians and physicists referred to as 'Hill's brigands'. He was Professor of Physiology, Manchester University, from 1920 to 1923 and Professor of Physiology, University College London from 1923 to 1925. He was made a Fellow of the Royal Society in 1918 and, with Otto Meyerhof, was the recipient in 1922 of the Nobel Prize in Physiology and Medicine. In 1926 he was made Foulerton Research Professor of the Royal Society at University College London. He died in 1977.

Patrick M. S. Blackett was born in London in 1897. A member of Magdalene College, Cambridge, he obtained his M.A. in physics, and then served as a Naval Officer before and during the First World War. He carried out research in Nuclear and Atomic Physics with Rutherford. From 1933 to 1937 he was Professor of Physics at Birbeck College, London, and in 1937 he became Professor of Physics at the University of Manchester. After the war, in 1948, he received the Nobel Prize in Physics. He died in 1974.

Henry Egerton Wimperis, who was born in 1876, received his university education at Gonville and Caius College, Cambridge, completing the Mechanical Science Tripos. He worked for a while with Southern Railways and with Armstrong Whitworth. He was inventor of the Wimperis acceler-ometer in 1909, of the gyroturn indicator in 1910 and of the course setting sight for aircraft in 1917. Between 1915 and 1925 he served in the Royal Naval Air Service and the Royal Air Force. He was Director of Scientific Research, Air Ministry, from 1925 to 1937 and President of the Royal Aeronautical Society from 1936 to 1938. He died in 1965.

Albert Percival Rowe was born in March 1898 at Bromyard, Hereford. He was trained as a physicist and associated with defence science from 1922 onwards. When he was made Secretary of the Committee, he was Personal Assistant to H. E. Wimperis at the Air Ministry. He was later, in 1938, to succeed Watson-Watt as Chief Superintendent Telecommunications Research Establishment. He is credited with having originated operational research. After the war, in 1946, he became Deputy Controller of Research and Development at the Admiralty and, from 1948 to 1958, he was Vice Chancellor of the University of Adelaide, Australia. He died in 1976.

Just before the first meeting of the Committee, H. E. Wimperis, who was a member of the Radio Research Board of the DSIR (Department of Scientific and Industrial Research), contacted Robert Watson-Watt, Superintendent of the Radio Research Station, Slough, and had an informal discussion with him on the possibility of 'death rays'. The discussion took place at the Air Ministry on the 18th January and as a result Watson-Watt* and his assistant, Arnold F.

---

* Robert Alexander Watson Watt (1892–1973): when he was knighted in 1942, he changed his surname to 'Watson-Watt'.

Wilkins, calculated that a 'death ray' was not feasible, but that it should be possible to detect the presence of an aircraft by irradiating it with radio waves of appropriate wavelength.

There rapidly took place, as recounted in Chapter 5, the successful testing of the Watson-Watt–Wilkins calculations in what is known as the Daventry Experiment, and, by the 31st May 1935, a simple radar transmitter and receiver were working at Orfordness. From then onwards, steered by Robert Watson-Watt, there arose the cornerstone of British radar effort, the early warning Chain Home system. This protected, at first, the Thames Estuary, but was extended to cover the whole coastline. With the purpose of ensuring that incoming low-flying aircraft were detected and that one's own fighter aircraft made contact with enemy bombers, there grew naturally from the Chain Home system the CHL (Chain Home Low), the GCI (Ground Control of Interception) and AI (Aircraft Interception) systems which are discussed in Chapter 5.

Outside of this main stream of advancement, important work was carried out in gun laying radar for the Army and in long range and fire control sets by the Signal School, Royal Naval Barracks, Portsmouth, for the Navy. Between July 1935* and August 1939, when staff were transferred to the control of the Ministry of Supply and external assistance brought in, the development work on radar was carried out entirely by War Department scientists and technicians who constituted an Army Cell at Bawdsey. Among these were Dr E. T. Paris, P. E. Pollard, H. S. Young, C. S. Slow and W. S. Butement. The MB (mobile base) set, an early warning equipment, was developed and initial tests and modifications completed before its further development was handed over to the Air Ministry. This set, which was initially designed for 23 MHz working, operated at about 40 MHz and was originally intended for use in overseas states and mobile theatres of war where the overall responsibility for defence would rest with the Army. As the MRU (Mobile Radio Unit) or AMES† Type 4, it gave excellent service throughout the war, particularly in the North African campaigns. The sets most associated with the Army group were the series of Gun Laying (GL) sets.

The objective in designing the first Gun Laying radar, GL1, was to provide some degree of early warning information to enable searchlights and predictors to be 'laid' on targets. In the pre-war, and indeed early war years, aircraft speeds for bombers were slow so it was quite in order to consider the radar as an early warning device for visual aids [4]. GL1 had a high accuracy in range, no great accuracy in bearing and gave no measurement of angle of elevation. It operated at 50 MHz with a transmitter peak pulse power of 50 kW. The range of the target was read continuously, but bearings were obtained only intermittently, about once per half-minute. The DF system used [4] consisted of a pair of horizontal half-wave dipoles, spaced 1λ apart between

---

* When Colonel J. P. G. Worledge submitted his report on the 'The proposed radio method of aeroplane detection and its prospects' [3].

† AMES: Air Ministry Experimental Set.

centres and in phase-opposition so that an extremely sharp minimum was obtained in the broadside direction, the direction of the incoming signal (resonant reflectors were placed behind each dipole). Contracts for the production of the GL1 were placed with Metropolitan-Vickers and A. C. Cossor, the former to supply the transmitter and the latter the receiver and DF system. The Bawdsey staff, who had designed the first experimental GL1 set, co-operated closely in the development work with these firms. 59 equipments had been delivered by the end of 1939. The set was used during the night raids on London in 1940. The usefulness of the equipment in action may now be a matter of opinion, but the following figures* given by Brigadier Sayer [6] show an improvement in 1941 when height-finding facilities and other advancements became available.

September, October 1940:	
Rounds fired	260 000
Aircraft destroyed	14
Rounds per bird	18 500
Year 1941:	
Rounds per bird	4 100

The GL2 incorporated height-finding facilities, and also a method of magnifying the range display so as to improve accuracy when following targets. The first prototypes were delivered in the summer of 1940. The method of elevation finding was similar to that used in the Chain Home system (see Section 5.6.5) where the signals received by two antennae at different heights above an assumed perfectly conducting earth were compared in a goniometer. An important modification was made to the GL1 set by L. H. Bedford of A. C. Cossor Ltd, in the form of an EF (elevation finding) attachment. Similar to the height-finding antennae for the GL2, the attachment, which was produced and ready for field use within 6 weeks, enabled GL1 sets to be afforded a height-finding facility before the GL2 could be introduced into service. Some 410 of the EF attachments were produced.

The GL2 was probably as accurate a gun laying device as was possible at the wavelengths used. The next stage was a move to microwave frequencies and this occurred in the GL3, which still, however, required manual following of targets; automatic following was experimented with from 1942 onwards and in January 1943 three experimental models (AF1) were being made [7]. The AF1 did not see service until after the war, when it was designated the Radar A.A. No. 3 Mk7. The availability of the auto-following American SCR 584 set may have taken the urgency out of the placement of contracts for the Mk7.

When the GL1, GL2 and GL3 were being designed, the technique of using guns to shoot down aircraft had to be fully understood. In this regard, two reports from A. C. Cossor Ltd, 'Shooting at aircraft' [8] and 'Study of AF1' [9], which were produced during the war, are of interest, as is a paper by Beeching [10].

---

* L. H. Bedford [5] comments: 'to which you can apply as many pinches of salt as you think fit'.

Two operational problems [4] which occurred with metric GL equipment are worth noting. Polythene, which had only become available for cable insulation [11] in 1938, was used to sheath the coaxial transmission lines from the antennae to the receivers. This cable was 'tropicalised' by encasing it in vulcanised India rubber and anti-pest covering. Unfortunately, sulphur compounds from the pest coverings penetrated into the braided screen of the cable, causing poor contacts. E. C. Slow has related [4] how this situation was temporarily remedied by a drill of stroking the cable each morning. Propeller modulation* of the amplitude of the returned signal on occasion interacted with the 50 Hz switching and strobe system in the receiver, causing a beat frequency which, when it passed through zero, caused an apparent abrupt change in aircraft position.

The Admiralty and the Signal School of the Royal Navy were acquainted with the activities of the Tizard Committee as early as March 1935 [12]. If progress in radar development by the Royal Navy from 1935 until 1938, when the first shipborne sets were turned out, should appear slow, the principal cause was the small number of staff available for research and development. It should be realised also that, akin to airborne radar, shipborne systems posed their own special problems [13]. HM Signal School was no way backward in radio science. In 1930, for instance, extensive and successful VHF trials at 43 MHz had been carried out and there had been ongoing research into thermionic valves. The use of silica envelopes for valves allowed high anode dissipation without fear of cracking the valve envelope, and such valves had been manufactured by the Navy since about 1920 [14]. The Naval NT46 silica triode valves obtained from the Signal School in the early months of 1935 were an essential factor in the successful operation of Dr E. G. Bowen's first transmitter at Orfordness. At a later stage, in September 1938, there was a very large demand on the Signal School for silica valves, when the Munich Crisis made it necessary to keep the Thames Estuary Chain in continuous operation.

The following tentative requirements were laid down in 1935 for a naval radar set:

Aircraft: Warning of approach 60 miles
               Precise location 10 miles

Ships: Warning of approach 10 miles
               Precise location 5 miles

Approval for work to proceed on development was given on the 30th September 1935 [15], and on the 17th October 1935 the set under development was named the 79X [12]. It was decided to operate initially at a wavelength of 0·5 m and then to explore the use of higher and lower frequencies. This was done in 1936 and 1937 and then in March 1938 it was decided to build a set

* Propeller modulation effects can still occur in present-day radio-navigational and aircraft communication systems.

working on 7 m (43 MHz); this would facilitate large transmitter power while still allowing an antenna of moderate dimensions. (It is worth recording that in October 1936 an experimental set, operating at 75 MHz with wire antennae strung between the masts, was installed on board the Signal School tender *Saltburn* and an aircraft at 500 ft and 17 nautical miles range was detected.) In August 1938, a set designated 79Y was fitted in *HMS Sheffield* and underwent trials in September 1938; another 79Y was fitted in *HMS Rodney* in October 1938 and underwent trials in January 1939. Transmitter power output of the 79Y was 15 kW to 20 kW. One antenna was used for transmitting and another for receiving, the two rotating in synchronism. Each antenna consisted of two horizontal half-wave dipoles spaced one half-wavelength above the other, and each with a tuned reflector a distance $\lambda/5$ behind. The power gain of the antenna was approximately 5 and horizontal beamwidth was about 70° (bearing accuracy was about 5°, the bearing being obtained by rotating the antenna for maximum signal). During trials on *HMS Sheffield*, aircraft were detected at 30 nautical miles at 3000 ft, at 48 nautical miles at 7000 ft, and at 53 nautical miles at 10 000 ft.

An improved version of the set, type 79Z, was installed in *HMS Curlew* in September 1939. With new thoriated filament silica valves, it had a peak power output of 70 kW; pulse width was 8 µs to 30 µs and PRF was 50/s. 40 of these sets were ordered. The 79Y and 79Z sets provided essentially only long-range warning of the approach of aircraft. The type 279M, produced in 1940, was a type 79 with a ranging unit RBL10 which allowed, in addition to long range warning, accurate ranging of aircraft (±50 yards) between the ranges 2000 yards and 14 000 yards. The power supply requirements on board ship for the set were provided by two DC motor-driven alternators. The alternators were 230 V, 50 Hz single phase. One machine, rated at 5 kVA, supplied power to the masthead antenna rotating gear, while the other, rated at 14 kVA, supplied all the set's power requirements, including filament power and HT power for the transmitter.

A specification was written in 1939 for a long-range surveillance set which would include an accurate ranging facility. This set, type 281, was produced in prototype form by Signal School workshops in June 1940 and became the principal wartime search radar for heavy ships. It operated at 90 MHz and delivered over 1 MW peak power at a short pulse length of about 2 µs and a PRF of 50/s; at a pulse width of 15 µs, it delivered about 350 kW. Range against aircraft was typically 110 nautical miles at 15 000 ft, and 22 000 yards against large ships. 59 of this model were eventually fitted, the first ships to be fitted being *HMS Dido* in October 1940 and *HMS Prince of Wales* in January 1941.

A development which deserves mention is that of the series of fire-control radars which emerged from the Naval group working under John F. Coales. Experiments had been going on at the Signal School on 25 cm, but when Coales took over this work at the end of 1937 he decided to concentrate on 50 cm operation. H. C. Calpine built a transmitter using W.E. 316A door knob valves, but peak transmitter power of only about 500 W was obtained. There was also co-operation with the Research Laboratory of GEC at Wembley and the E1046 copper-anode triode was produced. A pair of these in push–pull enabled powers of 15 kW to be obtained. A later enlarged development of this valve allowed in mid-1942 a peak power output of 200 kW to be achieved. Fitment to ships of the 282/283/284/285 series of 600 MHz sets began in mid-1940; in June 1940 there were 900 50 cm gunnery sets on order. For an analysis of the problems of naval fire control radars, reference may be made to a paper of Coales, Calpine and Watson [16]. Friedman [17] provides a very useful synopsis of British naval radar from the pre-war type 79 to the present day.

In February 1939, Britain made an official disclosure of her radar work to countries of her commonwealth [18]. Representatives from Australia, Canada, New Zealand and the Union of South Africa visited radar stations and research laboratories in Britain. No attempt will be made here to record or assess the role of the Commonwealth* in the British war effort, but mention will be made of the Canadian production of the GL3 radar, referred to as GL3C, some 665 of which were produced; the GL3C was available in quantity before its British counterpart, the GL2. The story of this and other radar developments in Canada, which were the achievements of the Radio Branch of the National Research Council of Canada, are told in detail by W. E. K. Middleton [21]. W. Eggleston [22] likewise relates the Canadian radar story, but against the wider background of overall Canadian scientific wartime effort. The first radar experiment carried out in Canada was undertaken by Dr J. T. Henderson, head of the Radio Section of the National Research Council. In March 1939, using a Western Electric altimeter, which he had specifically purchased for the purpose and then modified so it would emit short pulses, he observed on a cathode-ray tube display echoes from nearby aircraft and buildings.

A major factor in the blossoming of technological war effort in Canada was the visit there in mid-August 1940 of Henry Tizard, when, with three other members of the 'Tizard Mission", he stopped over for a week before continuing on to the United States. He met Dr C. J. Mackenzie, President of the National Research Council, on his arrival and this was followed by meetings with service personnel, with politicians, including the Canadian Prime Minister, and with scientists. The Tizard Mission, or, to give it its formal title, the 'British Technical and Scientific Mission to the United States', arose from the foresight and perception of Henry Tizard; that Tizard himself was asked to head the Mission was fortuitous. The Mission set up a channel which allowed an interchange of knowledge and resources throughout the war between America and Britain.

An immediate result of the visit to the United States was a disclosure by the British of the resonant-cavity magnetron, of their airborne radar work, of the

* Original work was carried out in Canada, Australia [19] and South Africa [20].

proximity fuse and of the jet engine. From the point of view of British radar, the following statement by Otto J. Scott [23], based on an observation attributed to Sir John Cockcroft, puts one favourable aspect of the Mission into focus:

As the Battle of Britain began, the boffins were dismayed to learn that if every machinist in Great Britain was taken off his other work and assigned exclusively to making magnetrons, the nation could still produce only ten thousand a year. It was obvious they could not spare every machinist, equally obvious, ten thousand magnetrons were insufficient.

There were some in Britain who initially did not approve of the Mission and who considered that the United States had nothing to offer Britain in the radio and radar fields. These included Watson-Watt and Sir George Lee, the then Director of Communication Development, Air Ministry [24].

Long before the Mission proper, in November 1939, Tizard proposed that Professor A. V. Hill should visit the United States to set up a liaison with American scientists. Professor Hill went to Washington in February 1940 and was stationed there as a temporary Scientific Attaché. He stayed until June 1940 and discreetly assessed the state of military oriented research there. He also visitied the National Research Council in Canada. Hill very effectively laid the foundation for the Tizard Mission and advised on the form of diplomatic strategy to be employed.

On the 8th July 1940, Lord Lothian, the British Ambassador in Washington, in a letter to Roosevelt, wrote [25]:

The British Government have informed me that they would greatly appreciate an immediate and general interchange of recent technical information with the United States, particularly in the ultra short wave radio field.

A favourable reply from Roosevelt was forwarded to the British Cabinet on the 22nd July and within days authorisation for the Mission was granted. Tizard himself, with authority from Winston Churchill to make a full disclosure of British achievements, arrived in Washington on the 30th August 1940 and did not leave the United States until the 5th October 1940, when he handed over the charge of the Mission to Professor Cockcroft. The Mission included Colonel F. C. Wallace, Army Representative, Captain H. W. Faulkner, Navy Representative, Group Captain F. L. Pearce, Royal Air Force Representative, Professor J. D. Cockcroft, Dr E. G. Bowen and W. E. Woodward Nutt, Ministry of Aircraft Production, who acted as Secretary to the Mission. Also attached to the Mission were Professor R. H. Fowler, the DSIR's (Department of Scientific and Industrial Research) liaison officer in Canada and the United States, and Colonel H. F. G. Letson, the Canadian Military Attaché in Washington.

The sample 10 cm cavity magnetrons brought over by sea from Britain by Dr E. G. Bowen were shown to the National Defense Research Committee members at the Loomis laboratory in Tuxedo Park on the 30th September. On the 6th October 1940, the magnetron was tested for the first time in the United States at the Bell Laboratories, Whippany, New Jersey; it gave an estimated peak power output of 6·4 kW*. The valve was X-rayed and, with the help of the X-ray photographs and other drawings, work began forthwith to have a copy made.

One important item gained in turn by the British was access to the research work on silicon crystal rectifiers which had been conducted by Bell Laboratories. These crystal diodes, on which work had also been carried out in Britain, proved excellent mixers in microwave receivers.

4.2 Beginning of radar in Germany

Germany can, as was seen in Chapter 3, claim a certain precedence in radar development in that in 1904 Christian Hülsmeyer constructed a workable shipborne anti-collision device; the efforts of Hans Dominik in 1916, and the proposals of the Austrian Henrich Löwy in 1922 for an aircraft altimeter, likewise merit regard.

German (Telefunken) experimental radar set, 1935.

---

* This figure is quoted by Guerlac [26]. Dr E. G. Bowen [27], who was present at the demonstration, mentions a figure of about 20 kW.

None of these events directly influenced the development of German radar, whose beginnings may be traced [1, 2] to the year 1933 and to Dr Rudolf Kühnold who was then Chief of the German Navy's Signals Research division. For the incentive to undertake this development one could, perhaps, regress a few years to 1929, when Dr Kühnold was engaged in research with echo sounders for measuring depth and for measuring the distance of ships from each other. He developed a sonic device, operating at about 30 kHz, which simultaneously measured the range and bearing of targets. It was intended that the instrument be used both above and below water. In 1933, a policy decision was made by the Navy to develop a radio-detection device.

A wavelength of 13·5 cm (2200 MHz) was chosen for the proposed transmitter and receiver. Barkhausen-Kürz valves (see Appendix B) were used for both transmission and reception, each valve being placed at the focal point of an 80 cm diameter parabolic mirror. The transmitter sent out a

German medium-range 125 MHz early-warning set, 'Freya'.

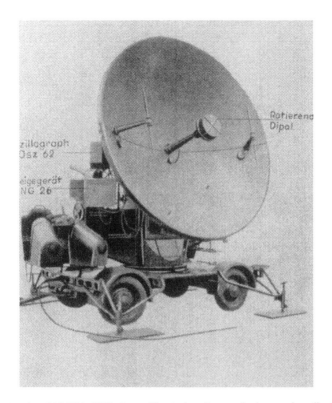

German Gun-Laying 560 MHz 'Würzburg C' set showing conical scanning dipole and IFF attachments.

continuous wave signal, which had been modulated by an audio frequency, but the power radiated was only about 100 mW. Both antenna reflectors were positioned next to each other and directed towards large metal objects such as the hull of a vessel. To increase the radiated power, a magnetron valve, giving an output of 80 W, was acquired from the Philips factory in Holland. This proved unstable and was replaced by a magnetron which delivered 40 W. This valve, modulated by 1000 Hz and feeding into a Yagi antenna, constituted the transmitter, while a Barkausen-Kürz triode fed from a simple dipole formed the heart of the receiver. The wavelength now used was 48 cm (630 MHz). At this junction, an incident occurred which will be dealt with at greater length below. Dr Kühnold approached Dr Wilhelm Runge of Telefunken and requested that Telefunken take over their radar research. Dr Runge's reply was interpreted as an official refusal by Telefunken, and thus, in January 1934, Dr Kühnold and his assistant, Dr Schultes, set up the form of GEMA (Gesellschaft für Electro-akustische und Mechanische Apparate).

GEMA developed the 48 cm apparatus and used a Yagi antenna array for the receiver which, if required, could be replaced by a 1·5 m parabolic mirror.

In Kiel harbour on the 20th March 1934, echoes were received from the warship *Hessen* as she lay at anchor in the harbour some 600 m away. This result was achieved with an output of 40 W from the transmitter and, even with this low output, sufficient energy spilled over directly into the receiver to overload it. This experience led eventually to the change over to pulse operation. The next notable event in the GEMA development occurred on the 24th October 1934, but in the meantime there were some relevant happenings.

The firm of Julis (*sic*) Pintsch of Berlin had developed, in the early summer of 1933, a transmitter and receiver operating on a wavelength of 13·5 cm [2]. The transmitter used a Barkhausen-Kürz valve which developed some 100 mW of power. With an increased power of 300 mW and using a modulation frequency of 1000 Hz, short ranges were obtained, using the research vessel *Welle* as a target, on the 15th May 1934. In addition, successful radiotelephonic experiments were carried out with the transmitter and receiver separated by 43 km across the stretch of sea between Heligoland and Wangerooze. This was creating competition for the GEMA firm who improved the shielding between transmitter and receiver on their 48 cm equipment. Comparative tests were carried out with the Pintsch and GEMA sets in the course of which it was demonstrated that reflection from targets depended on target aspect, and particularly so in the case of the higher frequency Pintsch equipment.

To return to the experiment of the 24th October 1934; the transmitter and receiver were separated by a distance of 200 m and a magnetron was used for the transmitter and positioned at the focus of a parabolic reflector which was five wavelengths in diameter. A 1000 Hz modulating tone was provided but, in addition, a sub-carrier of 1·6 MHz was employed. At the receiver, which was a superheterodyne type, the intermediate frequency was 1·6 MHz and the indication of signal was by means of a valve voltmeter. A range of 12 km was achieved with the research ship *Welle* and, in addition, a Junkers W34 aircraft passing overhead was followed to an estimated distance of 700 m at a height of 200 m. A representative of the TVA (Torpedo-Versuchsanstalt, i.e. Torpedo Research Institute) was present and, as a result, a subvention of 70 000 Reich-Marks was obtained for further development to be carried out in secret and with the following guidelines being observed [2].

1 Introduction of a pulse method similar to the ionospheric measurement technique of Bright and Howell;* according to a proposal of Dr Th. Schultes, P. G. Erbeloh and Von Willisen, this should allow a higher power capability in the transmitter through a short-time overloading of the transmitting valve.

2 Introduction of a blocking of the receiver while the transmitting impulse was going out. Thereby the unwanted direct signal from the transmitter into the

---

* These two names are quoted on p. 9 of Reference 2. It appears to be a transcription error, as no reference to their work can be found.

receiver should be eliminated. The unstable retarded-field audion valve should be replaced by a triode.

3 Development of a method to measure distance:

   (*a*) by using a phase delay procedure with indication on an instrument, dynamometer or something similar;

   (*b*) by indicating phase delay or transit time on a Braun tube,* with the insertion of a calibrated phase displacement on the time-base of the Braun tube.

4 Replacement of the unstable retarded field audion by a triode with feedback control.

5 Replacement of the transmitter parabolic mirror antenna by a Christmas-tree† radiator.

6 Development of a direction-finding method through the already familiar A–N procedure.‡

At the beginning of May 1935, a new GEMA transmitter was ready for testing. The transmitter provided 2 μs pulses with a peak pulse power of 800 W and a pulse recurrence frequency of 2000 Hz. A special blocking device [2] immobilised the receiver when the transmitter pulsed. The display on the cathode-ray tube was a form of A-display (see Section 2.9). The transmitting antenna was a 'Christmas-tree' type, with ten pairs of dipoles stacked in front of a reflector, and the receiving antenna was a smaller version of this; the shielding between the two arrays was so good that they could be mounted side by side. Although in tests signals were returned from 15 km away, the ship *Welle* could not be detected beyond 3 km. Further development of the set took place and, afterwards, the cruiser *Königsberg* was picked up at 8 km, while echoes from land 20 km away were obtained. On the 26th September 1935, a demonstration ws given in the presence of Admiral Erich Raeder, Comman-der-in-Chief of the Navy, and then, before the set was returned to Berlin, it was installed on the research vessel *Welle*. From the ship it was possible to detect vessels at sea at ranges up to 7 km, while the coastline could be picked up when 20 km distant. This event was significant in that it was the first installation of a radar set on a ship and also the proof that such installations could be successful.

At this particular stage of development, GEMA had discovered that the orientation of a target could appreciably affect the amount of signal reflected back from it, but that this dependence was less critical when longer wavelengths were used. Hence a decision was made to develop an equipment working on a wavelength of 2 m. An interesting fact is that some radio-

---

* Cathode-ray tube.

† Array of stacked dipoles.

‡ This refers to the aural radio range navigational aid. Spatially divergent, partly overlapping and alternately switched antenna radiation patterns were used. One lobe transmitted the Morse code letter A (· —), the other the letter N (— ·). In the equisignal zone, where the patterns overlapped, the interlocking of the Morse characters produced a continuous tone in a receiving set.

scientists erroneously considered that at wavelengths below about 20 cm, deflection only by targets, and not reflection, would occur [3]. The 50 cm GEMA set was improved. The magnetron in the transmitter was replaced by a triode and the wavelength, on the suggestion of Von Willisen, was increased to 82 cm. This 82 cm set was the forerunner of the 'Seetakt' [1] series of radar sets.

Reference has been made to Dr. Künhold's approach to Dr W. T. Runge of Telefunken in 1934 for assistance in developing radar. The background to the incident is as follows [4, 5]. Dr Runge had taken over the radio receiver laboratory of Telefunken in 1924 and had introduced testing techniques and generally built it up. The laboratory in the early 1930s was one of four sections which worked with high frequencies over 30 MHz. These sections or departments worked almost completely independently of each other. The other three departments were:

(a) A television department which used frequencies of the order of 50 MHz as carriers of TV signals.
(b) The 'Physical Laboratory' under the direction of Professor A. Meissner (whose name is associated with feedback oscillators). The laboratory experimented with Barkhausen–Kürz valves operating at about 500 MHz.
(c) Thermionic valve development department.

Dr Runge was himself interested in decimeter waves, particularly for radio-relay or point-to-point working. At this time the Telefunken firm was still affected by the financial depression, and the management, which was itself somewhat disorganised, showed no interest in these projects. There was certainly no Company interest in the radio detection of objects, and no one in Telefunken is known to have been carrying out experiments in this area. When Dr Künhold visited him early in 1934 and proposed that the Navy and Telefunken embark on a joint project together, he overestimated Dr Runge's position. Dr Runge had neither the staff nor the means to pursue such research. The head of the department for the development of equipment was Dr Bohm; he was Jewish, and understandably would not, in the political climate of that time, become involved. The management committee, to whom the request should have been addressed, would most likely have refused also. Dr Runge's refusal may well have pleased Dr Künhold, who now had freedom to found his own company, GEMA. Later, in 1936, when Dr Runge had become involved in radar experiments, Dr Künhold tried to sue him because he, Künhold, considered that radar was his sole invention.

Dr Runge forgot about these proposals, but then in his notes for November 1934 there appeared the statement 'studied ship effect'. This 'ship effect' was the beat phenomenon observed when a ship passed through a decimeter radio beam. Telefunken had, in fact, been experimenting with such beams in the Kiel Fjord as part of a navigational project. In the summer of 1935, Dr Runge had the idea of measuring the radiated power reflected from an aircraft. To accomplish this he placed a transmitting antenna on the ground and fed it with 0·5 W of a modulated signal. Beside it he laid an identical antenna which was connected to a diode which, in turn, was connected to an audio-frequency amplifier and a voltmeter. When a Ju52 aircraft approached and then flew overhead, the meter moved back and forth. (The close resemblance of this experiment to the manner in which reflections from aircraft were observed by Lawrence Hyland and Leo Young in the United States in 1930, and by Wilkins in the Daventry Experiment in February 1935, will be apparent.) Dr Runge was very impressed and he reported the result to the manager for research and development, who was completely indifferent. From then on, however, Dr Runge became interested in radar and particularly in fire-control radar.* Not until 1938 did he receive financial support for his ideas by which time he had constructed a parabolic reflector antenna 3 m in diameter, and fabricated from screen-wire, and had developed a method of conical scanning.† In the summer of 1938 a small incident occurred which is indicative of the attitude to radar in certain quarters at that time. The embryonic fire-control set was being shown to General Udet of the Luftwaffe, when he turned to Dr Runge and remarked: 'You know, if you introduce that in the armed forces then flying won't be fun anymore'.

Matters changed after the managing Director of Telefunken accompanied Dr Künhold on a visit to the Navy and Dr Künhold criticised Telefunken for its lack of technical initiative. This criticism included a disapproval of their use of a frequency of 600 MHz, and this angered the Manager for Research and Development, who was an old pioneer of valve construction, to such an extent that he undertook to develop, within four weeks, a valve that would develop 10 kW pulse power on this frequency. The valve was manufactured.

In February 1936 the Gema 2 m wavelength set was completed. The actual wavelength was 1·8 m (165 MHz). The transmitter, using an output triode valve developed by GEMA, delivered a peak pulse power of 8 kW. On test, the equipment failed to pick up a 1500 tonne steamer, while detecting an aircraft at a range of 28 km. The bad result with the ship was solely due to the installation height of the antenna which produced an elevated radiation pattern. Because of these results, the set was assigned for use against airborne targets. It was sent back to Berlin where the wavelength was altered to 2·4 m (125 MHz) and the antenna array redesigned, the Yagi antennae being replaced by a rotatable array of dipoles mounted in front of a wire mesh reflector. This set was the originator of the 'Freya' series of aircraft warning sets.

Perhaps a significant factor in the story of Telefunken's association with radar was that the Luftwaffe became envious of the Navy having its Freya

---

* Radar sets which provide the requisite information to direct the fire of guns on to targets.
† Conical scanning is used in the automatic following of targets. The antenna system describes a rapid conical motion and the error signals that are produced by the displacement of a target from the axis of rotation are then used to keep the antenna pointing at the target.

equipment. Not wishing to take equipment from the Navy, they ordered 600 MHz sets from Telefunken in April 1939. Thus, by April 1940, the first 'Würzburg' sets had been built and purchased.

The Lorenz firm experimented with radar devices in 1935. Independently of Telefunken they also chose the decimeter waveband and so could use the valve and antenna technology of the beamed radio systems which had been developed by the firm. Their first laboratory radar set operated at 430 MHz with a peak power output of 400 W and a PRF of 10 kHz. This was eventually developed into the Anti-Aircraft, or 'Flak'* radar set A2-Gerät.

One could summarise the development of radar in Germany before the war by stating that, except for the Navy and GEMA, no real interest was taken in radar until 1939. The firm of Telefunken with its particular expertise and its resources was, in the early 1930s, the most suitable in Germany for the development of radar, but it was not until after 1938 that they got involved. The firms which were particularly associated with radar at the beginning of the war were GEMA, Lorenz and Telefunken.

The early development of radar in Germany must be seen against a background of an offensive military policy in which such thinking as was behind the British Chain Home system had no part. Furthermore, at the outbreak of the war, research was seriously curtailed by a decree from Adolf Hitler, which prohibited all projects which would not lead to tangible results within a few months [1]. This adversely affected the development of centimetric equipment, broadband antennae and navigational equipment [3].

The accomplishments of German radar technology before 1945 are outlined, but by no means exhaustively covered, in the bibliography. For an explanation of the coding or classification of sets as used in the three services, Trenkle [1] and Sieche [6] may be referred to.

Below are listed some details of the principal German ground radar systems.

Freya

Manufacturers	Gema
Function	Medium range early warning. Ground controlled interception was possible. Somewhat equivalent to British CHL set.
Frequency	125 MHz, nominal
Polarisation	Vertical
Transmitter peak power	20 kW
Azimuth coverage	360°
Maximum range	200 km
PRF	500/s
Pulse width	2–3 µs
D/F accuracy	0·5°
Height finding	None

* Flak = *Fl*uger*a*bweh*r*kanonne. The Flak arm of the Luftwaffe was set up in 1935.

Mammut

Manufacturer	Telefunken and Experimental Signals Kommando
Function	Long range early warning
Frequency	120–150 MHz
Polarisation	Vertical
Transmitter peak power	200 kW
Azimuth coverage	Static antenna; electrical swinging of beam through 100° by means of phase-change system; either front or back array useable, hence two blank sectors each 80° wide.
Maximum range	300 km
D/F accuracy	0·5°
Height finding	None

Wassermann

Manufacturer	Siemens and Gema
Function	Long range early warning with height finding. Antenna mounted on cylindrical steel tower 120 ft to 195 ft high.
Frequency	120–150 MHz
Polarisation	Horizontal
Transmitter peak power	100 kW
Azimuth range	300 km
Height finding	Measurable to within ±1000 ft

Jagdschloss

Manufacturer	Jointly by Gema, Siemens and Lorenz
Function	Early warning radar against low flying and medium flying aircraft
Frequency	129–165 MHz
Polarisation	Horizontal
Transmitter peak power	30 kW
Azimuth coverage	360°
Maximum range	80 km
PRF	500/s
Pulse width	1 μs
Height finding	None

Elefant–Russel

Function	Long range warning system. Somewhat equivalent to British CH. (The Elefant part of the combination used a high standing antenna mounted between steel towers, 310 ft high, for transmission and reception. There was a front and a back array and transmitter and receiver could be

switched between arrays. The Russel was the DF component of the combination and its receiver was locked to the Elefant transmitter.)

Frequency	20–40 MHz
Polarisation	Vertical
Transmitter peak power	380 kW
Azimuth coverage	Elefant: Floodlit 120° arc
	Russel: Rotatable through 360° (4 min/rev)
Maximum range	450 km
PRF	25/s
Pulse width	10 μs

Würzburg

Manufacturer	Telefunken
Function	A Gun Laying set which was also used in the ground controlled interception role, and some-times in an early warning role (over 4000 were manufactured).
Frequency	560 MHz
Transmitter peak power	8 kW
Maximum range	30 km
PRF	3750/s
Pulse width	2 μs
D/F accuracy	±0·75°

Würzburg–Riese (Giant Würzburg)

Manufacturer	Telefunken
Frequency	560 MHz
Transmitter peak power	8 kW (note: Antenna paraboloid dish had a diameter of 22 ft, over twice that of the ordinary Würzburg)
Maximum range	60 km to 80 km
PRF	1875/s
Pulse width	2 μs
D/F accuracy	±0·2°
Elevation accuracy	±0·25°

The following quotation concerning the Würzburg (Fu MG 62A) appears in Hoffmann-Heyden's book on Flakartillerie [7].

The TRE Report 6/R/25 'Final technical report on the German RDF equipment captured at Bruneval on February 28, 1942' brings to light two very interesting points. The first is that the Germans were using intermediate frequency bandwidths of 1/t, which is shown to be optimum in the report. The second point of interest is the statement by TRE ' . . .

this shows that the receiver gain was usually operated at a high level, giving about 1 cm of noise on the table. Owing to the high recurrence frequency used (3750 pulses per second), it would be possible to see signals below the noise level . . .'. This 'seeing below the noise' is equivalent to saying that the visibility factor as described in this report is less than 1. The actual calculated value is 0·78. This is of especial interest when it is noted that this German equipment was designed at least as early as 1939.

4.3 Beginning of radar in the United States

In Section 3.8 it was mentioned that an event occurred on the 24th June 1930, when Leo Young and Lawrence Hyland of the Naval Research laboratory were testing the directional properties of an aircraft HF antenna system, and that this occurrence laid the foundation for radar research in the United States.

Hyland was at the aircraft which was parked on the compass rose* of the Air Station at Bolling Field and about two miles distant from a ground beacon at the Naval Research Laboratory that transmitted on 32·8 MHz using horizontal polarisation. The behaviour of a 15 ft fore-and-aft wire antenna on the aircraft was being observed [1–3]. The polar pattern of the wire antenna would ideally be a figure-of-eight, with two maxima and two minima. The beacon's signal was received in the aircraft, and as the aircraft turned on the compass rose, Hyland observed the bearings at which the maxima and minima of the signal occurred. He noticed that a perfect minimum position could be seriously disturbed by aircraft passing overhead and along a line between his position and that of the beacon station. He was deeply impressed by this and reported it. The next day the phenomenon was again observed in the presence of C. B. Mirick, Ross Gunn and Lieutenant Rowe. The timetable for the following significant events was:

26th June 1930	Hyland reported to the Honorable Frank Knox, Secretary of the Navy, through Commander E. D. Almy, Assistant Director of the Laboratory.
27th June 1930	Mirick reported to Dr Albert Hoyt Taylor, Superintendent of the Radio Division of the Naval Research Laboratory.

A small portable super-regenerative receiver was constructed and successful confirmatory experiments were carried out at distances up to 10 miles from the transmitter.

19th August 1930	Hyland composed a memorandum to Commander Almy describing the phenomenon.

* Generally a special concrete circular area at airfields on which an aircraft can be turned when 'swinging' or calibrating the aircraft's magnetic compass.

Dr Robert M. Page beside 200 MHz antenna array which was used in USS LEARY tests, 1937.

The scope of the experiments was extended with further tests being carried out on 32·8 MHz, and on other frequencies including 65 MHz.

5th November 1930 H. Taylor forwarded a detailed memorandum consisting of eleven pages with diagrams to the Chief of the Bureau of Engineering. Indeed, on the 10th December he provided a demonstration of the beat phenomenon for representatives of the Signal Corps, the Coast Artillery and the Air Corps.

The response to the memorandum of the 5th November was in two stages. On the 25th November 1930 the problem labelled B1-1 was formulated; this concerned the investigation of high and super* frequency antenna systems of a type suitable for naval use. The second stage was the formulation of problem W5-2, for which formal authorisation came on 19th January 1931. This gave permission to 'investigate use of radio to detect the presence of enemy vessels and aircraft' [3, 4].

Between 1931 and 1934, all experimentation by the Naval Research Laboratory was concerned with continuous-wave or Doppler methods of detection [1, 5]. During this period, Taylor could not foresee the use of the

---

* Super frequencies corresponded to the present-day VHF band.

Model XAF 200 MHz equipment on USS NEW YORK, installed December 1938.

technique on board ship, yet he considered it of value to the US Army. He drafted a letter for the Secretary of the Navy which was sent to the Secretary of War on the 9th January 1932. The letter suggested that a system of transmitters and associated receivers could be used to detect the passage of hostile aircraft into a designated area (within months, the Army initiated its own programme, but not just as a result of this letter) [4]. At the Naval Research Laboratory progress was slow because of the limited staff and financial resources then available.

Carl R. Englund, Arthur B. Crawford and William W. Mumford of Bell Telephone Laboratories, on the 12th January 1933, reported openly at a meeting of the Institute of Radio Engineers in Washington DC on their work on the propagation of ultra-short waves (64 MHz to 81 MHz, approximately) including reflection from moving bodies. Their paper appeared in the

NRL 28 MHz radar transmitter used extensively during 1936 for tests and demonstrations: peak pulse power 3 kW to 5 kW.

Proceedings for March 1933 [6] and the relevant section is at the end of this account [7]. As a result of this disclosure, the Director of the Naval Research Laboratory took action and the Bureau of Engineering filed a patent application on the beat method on the 13th June 1933 in the names of A. H. Taylor, L. C. Young and L. A. Hyland [1]. The patent, No. 1 981 884, for a 'system for detecting objects by radio' was issued on the 27th November 1934.

The Naval Research Laboratory assigned priority 'A' or 'Urgent' to the project in early 1934. A demonstration of the continuous-wave aircraft-detection system for the benefit of the Naval Appreciations Subcomitee took place at the Laboratory in mid-February 1934. To help prepare the equipment for the demonstration, Robert M. Page, who had been working with the Laboratory since 1927 and had the reputation there of being an inventor, was assigned to the project team. This assignment in effect heralded both the use of

NRL XAF set with the RCA CXAM set (at right) undergoing comparison tests at the NRL, 1940.

pulse techniques for detection by the Naval Research Laboratory and a proper development programme for radar in the United States.

Robert Page had been thinking about the possibilities of the beat method from mid-February until mid-March 1934 when Leo Young, on 14th March, came into his laboratory with a definite suggestion to use pulses [8], Leo Young was confident that short-power pulses could be generated because of his work on the suppression of transmitter 'key-clicks'; during high-frequency morse transmissions, keying the transmitter produced very large, sharp pulses which could be observed on a cathode-ray oscilloscope. (It is worth mentioning also that the idea for the circular time base, used in their first experimental pulse radar, came from their work on sonar.)

Dr Robert M. Page was a member of the staff of the Naval Research Laboratory from June 1927 until he retired as its Director of Research in December 1966. He had a talent for invention* and to him, more than to any other one person, must be credited the development of pulse radar in the United States. Between March 1934 and December 1934, Page assembled, almost single handed, an experimental pulse radar set. A transmitter frequency of 60 MHz was chosen because this frequency had been used in the continuous-wave beat method equipment. The transmitter, which used two Raytheon RK-20 valves (nominal 50 W) in push–pull, was keyed by an

---

* He received a total of 65 patents during his working life [4].

US Army SCR-268 set on Signal Hill, Greenland Base Command.

asymmetric multivibrator and pulses of somewhat less then 10 μs spaced at intervals of about 100 μs resulted, giving an estimated peak pulse power of over 100 W. A high gain wideband communciation receiver was borrowed from the receiver section of the Laboratory and was modified, principally by loading its anode circuits with resistance to increase bandwidth. In addition, all resistance was removed from the grid circuits of the amplifying valves, so as to prevent 'blocking' or paralysis of the receiver by the transmitted pulses. This superheterodyne receiver had four stages of IF amplification operating at 10 MHz. A horizontal dipole with tuned reflector was used for transmission and a similar antenna, spaced some 10 wavelengths away, for reception; both antennae pointed up the Potomac River.

The complete system did detect aircraft as they flew up and down the line of the river, but, because of ringing in the receiver for durations of from 30 μs to 40 μs, the indications on the circular time-base were not very definite. Observing this ringing, which had considerably increased the apparent width of the received pulse, it became clear to Page that, while he had overcome the danger of the receiver being paralysed by direct signals from the transmitter, he had now to design a receiver which would have sufficient bandwidth to cope with short pulses. He worked on the design and fabrication of a new receiver throughout 1935. In the design he was assisted by a paper [9] written by René Mesny on 'Time constants, build-up time and decrements', which enabled him to determine the overall bandwidth of several amplifying stages in cascade.

The design of this new receiver was helped also by the availability of the new RCA 'acorn' [10] type 524 very-high-frequency pentode valve. Regenerative feedback was guarded against by using two intermediate frequencies,* by restricting the voltage gain at any one frequency to 1000, and by scrupulous attention to shielding, filtering and common-point grounding [11]. The cathode-ray tube indicator used a linear timebase sweep instead of a circular one.

The transmitter was self-keying using a 'squegging'† circuit and was built by Robert C. Guthrie. The squegging oscillator had been used in the laboratory by Page as a low-power signal generator in 1934, but its use at high power as a pulse transmitter had been suggested by La Verne Philpott in 1935 [12]. A frequency of 28·2 MHz was chosen for the system as there was already available at the Naval Research Laboratory for communication experiments a horizontally polarised 4 × 4 wavelength curtain array (dipole elements stacked 6 vertically, 8 horizontally), with resonant reflectors,‡ suspended between two 250 ft towers.

This antenna was used for transmission, while a single half-wave dipole with resonant reflector was used for reception. The transmitter was capable of giving a peak pulse power of over 3 kW, with a pulse length of 5 μs to 7 μs and a repetition rate of about 1800/s. The equipment was ready for testing on Tuesday 28th April 1936, and, when switched on, picked up an aircraft which happened to enter the beam. The range was about 2½ miles but, with the anode voltage of the output valves raised from 5 kV to 7·5 kV, an aircraft on a scheduled flight was tracked out to over 8 miles on the 30th April. Tests were carried out throughout May and early June and, with anode voltage raised to 15 kV, ranges of 25 miles, the limit of the display, were obtained.

The equipment was used extensively and successfully for demonstrations throughout the remainder of 1936, even while other higher frequency equipments were being developed. The demonstrations were arranged principally for influential Naval officers who might be passing through Washington. (It should be remarked that Dr A. Hoyt Taylor had in 1935 succeeded in obtaining more money from Congress for the radar project and that this, among other things, resulted in Robert C. Guthrie being assigned to Robert Page in November 1935). On the 12th June 1936 Admiral Harold G. Bowen, head of the Bureau of Engineering, instructed the Naval Research Laboratory to place the research programme under a security classification of 'secret'.

During 1936, Robert Guthrie carried out experiments at both 50 MHz and 80 MHz. When transmitting on 50 MHz, he used an existing 16·8 MHz curtain

---

* Three stages of amplification at 35 MHz and a single stage of amplification at 25 MHz.
† Squegging is a self-quenching process in an oscillator arising from auto-biasing; oscillations build up, reach a maximum, decay and after a quiescent period the cycle repeats itself. See Reference 34, Chapter 5.
‡ Array was beamed towards London, England.

array, thus employing a harmonic mode, and in July 1936 he obtained very good results at 15 miles. The output stage of the transmitter employed a pair of Gammatron 354 valves* in push–pull. Having discovered that these valves could operate efficiently up to 80 MHz, he modified the transmitter and built a new 80 MHz antenna array which was suspended between two towers. It was a vertically polarised Sterba† array of radiators and parasitic reflectors and its beam pointed southwards down the Potomac River. This antenna was completed in September 1936. (The supporting cables for the array were First World War suplus aircraft stranded-steel control cables which rusted within a year.) Duplexing, which had been developed for a 200 MHz radar and which is discussed below, was used, thus permitting the use of a single antenna for both transmitting and receiving. The receiver used with the 80 MHz equipment was the same as that used in the 50 MHz experiments, namely Page's 28·2 MHz set, modified by the use of a suitable heterodyne stage. The 80 MHz equipment was first put into operation on the 5th November 1936. By the 24th November 1936, aircraft out to a range of 36 miles were being observed. When successful working at 200 MHz was assured, no further developments at 80 MHz were undertaken.

To another member of Page's small team, Arthur A. Varela, goes the credit for the original design of a 200 MHz radar set, which became the forerunner of the XAF and CXAM radars. Development began on the 8th May 1936 and the equipment was working by the 22nd July 1936. Robert Page's receiver was still used, together with a 200 MHz to 28 MHz convertor. The transmitter used two valves in push–pull in its output stage. Yagi antennae were employed. The set was notable for its use of an antenna duplexer. The suggestion for the use of a duplexer originally came from A. Hoyt Taylor, but the method of accomplishing it was worked out by Page. His first duplexer [14] was based on the impedance inverting property of a resonant quarter wavelength section of transmission line. When the transmitter pulsed, grid current flowed in the grid-cathode circuit of the receiver input valve, thus greatly lowering the grid-cathode resistance. A quarter of a wavelength away, this low resistance was transformed into a very high resistance, thus effectively blocking the further flow of energy from the transmitter to the receiver and allowing it to be channelled totally into the antenna.

Page tried this basic duplexer on Guthrie's 80 MHz curtain array, where it worked effectively. In the case of the 200 MHz radar, when the power was increased the receiver input valves began to burn out. This was because the action of the duplexer depended on the receiver input circuit presenting a virtual short-circuit to its feeder line under conditions of grid current. It was not a true short-circuit and so the duplexing action was imperfect. This problem was circumvented by placing a neon gas discharge tube across the

* Rated at 100 W continuous-wave operation.
† The Sterba antenna curtain was an A.T. & T. system [13] developed by E. G. Sterba.

receiver input. Special glass-enclosed spark gaps were later developed by a glass blower, a Mr Kueck [12].

In August 1936, the 200 MHz radar was used to ascertain the relative effectiveness of using horizontal, vertical or circular polarisation and also to observe the variations in echo intensity caused by changes in a target aircraft's attitude.

In search of more transmitter output, Page developed a ring-oscillator, initially using four valves. This is illustrated schematically in Fig. 4.1. The ring oscillator is essentially a multiple push–pull tuned-grid tuned-anode arrangement. Page later regarded it as the decimeter counterpart of the microwave multi-cavity magnetron and indeed, by 1941, 1 MW of pulse power at 200 MHz and 50 kW at 600 MHz were achieved with the circuit. The advantage of the ring arrangement over a parallel or a push–pull parallel configuration is that in the latter, the valve capacitances are effectively in parallel and add, thus militating against high-frequency operation, whereas in the former, valve capacitances are in series. With an even number of valves in a ring, grids of adjacent valves assume opposite polarities around the ring. The tuned circuits are lengths of parallel transmission lines with adjustable shorting bars. At any instant, the radio-frequency currents in the shorting bars will all be in the same direction; the shorting bars form an effectively closed circuit from which radio-frequency energy may be picked up by a suitable coupling coil.

The next development of note was the installation of a 200 MHz set on board a ship. A. Hoyt Taylor, aware of the interest of Vice-Admiral A. J. Hepburn, Commander-in-Chief of the United States Fleet, in some practical test of

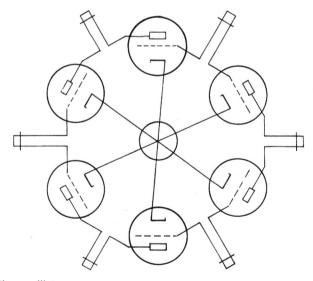

Fig. 4.1   *Ring oscillator*

radar, gave instructions that a trial of Robert Page's 200 MHz equipment be carried out at sea. The destroyer *USS Leary* was made available and the 200 MHz equipment, modified to withstand conditions at sea, was installed on board. The tests were undertaken in April 1937. Several Yagi and planar antennae were tried out; these were mounted on the 5 inch guns and so could be rotated. Ranges of 20 miles against aircraft were obtained. These tests highlighted the fact that the transmitter power output was inadequate. This deficiency was overcome by using 6 Eimac 100TH valves in a ring oscillator, with the anodes driven at 10 kV to 15 kV. The Eimac 100TH was a product of the Eitel-McCullough Company and had been purposely designed to withstand the often exacting demands of radio amateurs. Other modifications were carried out, including the redesigning of Page's receiver and the use of a rotating antenna. The new radar was completed on the 17th February 1938 and was demonstrated to numerous officials during 1938.

The Commander-in-Chief of the United States Fleet had requested that 'radio detection and ranging' equipment be provided in the Fleet; a conference, at which representatives of Naval Operations and the Bureaux of Ordnance, Aeronautics and Construction and Repair were present, was held in February 1938 to consider the matter. A decision was made to instal a sea-going model of a radar set on board a major ship at the earliest possible date. An allocation of 25 000 dollars was made from the Bureau of Engineering to the National Research Laboratory to carry out the work. Development began in March 1938 and in December 1938 the Model XAF was installed on board the *USS New York*. The overall dimensions of its planar antenna array were 20·5 ft × 23·5 ft. The XAF provided the capabilities of air and surface detection of targets, of navigation, of spotting the fall of shot and or tracking the flight of projectiles; aircraft were detected at ranges of 100 nautical miles, surface ships at 15 nautical miles, birds in flight at 5 nautical miles, navigational buoys at 4 nautical miles and mountains at 70 nautical miles.

There had been some radar experimentation in progress at RCA (Radio Corporation of America) since 1932 and the Bureau of Engineering were aware of this. In 1937 the Bureau informed RCA of the work in progress at the Naval Research Laboratory and contracted with them for the manufacture of a 400 MHz set. This equipment, named CXZ, was installed in the battleship the *USS Texas* and tested in the Carribean in January 1939 in conjunction with the tests of the XAF. It did not perform well. Nevertheless, when quantity production of the XAF design was decided upon, RCA obtained the contract. A preproduction model of this new set, the CXAM, became available in November 1939. The first of 20 sets was then delivered in May 1940. The sets were installed on battleships *California, Texas, Pennsylvania, West Virginia, North Carolina* and *Washington*, on aircraft carriers *Yorktown, Lexington, Saratoga, Ranger, Enterprise* and *Wasp*, on 5 heavy cruisers, 2 light cruisers, and the seaplane tender *Curtis*.

The CXAM performed well during the Second World War. Its development could be regarded as bringing the first stage of development of radar in the US Navy to a close. The setting up of the NDRC (National Defense Research Committee) and of the Radiation Laboratory, and the introduction of the cavity magnetron by the Tizard mission from Britain, all occurred in 1940, a little more than a year before America's entry into the Second World War. These events heralded a completely new degree of growth in the United States' radar programme. Before the story of the early development of radar in the Navy is brought to a close, there are two other events worth mentioning.

In October 1931, a Lieutenant J. Wenger of the Navy forwarded to the Bureau of Engineering a confidential memorandum, the subject matter of which was 'Radio bearing indicator', in which he proposed the use of microwaves in a type of 'radio searchlight' [1]. It was proposed that the device could be used for gunnery, navigation and for obstacle detection when flying. The memorandum was examined by Naval Research Laboratory engineers who considered the proposal useful but, as yet, not feasible. The idea appeared again in 1933 when Lieutenant W. S. Parsons considered the possibility of using microwaves for fire control. Memoranda drawn up by him were sent to the Bureau of Ordnance via the Bureau of Engineering, but no sanction was given to investigate the problem. A more comprehensive memorandum on the uses of microwaves was forwarded to the special Board of Ordnance in March 1934, but again without concrete results. However, as a result of a conference held at the Naval Research Laboratory, at which the representative of the Bureau of Aeronautics, a Lieutenant A. Smith, discussed the problem of measuring absolute height and ground speed, it was decided that the Naval Research Laboratory should investigate the possibility of using microwaves. Work was begun by W. J. Cahill and L. R. Philpott and continued through 1934. Westinghouse split-anode magnetrons, operating at 9·2 cm, were employed as transmitters. In 1935, Ross Gunn, the Technical Assistant to the Director of the Laboratory, lent his support to microwave development. Research continued, all on continuous wave methods, with a small team of three to five men and concurrently with Robert Page's 200 MHz tests on the *USS Leary* two systems were tried out, one operating at 1200 MHz and the other at 500 MHz. The former system provided ranges of a little over a mile, while the latter, because of bad weather, gave no results.

In 1937 disclosures of pulse radar development at the Naval Research Laboratory were made both to RCA and to the Bell Telephone Laboratories. Engineers from Bell Laboratories visited the Naval Research Laboratory in July and November 1937. Early in 1938 a small group, under the leadership of W. C. Timus, was set up at the Whippany Radio Laboratory, New Jersey, to do exploratory work on radar at their parent company's (AT & T) expense [15]. A 700 MHz fire control set was developed. This was demonstrated to representatives of both the Navy and the Army in July 1939. The Navy signed a contract for a production model to be built. This set, developed by the Bell

Laboratories and manufactured by Western Electric, became known as the CXAS. The first installation of a CXAS was on board the cruiser *Wichita* in June 1941. It was known as FA or Mark 1 in production and was but the first of a long series of fire control radars developed by Bell Laboratories.

4.3.1 United States Army Signal Corps

The eventual development of radar by the Signal Corps can be viewed as a quite logical culmination of experiments in sound location and thermal detection carried out by the US Army since the First World War. Although the Signal Corps spearheaded developments within the Army, interest in methods of detection was shown also by the Ordnance Corps and even by the Corps of Engineers.

The Signal Corps Radio Laboratories were established at Camp Alfred Vail, New Jersey, in March 1918. The Camp changed its name to Fort Monmouth in 1925. During the latter part of the First World War, the Laboratories were very active in developing radio direction finding equipment and aircraft radio. After the First World War, aviation research shifted to Wright Field, Dayton, Ohio and in 1929 three Signal Corps Laboratories were transferred to Fort Monmouth from Washington DC. There was a transference also to Fort Monmouth in 1930 of the Signal Corps Subaqueous Sound Ranging Laboratory. The consolidated organisation at Fort Monmouth was then referred to as the Signal Corps Laboratories.

Fig. 4.2 is a map [16] outlining the location of the Signal Corps Laboratories and also the area in New Jersey used by the Signal Corps when evaluating their detection equipment.

Major William Blair was Director of the Laboratories from June 1930 until October 1938. His background was, briefly, that he had, for his Ph.D., studied the properties of electromagnetic waves between 860 MHz and 3 GHz including their reflection from various surfaces; he had also been involved in the development of a radiosonde, a miniature weather station with radio transmitter borne by a balloon, and in work with sound locator systems.

In February 1931 Major Blair assigned Project 88 to the Laboratories' Sound and Light Section. The project's title was 'Position finding by means of light' where the term 'light' included the radio and infra-red parts of the electromagnetic spectrum. Techniques involving the reflection of infra-red waves were tried at first and then, in 1933, these were abandoned in favour of research on the heat radiated from targets and on reflected radio waves. With the latter in mind, the Signal Corps Laboratories undertook a systematic survey of what was then known, in all countries, about methods of generating, modulating and detecting radio waves of less than 1 m wavelength. Barkhausen–Kürz (see Appendix B) oscillators and split-anode magnetrons were obtained and used in experiments. With a Barkhausen–Kürz transmitter, which delivered 5 W at 75 cm wavelength into a Yagi beam antenna, signals were picked up at a range of 85 miles. Again, with a 9 cm wavelength

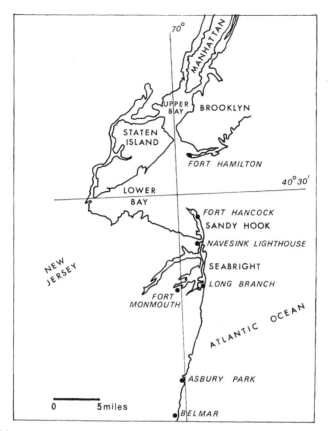

Fig. 4.2   *Map: experimental area used by the Signal Corps Laboratories*

magnetron from Westinghouse and with transmitter and receiver 50 ft apart, a Ford truck was detected at 250 ft using the continuous-wave beat method.

In his annual report in July 1934, Major Blair indicated that consideration was being given to a pulse method [17]:

> It appears that a new approach to the problem is essential. Consideration is now being given to the scheme of projecting an interrupted sequence of trains of oscillations against the target and attempting to detect the echoes during the interstices between the projections. No apparatus for this purpose has yet been built.

The construction of a demonstration model of a pulse detector began at the Signal Corps Laboratories early in 1936 with a transmitter operating at 133 MHz; this frequency was later changed to 110 MHz. These frequencies were chosen simply because sufficient power was not obtainable from the microwave valves then available. It is pertinent to point out that microwave propagation tests continued into 1935 with the use of German built Hollmann valves,* RCA magnetrons operating at 9 cm, and RCA acorn valves operating at 45 cm in a Hartley circuit. The plan was to place the transmitter on the New Jersey shore and the receiver on a launch that would sail into the waters of New York Harbour. Again, considerable and quite successful work on heat detection progressed within the Army until 1937. Heat detection was employed with radio detection in an early form of the SCR-268 radar. A specification (see Appendix J) for a 'Detector for use against aircraft (heat or radio)', which was drawn up by the Chief of Coast Artillery and dated 1st February 1936, was forwarded to the Chief Signal Officer[18]. The specification underlines the thinking at that time in Army circles. The work in heat detection did not go to waste and in limited fields was continued through the Second World War.

Returning to the breadboard model equipment produced at the end of June 1936, we find that the transmitter employed two triodes in its output circuit and gave a rated average power output of 75 W. The triodes, RCA 834s, were operated in push–pull with parallel wire lines as tank circuits. A 20 kHz master oscillator triggered the keying circuit, a multivibrator which produced pulses of 2 or 3 microseconds, and also synchronised the sweep circuit of a commercial RCA oscilloscope. In addition, the master oscillator provided a quenching voltage to the super-regenerative receiver and it was intended that the receiver would recover its sensitivity before the return of the echo pulse; a phase-shift circuit was used to ensure that the receiver was quenched at the precise moment of pulse transmission. This scheme never worked satisfactorily and eventually a superheterodyne receiver with low-Q circuits was employed. These circuits allowed sharp damping of the receiver at the moment of transmission, but also a quick recovery without 'ringing' in time to respond to echo signals. In addition, the transmitter and receiver were separated and each had its own antenna. The multivibrator keyer was modified to produce a 300 V pulse capable of modulating a 100 W transmitter.

While this work was in progress, Lt. Col. Louis B. Bender of the Office of the Chief Signal Officer, appraised the work being carried out at the Naval Research Laboratory and at the research laboratories of GE, RCA and Bell Telephone Laboratories.

The first successful tests with pulse radar occurred on the 14th and 15th December 1936 near Princeton Junction, New Jersey. The system used was as described above and employed a 75 W transmitter and a superheterodyne receiver. The transmitter and receiver were one mile apart, with the transmitter using a Yagi antenna and the receiver a dipole antenna. A letter from Lt. Col. Blair to Lt. Col. Bender on the 15th December, 1936, read:

> You will be interested to know that yesterday, December 14, we succeeded in locating an airplane by the pulse method over a range of approximately 0 to 7 miles. [19]

---

* A retarding field valve, see Appendix B.

The next noteworthy event was a demonstration at Fort Monmouth on the 18th and 19th May 1937. After the tests of 1936, there were concentrated efforts to improve both the gain and the directivity of the antenna systems. Parabolic cylinders of wire netting were tried but these were discarded in favour of arrays of half-wave horizontal dipoles with matching parasitic reflecting elements mounted one-quarter wavelength behind the dipoles. The arrays were fixed in wooden frames. For the demonstration of May 1937, the transmitting antenna consisted of an array of dipoles stacked 5 vertically and 2 horizontally. For the azimuth receiver, an array stacked 2 vertically and 6 horizontally was used, while an array of 7 dipoles stacked vertically and 2 horizontally was used for the elevation receiver. The antenna arrays were mounted on a sound locator chassis. (It is of interest to mention that at the 23rd meeting of the British Committee for Scientific Study of Air Defence held on 31st October 1936, at which RDF Beam Technique was discussed, the use of sound-locator-type mountings was proposed [20].) The transmitter operated on 110 MHz and gave an output of from 5 kW to 10 kW peak pulse power with a pulse width of about 5 µs and a PRF of 8000/s. RCA 806 valves were used in the transmitter. The radio detector system was used in conjunction with a thermal detector and a searchlight; the radio detector with a maximum possible range of 20 500 yards controlled the pointing of the thermal detector which, in turn, controlled the searchlight.

A 240 MHz system was also demonstrated. The demonstrations, which were very successful, were carried out during both day-time and night-time and were witnessed by the Chief Signal Officer, the Chief of Coast Artillery, the Assistant Chief of the Air Corps and by representatives of the Ordnance Department and the Corps of Engineers. The operational results of the two-day exercise could be summarised by saying that aircraft coming in at night at any altitude above 6000 ft were successfully illuminated on 4 approaches out of 7. The demonstration was repeated a week later on the 26th May for the benefit of Harry A. Woodring, Secretary of War, Major General W. Westover, Chief of the Air Corps, and several senators and representatives from the Congressional Committees on Military Affairs. At this demonstration the thermal locator, which a week earlier could not be used because of bad weather, was now fully utilised, and in one series of tests a score of 13 out of 15 was obtained. The real success was that Harry H. Woodring was so impressed that he wrote a congratulatory letter on the 2nd June to the Chief Signal Officer and also informed the President, so that by the 26th June 1937 the Chief Signal Officer was in a position to enquire of Lt. Col. Blair, Signal Corps Laboratories director 'whether or not $200,000 will be sufficient for estimates for the research during the ensuing fiscal year' [21].

It can be said that the equipment which was used at the May 1937 demonstration was later developed into what became the SCR-268 [22], the first US Army radar. Also, on the 3rd June, the Chief of the Air Corps proposed an operational specification for a long-range detector and tracking equipment. The requirement in height and range was 40 000 ft and 50 miles, respectively, the latter being later increased to 120 miles. From this request of the Chief of the Air Corps in 1937, there eventually evolved the SCR-270 and the SCR-271 radars, the former for mobile use and the latter a fixed-station equivalent. As the Second World War loomed, the SCR-270 and SCR-271 were given a high priority in their development and in fact went into field use before the standardised SCR-268.

Below are listed the salient characteristics of the SCR-268 set [23]:

Frequency	205 MHz
PRF	4098/s
Pulse width	3 μs to 9 μs
Peak pulse power	>50 kW (75 kW typical)
Power consumption	15 kVA
Design range	40 000 yds
Range precision	±200 yds
Antennae	
Polarisation	Horizontal
Transmitting antenna	4 × 4 array of $\lambda/2$ dipoles with reflector elements spaced $\lambda/4$ behind. Beamwidth (3dB points) 10°, azimuth and elevation.* Gain $\doteq$ 26dB
Azimuth receiving antenna	6 wide × 4 high $\lambda/2$ dipole array with reflectors. Beamwidth (3dB points) 12°. Electronic lobe-switching of antenna pattern for angular discrimination
Elevation receiving antenna	2 wide × 6 high $\lambda/2$ dipole array with reflectors. Beamwidth (3dB points) 9°. Electronic lobe-switching of antenna pattern for angular discrimination
Angular accuracies	Approx. ±1°. Serious errors occurred in reading elevation at angles <10° because of ground reflection.
Antenna side-lobe suppression	30dB
Transmitter	16 100TS valves in ring oscillator configuration
Modulator	8 type 304TL valves in parallel, when conducting, discharged a capacitor, charged to 15 kV, through the transmitter valves. 10 A at 12 kV passed through anode circuit of transmitter.

---

* The value of 24° as quoted by Ridenour (24) is a more likely value.

Receiver	Two identical units, one each for azimuth and elevation arrays. Each receiver comprised: 2RF amplifiers (type 954 acorn pentodes); a Hartley circuit local oscillator (type 955 acorn triode) with buffer stage (type 954 valve); common control grid pentode mixing (type 954 valve); 4 stages of IF amplification at 19·5 MHz (6SJ7 triodes), voltage gain per stage $\doteq$ 12; triode detector (6SJ7); video amplifier of gain $\doteq$ 2·5 (6SJ7). Maximum output of video amplifier was 10 V peak-to-peak into a 500 Ω output circuit
Displays	
Range operator	A scope (see Fig. 2.14) with movable marker
Azimuth operator	K scope (a form of A scope, with two slightly out of phase time-base sweeps; this allows the amplitudes of the signals from two separate antennae to be displayed side-by-side)
Elevation operator	K scope

The SCR-270 provided a peak pulse power of 100 kW at 106 MHz. Pulse widths from 10 μs to 25 μs at a PRF of 62/s were used. The antenna, a stacked array of dipoles with reflector, gave, typically, a beamwidth of 28° in azimuth and 10° in elevation and provided a gain of about 21 dB. Azimuth accuracy of about ±4° was achievable and height finding, using the method of ground reflection and comparison of signals (see Section 5.6.5), was possible. Ranges of 120 miles against large aircraft were readily achieved.

Some facts concerning the production of the SCR-268 and SCR-270 systems may be of interest. The SCR-268 was developed in three stages. The SCR-268-T1, which included a thermal detector and employed mechanical tilting of the receiving antennae to achieve an overlapping of lobes, was service tested at Fort Monroe during November and December 1938. An improved set, the SCR-268-T2, was available in May 1939, but it was never officially service tested. In the meantime, further modifications to the design were carried out, including the use of a single mount for the antennae and the incorporation of electronic lobe switching. This new equipment, the SCR-268-T3, was service tested in April 1940 by a subcommittee of the Coast Artillery Board, following which 18 of them were hastily built by the Signal Corps Laboratories for training purposes. In August 1940, a contract for the production of 212 standardised SCR-268 sets was signed with the Western Electric Company and by the 30th June 1941 a total of 85 sets, both Signal

Corps and Western Electric, had been delivered and accepted. The order with Western Electric was extended in June 1941 to a total of 520 sets. The SCR-268 remained the standard SLC (searchlight-control) radar of the Army until January 1944, when it was declared obsolescent. In August 1940, a contract was signed with Westinghouse for the delivery of 73 mobile SCR-270 units and of 21 fixed SCR-271 units. A total of 794 sets, both fixed and mobile, were produced between 1939 and 1944; this figure includes 5 pre-production SCR-270 sets made by the Signal Corps.

4.3.2 Aircraft Warning Service
Brief reference will be made to the Aircraft Warning Service (AWS). As part of a policy for the defence of the continental United States, there existed, before the advent of radar, the concept of civilian observers, information centres and air-raid warning plans [25] (this was very much akin to the British Royal Observer Corps organisation).

The Chief Signal Officer was instructed on the 10th March 1939 to prepare a comprehensive plan for the organisation and operation of the AWS for the continental United States and for overseas areas [26]. The study, which appeared in February 1940, provided for a total of 23 long-range detector stations and 9 information centres within the United States itself. In March 1941, there was a transfer of responsibility for the network to the GHQ Air Forces. By the end of 1941, surveys had been completed for 46 sites in the United States proper, 14 in Alaska, 12 in Hawaii and 5 in Puerto Rico. A high priority was bestowed on the Panama Canal Zone and, by February 1941, it was equipped with 6 SCR-270 sets. Indeed, in July 1941, three mobile SCR-270 sets were set up as fixed installations in Iceland to help protect the lend–lease convoys. By early 1942, the AWS had a chain of SCR-270 and SCR-271 stations covering the entire East Coast from Maine to Key West, Florida, and the West Coast from Washington to San Diego in California.

4.3.3 Establishment of the NDRC
The development of radar independently by the Navy and the Signal Corps during the 1930s has been outlined and reference has been made to participation therein by some commercial companies. Undoubtedly, some cross-fertilisation occurred between the Navy and the Signal Corps over the period 1936 to 1940 [27]. The year 1940 represented a turning point in the story of United States radar. While it brought to an end the pre-magnetron era of radar growth, which is the area of concern here, it led into a mainstream of development which, during the following 5 years, advanced the technology of radar to the threshold of its present-day potential.

There are three significant and related factors which, in 1940, initiated this growth. They are:

(*a*) the formation of the NDRC (National Defense Research Committee);

(*b*) the setting up of the Radiation Laboratory at the Massachusetts Institute of Technology, and

(*c*) the visit of the Tizard Mission from Britain to the United States, which has already been discussed in Section 4.1.

The setting up of the NDRC was the outcome of an acute sense of awareness, that had arisen among civilian scientists in industry and academic institutions, that their country was entering a war situation in which scientific resources would have to be properly harnessed.

As far back as 1863, the National Academy of Sciences had been set up by an act of Congress so that the best scientific advice would be available to Government departments. In 1916, prior to America's entry into the First World War, a Council of National Defence, composed of Cabinet members, had been set up to co-ordinate industrial and scientific resources and, although inactive, the organisation had never been disbanded. Again, the NACA (National Advisory Committee for Aeronautics) which had been established by Congress in 1915 had been very effective, and in June 1939 President Roosevelt designated it as a consulting and research agency for the joint Army and Navy Aeronautical Board in the event of a national emergency. Dr Vannevar Bush [28] became a member of NACA in 1938. Dr Bush, who had also in 1938 become president of the Carnegie Institution of Washington DC, after a distinguished career in both education and industry, was the man principally responsible for the setting up of the NDRC. Aware of support from colleagues such as Dr Frank B. Jewet, President of the American Academy of Sciences and President of Bell Telephone Laboratories, and visualising a federal organisation with a structure similar to the NACA, he approached Harry Hopkins, adviser to President Roosevelt, and was granted an interview with the President. Roosevelt accepted the proposals made by Bush at the interview and, as a result, an executive order was issued on the 27th June 1940 for the creation of the NDRC.

One of the first acts of the NDRC was to establish in July 1940 a Microwave Committee under the direction of Alfred L. Loomis, a New York attorney and physicist who had been experimenting for many years with microwaves at his private laboratory in Tuxedo Park, New York. The first meeting of the Microwave Committee was on 13th July 1940, at Tuxedo Park. The stated object of the committee was 'so to organise and consolidate research, invention and development as to obtain the most effective military application of microwaves in the minimum time' [29].

It is pertinent to reflect briefly on the state of microwave technology in the United States in 1940. While theoretical studies of the propagation of electromagnetic waves in hollow pipes had been carried out since the 1890s, experimentation with waveguides for practical applications had taken place only from about 1932 onwards. By 1940, in the United States, experiments, using klystron sources operating at about 3 GHz, crystal detectors and matching devices, had been carried out. The various modes of propagation were understood and impedance calculations were facilitated by the use of the Smith [30] Chart. Southworth, who was one of the notable workers on waveguides at Bell Laboratories, gives an excellent review of waveguide development [31]. Only two valves were available which gave worthwhile outputs at wavelengths below 1 m; these were the kylstron and resnatron,* the latter having been developed by David Sloan and L. C. Marshall of the University of California. In August 1940, the resnatron was capable of providing an average power of 1 kW at a wavelength of 45 cm.

The visit of the Tizard Mission in September 1940 (see Section 4.1) resulted in the disclosure by the British of the resonant cavity magnetron, which had just been developed by Oliphant's group at Birmingham and which, at that time, was capable of producing a peak pulse power of 10 kW at 3 GHz. The availability of such a powerful and suitable microwave generator gave a new impetus to the microwave programme. The setting up of a central research and development laboratory under civilian control and on the lines of the British Air Ministry Research Establishment (AMRE) was mooted. Events moved rapidly and on the 18th October 1940 it was decided to set up the Radiation Laboratory at the Massachusetts Institute of Technology with L. A. Du Bridge as its technical Director. It was decided mutually between the British and the Americans that the Laboratory should concentrate initially on three projects with the following order of priority:

(a) a microwave (10 cm) AI (Aircraft Interception) system;
(b) a precision gun laying radar;
(c) a long-range aircraft navigation system which would not be dependent on signals sent out from the aircraft.

On the 11th November 1940, a preliminary meeting of Laboratory staff was held at which the allocation of research projects into seven divisions, such as receivers, antennae, took place. Work progressed swiftly and, by the end of December 1940, the first experimental radar system was set up on the roof of the Laboratory. On 10th March 1941, a 10 cm equipment was flown on board a B-18 aircraft and on that day also the Microwave Committee submitted its first report on the Radiation Laboratory to Dr Vannevar Bush. With a total staff of 140 people, it had successfully come to the end of its first phase of growth. The remainder of its history was rich in achievement and varied in detail. Guerlac's unpublished history of radar [1] devotes the greater part of its text to the work of the Laboratory.

4.4 Beginning of radar in France

France developed its own radar industry before the war and, indeed, quite significant developments were achieved in their naval radars during the years of occupation.

---

* Resnatron is an acronym for RESoNAtor-TRON, a cavity resonator tetrode; one cavity between control grid and cathode, the other between anode and accelerator grid.

There were some exchanges of information between Britain and France just prior to, and after, the beginning of the war in 1939. One meeting of importance was that of 7th November 1939 [1], which was chaired by Général de Division Jullien, Technical Inspector of Signals for the National Defence, at which 11 British and 22 French representatives were present. The British attendance include Air Marshal Sir Philip Joubert de la Ferté and Watson-Watt. The meeting discussed a very comprehensive plan for a network of stations which would cover the Northern and the North Eastern frontiers of France.

This liaison with the British is considered to have in no way influenced the main line of French radar development. The story of the liaison is well told in the British Air Ministry publication 'Radar in raid reporting' [2]. Six French officers were trained on an 8-week course in the summer of 1939 at Bawdsey and so had a thorough knowledge of the functioning of the Chain Home system. After the outbreak of war, other personnel were trained on GM (mobile gun laying) and GL (gun laying) sets and on the IFF MkI, and some had even been integrated into crews on radar watch-keeping duties. Although Guerlac believed otherwise [3], the British assumed that the Germans were well informed of their radar after the Armistice with France, as the following quotation from an Air Ministry publication shows:

> Nevertheless it was only safe to assume that practically all the information about our ground RDF stations and some in regard to aircraft equipment was in enemy hands after the fall of France. [4].

In retrospect, a significant result of the French–British co-operation before the fall of France in June 1940 was that French research on the magnetron was critical to the successful British development of the cavity magnetron, as will be related in Chapter 6.

One could very broadly summarise the evolution of French radar from 1934 onwards by saying that there were two separate, but concurrent, streams of development. One stream was a military one with which the name of Pierre David is associated, and the other was an initially commercial one, involving the affiliated companies CSF and SFR* and the name of Maurice Ponte.

Although Pierre David did, in 1938, advocate the use of pulses, the system known as the David System used a continuous wave beat-frequency method in a bistatic configuration, and, in its ultimate form before the outbreak of war in 1939, pairs of transmitting and receiving stations were combined into a chain, which functioned as an early warning system and even as a ground control of interception system. This type of system, which was used by the Japanese and was experimented with for a while by the Americans, was referred to very aptly by the French as a 'barrage électromagnétique' (electromagnetic barrier).

---

* CSF: Compagnie Générale de TSF (Télégraphie San Fils), and SFR: Société Française Radioélectrique.

Maurice Ponte experimented initially with continuous waves but later used pulses. While separate transmitting and receiving antennae were used, the configuration was monostatic, and the most common wavelength employed was 16 cm. The power generated was small so that ranges obtained were never comparable to those obtained with the David System. In 1939 SADIR (Société Anonyme des Industries Radioélectriques) combined the lower frequencies of the David technique and the pulse method of Ponte to produce a naval radar, operating first at 50 MHz and later at 150 MHz, with approximately 20 kW peak power output.

A brief account of the development of radar in France will be set out in a chronological order. One may go back as far as 1923 to discover the first recorded interest in electromagnetic detection. The story has been recorded by Captain J. Bion [5] who, at that time, was head of the Scientific Research Department of the General Staff of the Navy. The story is significant in that its central figure, Mr Brard, was a radio amateur. He seems to have been inspired by material he read in the amateur journal *l'Antenne* and also by the work of Professor Paul Langevin who had experimented in ultrasonics and underwater detection during the First World War [6]. Briefly, the method Brard employed was to connect a 10 W transmitter to a frame antenna which was rotatable. A vertical antenna was set up some 10 metres away, and it was connected to the ground by a 'ticker' or interrupter which was operated by a clockwork mechanism. The signal reradiated back from this vertical antenna was picked up by a receiver and this occurred only while the beam of the transmitting antenna irradiated the vertical antenna.

Two years later, in 1925, Pierre David proposed that aircraft might be located by detecting the radiations generated by their magnetos, an idea which had been tried by Major E. H. Armstrong of the United States Signal Corps during the First World War. General Gustave Ferrié supported the idea, but, although interesting results were obtained, the experiments were abandoned.

Camille Gutton and his assistant, Émile Pierret, at the University of Nancy tried, in 1927, to detect nearby objects using retarding field generators (see Appendix B) with parabolic reflectors, and operating at a wavelength of 16 cm.

On the 5th June 1928, Pierre David submitted a memorandum indicating that the detection of direct radiations from aircraft was not feasible because of the screening of magnetos, but that aircraft, if irradiated by a beam of high-frequency radio waves, might be detected by receiving the reflected energy. The scheme was considered feasible, and work on it was sanctioned, but the death of General Ferrié and lack of support from Henri Guton, who had been appointed the new director of the laboratory (Laboratoire de Télégraphie Militaire), resulted in experimental work not going ahead [7].

In 1932, SFR succeeded in using split-anode magnetrons, operating at wavelengths of 70 cm or longer, as transmitters. These delivered several watts of continuous power.

In 1933, Pierre David, after reading the article of Englund, Crawford and Mumford [8], was reminded of his proposals of 1928. He was now very clear about it feasibility. At his instigation, a letter was dispatched on the 12th January, 1934, by the President of the 'Comité d'Etudes et d'Expériences Physiques' to the Ministry of War, and results were obtained by the 22nd January when a Squadron Leader* Blondel was appointed to the project. Then, during June and July, apparatus was set up and successful experiments were carried out at Le Bourget. A frequency of 75 MHz was used with the transmitter and receiver separated by a distance of 5 km. The transmitter delivered 50 W and the receiver was a super-regenerative type. During the first experiment, an aircraft was detected up to a maximum height of 5000 metres. At this same period, during the summer of 1934, Maurice Ponte and Henri Gutton, with the object of maritime detection in mind, experimented with the following:

(a) A magnetron transmitter working on 80 cm with a power output of a few watts, and a regenerative receiver using a retarded field valve.

(b) A retarded field valve transmitter operating on 16 cm with a power output of a few tenths of a watt, and a regenerative or a super-regenerative receiver employing a retarded field valve.

In February 1935, the Navy experimented simultaneously at Toulon with SRF equipment working on 16 cm, and with the metric system of David's. The power of the former proved insufficient for the detection of some targets, so studies were continued using the metric system, which had detected aircraft passing over a base-line of 15 to 21 km. A plan of work was established by the 'Centre of Naval Studies', who foresaw the establishment of a 'barrage' between the Continent and Corsica.

On the 10th March 1935, Maurice Ponte wrote a confidential memorandum on the detection of mobile objects by very short waves. He envisaged wavelengths of 4 to 6 metres, power of several kilowatts, ranges of 100 km and the determination of the co-ordinates of a target in three dimensions.

On the 6th April 1935, 70 000 francs† were allocated by the Government to allow the construction of 3 transmitters and 6 receivers working on metric wavelengths, to implement a David type of 'barrage' system.

In January 1935 the owners of the newly built liner NORMANDIE requested SFR to install an obstacle detector system on board. As a preparatory exercise SFR installed 80 cm and 16 cm equipments on the cargo vessel *Oregon*. In experiments on the 5th and 6th April 1935 in Dunkirk harbour, the following results were obtained.

---

* French title: Chef d'Escadron.
† Approximately £1000.

80 cm: The direct signal from the transmitter into the receiver was too strong and did not allow targets to be picked up.

16 cm: The receiving antenna was 3 to 4 m away from the receiving antenna. The coastline was picked up at 9 km to 10 km and a small vessel was detected at 5 km. Aircraft, however, could not be detected.

A 16 cm set using parabolic mirror antennae was installed on the *Normandie*. Its transmitter gave a continuous-wave signal, modulated by 800 Hz*, and a reaction receiver was employed. Tests were carried out between the months of August and December 1935 and it was found that direct signals, or spill-over from the transmitter, as well as reflections from the superstructure of the ship, hindered the obtaining of good results. Interesting results were reported while passing between Le Havre and Southampton.

In manoeuvres during the summer of 1936, the DAT (La Défence Aérienne du Territoire) set up a 'ligne de guet électromagnétique' (electromagnetic look-out line) in the region of Gien (Loiret). 25 aircraft were picked up, but several escaped detection.

The David system was set up in the Reims-Argonne region for the June manoeuvres of 1937. Three configurations of stations were used, the most complex of which was referred to as 'maille en Z' and allowed the speed, direction and altitude of an aircraft to be determined to within 10%, 10° to 20° and 1000 metres, respectively.

Also in 1937, SFR continued their studies of decimetric waves and Henri Gutton succeeded in detecting a man at a range of 150 m. SFR also installed equipment on the vessels *Normandie* and *Ville D'Ys*. The equipment used magnetrons which delivered 10 W of power at 16 cm, and it was deemed suitable for the detection of icebergs. Following on from this, SFR at the beginning of 1938 installed an improved system at Le Havre. The equipment consisted of

Transmitter: Wavelength 16 cm, power output 10 W and pulse width of 6 μs
Receiver: Reaction type or superheterodyne type; cathode-ray tube indicator with a sweep of 1 cm/km. An audible indication at 800 Hz was also available

Two parabolic reflector antennae, of 12 cm focal length and 1 m diameter, were used.

The results obtained were as follows: ranges of 8 km with cargo vessels, 3 km with fishing boats, and a precision in the measurement of range of between 200 m and 300 m.

In October of 1938, Pierre David forwarded a letter to the 'Comité d'Expériences Physiques' stressing the necessity of using monostatic pulse equipment for aircraft detection.

---

* An article in *Wireless World* (Reference 61, Chapter 3) gives a modulating frequency of 7500 Hz. All other accounts give the more likely figure of 800 Hz.

In 1939 significant developments occurred. First, the David system operating at 30 MHz and using equipment manufactured by SADIR and LMT (Le Matériel Téléphonique) was set up to protect the ports of Cherbourg, Brest, Toulon and Bizerte in Tunisia and also the north-eastern approaches to Paris at Epernay and Troyes. Then SFR installed near Paris a set which operated at 100 MHz and delivered a peak power of 25 kW. In September 1939, SADIR installed a 50 MHz set with peak power of 12 kW at Carqueiranne, near Toulon, which gave a range of 60 km. This was followed before the end of the year by the installation of a set at nearby Cap Sicié. This set worked at 50 MHz and, with a peak power of 25 kW, detected aircraft at ranges varying from 60 km to 130 km. Towards the end of the year also, between October and December, SFR experimented with a 16 cm set at Pointe de Saint-Mathieu near Brest. There were virtually no results with aircraft, but excellent performance was obtained with ships.

At the beginning of 1940, a new requirement in aircraft detection arose. This was the detection of mine laying aircraft in the English channel, which necessitated the spotting of aircraft at an altitude of only 50 metres at ranges up to 30 km. For this purpose, a SADIR set as used at Carqueiranne was modified to operate at 150 MHz.

Between February and May 1940, LMT installed at Port-Cros, near Toulon, a pulse radar equipment which was remarkable for that time. It operated on a frequency of 48 MHz with a transmitter peak power output of 350 kW, pulse recurrence frequencies of 50, 250 and 1250 Hz, and pulse widths of 10 µs and 25 µs. At times echoes from aircraft went beyond the range of the display, which was 130 km, and echoes returned from Corsica, 210 km away, were strong enough to be a source of annoyance.

What must be considered an achievement was the construction in April 1940 by SFR of a magnetron with oxide-coated cathode which delivered a peak power of 500 W. Maurice Ponte brought a sample of this magnetron over to Britain on the 9th May 1940. Then, and in the midst of the German occupation, SFR developed their magnetron until in September 1940 it was delivering a peak power output of 4 kW.

The fall of France in June 1940 did not mean the cessation of work on French radar. In January 1941 it was decided to install a SADIR set (250 MHz, 15 kW) on the battleship *Richelieu*. The equipment, which had functioned in Bizerte, Tunisia, was transported by warship to Dakar and was installed and functioning on the *Richelieu* by the 25th May. Sets were also installed on the *Strasbourg* and *Colbert* at Toulon and on the *Jean-Bart* at Casablanca. In April 1942, the *Algerie* at Toulon was fitted with a new radar set produced at the LMT laboratories at Lyons. Operating at 150 MHz, it had a peak pulse power of 60 kW and a maximum range of 150 km.

In October 1942, an LMT set operating on 50 cm with a peak pulse power of 25 kW was installed on the peninsula of Saint-Mandrier for the defence of Toulon. It was able to detect a cruiser of 10 000 tonnes at a range of 20 km with a precision of 30 m. Angular precision in azimuth was one degree and this was obtained by antenna lobe-switching.

All radar equipment in Toulon was destroyed when it was invaded in November 1942.

4.5 Beginning of radar in Italy

The origins of Italian radar can justifiably be said to be traceable to the speech which Marconi made in 1922 before the joint meeting of the American Institute of Electrical Engineers and the Institute of Radio Engineers, which has been alluded to in Section 3.7. It is not because Marconi was of Italian birth that this is deemed to be so, but because he actually verified his prediction of 1922 before the Italian Authorities in both 1933 and 1935, as a direct result of which a Government research programme began, which led to the production of the first Italian radar.

The driving force behind Marconi's work on propagation at the higher radio frequencies, from 1924 onwards, seems to have been the desire to break down the then accepted barriers and to demonstrate the usefulness of these frequency bands. From 1924 to 1928, he concentrated his efforts on studying propagation within the 3 MHz to 20 MHz high-frequency band. In 1930 he achieved a 30 MHz radio-telephone link between Sardinia and Rome, a distance of 170 miles. Then, in 1931, he experimented with waves of 600 MHz using Barkhausen–Kürz oscillators (see Appendix B) in both the transmitter and the receiver [1, 2].

The Barkhausen–Kürz triode valves were arranged in a balanced push–pull circuit and the anode, grid and filament circuits were all tuned by Lecher lines. It took some time to develop a suitable valve. A common experience at that time with valves which were capable of giving continuous-wave Barkhausen–Kürz oscillations was that the grids melted after a few minutes of operation. Marconi eventually succeeded in developing a valve which dissipated 200 W at 600 MHz and had a life of approximately 500 hours. The Barkhausen–Kürz transmitter was capable of being frequency-modulated by simply modulating the negative potential applied to the anode of the valve. The dipole antenna, which was directly coupled to the grid circuit by means of an open Lecher line, was placed at the focus of a parabolic reflector. In like manner, the receiving antenna was connected to a pair of push–pull valves also connected in the Barkhausen–Kürz configuration. The push–pull pair acted as a form of regenerative detector and this was followed by two stages of audio amplification.

By the autumn of 1931, the transmission of good quality speech was possible between Santa Margherita and Levante, a distance of 18 miles, and demonstrations were given to the Italian Ministry of Communications. In April 1932, a duplex telephone system was demonstrated and then the Vatican

Italian 'Gufo' (EC3-ter) set on board LITTORIO, 1941.

Authorities expressed an interest in linking the Vatican with the Pope's summer residence at Castel Gandolfo, a distance of 15 miles. A telephone and teleprinter link operating at a wavelength of 90 cm was officially inaugurated by His Holiness Pope Pius XI in February 1933, and this constituted the first microwave telephone subscriber link ever to go into service [2]. During the testing of this installation, it was common practice to transmit a modulating tone to enable adjustments to be made to the signal levels in the system. This tone could be received on the adjacent receiving antenna. On a particular occasion, a rhythmic modulation of this monitored signal was observed and Marconi noticed that the variations occurred each time a steam roller, which was working on the road in front of the antennae, moved. Controlled experiments were then carried out. A man walking in front of the antennae, for instance, could be detected. The results achieved were demonstrated in 1933 to the Italian Military Authorities including General Professor Luigi Sacco. On that occasion, Marconi suggested that research should begin in the field of bistatic CW apparatus with the ultimate aim of proceeding to a monostatic pulse system when difficulties with the pulsing of high emission valves had been overcome [3].

During 1934, Marconi built in his laboratory at Cornegliano (Genova), two sets with parabolic reflector antennae for the detection of obstacles. With this equipment, it was possible to detect the presence of motor cars at a distance of about $1\frac{1}{2}$ miles. A second public demonstration of the radio-detection of objects, which marked the culmination of his work in this area, was carried out by Marconi on the 14th May 1935 at the Acquafredda near Rome. With Marconi at this demonstration were the head of the Government, Mussolini, General Arturo Giuliano, Inspector General of the Army, and General Luigi Sacco (whose title or appointment was 'Direttore del Reparto Studi Generale'). General Professor Luigi Sacco may rightly be considered the prime instigator of Italian radar. He had many years of pioneering work in radio-science to his credit, particularly in the fields of direction finding, antenna design and propagation [4]. After Marconi's first public demonstration, General Sacco, who believed in the importance of the radio-detection of objects, chose a young but promising radio scientist, Professor Ugo Tiberio of the Istituto Radio dell' Esercito a Roma, to help develop the technique. Professor Tiberio was requested to make a theoretical study of the detection of objects by high-frequency radio waves. After what was in effect an investigation of a basic radar range equation [5], he made a report to the Department of Defence and in 1936 he headed a research group at the 'Regio Istituto Elettrotecnico e della Communicazioni della Marina (RIEC)' (Royal Institution of Electrotechnology and Communication) at Leghorn (Livorno).

The choice of a Naval establishment for this research work was because the Navy was the best equipped technically of all three services. The task set for Tiberio was fundamentally that of modifying an ionospheric sounder which operated between 1 MHz and 14 MHz into a radar equipment which would operate above 100 MHz. Between 1936 and 1941, Professor Tiberio carried out his research work practically single-handed except for some help in 1937 from Dr Ricamo. A gauge of the support received from the authorities for this project is the fact that the yearly financial allotment from 1936 until 1941 was a mere 20 000 Italian lire [6].

The turning point in Government radar policy was the naval battle of Cape Matapan on the 28th March 1941, when the Italian Navy lost three cruisers of 10 000 tons, two destroyers and 2300 men. The battle was fought during the night with the British fleet using radar and firing from a distance of 3000 metres.

It is worth recording the achievements between 1936 and 1940 of the small RIEC radar research team which was headed by Tiberio.

(*a*) The production of a 2 metre equipment with linear frequency modulation, similar to that studied by Espenschied and Newhouse in the United States and by Matsuo in Japan (see Section 3.9.3). The equipment was first tested in 1936 and again in 1938, when a superheterodyne receiver was used. It proved generally unsuitable but was recommended for development as an aircraft altimeter.

(*b*) The development in conjunction with the SAFAR* Company of Milan, between 1938 and 1939, of a 2 metre continuous-wave radar. This was not successful as the receiver proved too noisy.

* SAFAR = Società Anonima Fabrricazione Apparechi Radiofonici.

(*c*) Two pulse equipments started in 1939 and completed in 1941. One (to be known as EC3) was a 1·5 metre equipment using a superheterodyne receiver and was the first Italian pulse radar. The second (to be known as EC3-bis) was also a pulse radar and operated on a wavelength of 70 cm. It used a super-regenerative receiver. This latter equipment gave satisfactory results against ships at a range of 12 km and at a range of 30 km against aircraft. A simple block schematic diagram of each of these experimental sets is given in L. Castioni's 'Storia dei Radiotelemetri Italiani' [7].

The designation EC (EC-1, EC-2, EC-3, etc.) given to all Italian built pre-war and wartime radars derived from the title of the Institute at Leghorn, the RIEC. A brief summary of the characteristics and history of each type is given by Tiberio [8] and Friedman [9]. Of the three EC-3 types (EC-3, EC-3 bis and EC-3 ter), the EC-3 ter went into production with the SAFAR Company and was popularly known as 'Gufo', the Italian word for owl.

The battle of Cape Matapan was a turning point in Italian radar development. After it, there developed a serious approach to radar in all three services, and, furthermore, there was a release of information by the Germans on their own radar developments [3]. The battle occurred on the 28th March 1941 and on the 20th April 1941 an EC-3 ter set installed on the ship *Carini* was used to demonstrate the abilities of radar to high ranking officers of the Navy and Army, who were surprised at the revelation. The radar that day detected ships of medium tonnage up to ranges of 12 km and aircraft up to ranges of between 32 km and 80 km. Immediately after this demonstration, an EC-3 bis was installed on the battleship *Littorio,* where it remained until the end of the war. It was, however, not a very satisfactory installation. Before the Armistice in September 1943, the production model of EC-3 ter was installed in three battleships of the *Littorio* class, in the cruisers *Montecuccoli, Eugenio di Savoia, Scipione Africano,* and in five destroyers [9]. Indeed, from information [10] very kindly furnished by L. Castioni, there appears to have been some 100 Gufo (EC-3 ter) equipments in service with the Navy on the 8th September 1943, with almost 40 additional ready and awaiting installation.

The indication of the presence of targets to the operator of the EC-3 bis was by means of an auditory 600 Hz signal into headphones, while in the EC-3 ter set, a type 'J' oscilloscope display was used. The J-scope display is a modification of the A-scope display (see Chapter 2) in which the electron beam producing the display sweeps a circle with a uniform angular velocity, thus producing a larger sweep and effecting a more accurate range measurement. A broad specification of the EC-3 ter (Gufo) is as follows [7]. Minor variations to this specification will be found in the literature.

| Antenna | Two wire mesh horns, or truncated pyramids, each containing equiphase dipoles. The horns, one for transmission and the other for reception, were placed one above the other and could be rotated at 3 |

	rev/min. Beam dimensions were 6° horizontal by 12° vertical
Polarisation	Vertical or horizontal
Frequency	500 MHz
Pulse width	4 μs
PRF	500 Hz
Peak pulse power	10 kW
Maximum range	25 km, against ships
	80 km, against aircraft

An interesting feature of the transmitter was the two triode valves, which in push–pull configuration achieved the total peak power of 10 kW. These triodes were a type 1628, built at the FIVRE* S.p.A. at Florence (Firenze). The pair of triodes was connected to a special cavity resonator [11, 12], from which the feed for a dipole which energised a wire mesh antenna was taken.

Apart from the dedicated work of a few able scientists and engineers, the whole Italian wartime radar programme seems to have been a matter of 'too little, too late'. There were difficulties encountered in the training of radar operators and in the availability of adequate installation and maintenance personnel.

The delays experienced in procurement are highlighted by the following [13] which concerns an order placed with SAFAR for the delivery of a batch of EC-3 sets. The draft of the Contract (No. 0122) for the production of 50 sets (type EC-3 bis), which was delivered by SAFAR to the Italian Navy, is dated 9th May 1942 (this document contains descriptions, drawings and testing norms for the equipment). This was followed by a further two documents from SAFAR which were modifications to the original. One, dated 2nd December 1942, was for the supply of 26 receivers type EC-3 ter, and the other, dated 13th January 1943, was for the supply of 15 oscillograph display panels type EC-3 ter in place of the acoustic units.

Others, apart from Professor Ugo Tiberio, were involved in radar research before the war. One of these was Agostino Del Vecchio, who submitted on the 17th November 1938 a patent for a pulse radio detector using a magnetron and operating between 50 cm and 8 cm [14].

Simple but effective radar counter measures, interception and jamming, were undertaken during the Second World War. 200 MHz equipment, based in Sicily, was used against British radar stations in Malta [4].

4.6 Beginning of radar in Japan

No specific time or event, to which the birth of radar in Japan may be attributed, will be put forward here. Roger I. Wilkinson [1, 2] states that radar in Japan dates back to 1936, when Professor Kinjiro Okabe of Osaka University, who worked with Professor Yagi, demonstrated how the presence

* FIVRE = Fabbrica Italiana Valvole Radio Elettriche.

of passing aircraft and other moving objects could be detected by means of a continuous wave Doppler technique. Professor Yagi claimed that the idea for a Doppler radar arose following a trip he made to Germany before the First World War, when he became interested in supersonic signalling; his experiments in sound led him to a study of the effects of the presence of foreign objects in radio fields. The experiments of 1936 took place as a result of an interest displayed in 1935 by the Japanese authorities in the detection of distant objects.

We find in a book [3], published in Japan in 1944, a statement that 15 or 16 years previously it had been discussed whether or not it was possible to locate the position of an aircraft by transmitting electromagnetic waves and detecting the signals reflected back by the aircraft. There is abundant evidence that the Japanese were well versed in radio science. As far back as 1909, the Imperial Japanese Navy had carried out research on the use of Bellini-Tosi direction finders [4]; it produced its own Type-10 Direction Finder in 1921. Several medium frequency and high frequency ship and shore direction finders were produced by the Navy between 1928 and 1942, and in 1940 an ADF (automatic direction finder) for aircraft was constructed. At the Electro-Technical Laboratory, Ministry of Communication, radio direction finders had been used in the study of HF propagation from 1926 onwards [5]. Again, we find that Yoshihiro Asami of Hokkaido Imperial University went, in the 1920s, to Great Britain to study under Professor Fleming and then to Germany to study under Barkhausen [6]. At the close of the war in 1945, Asami was carrying out research on slot antennae. In an area closely related to radar, that of

Japanese Navy No. 12 1.5 m/2 m mobile early warning set.

radionavigational aids, we find that Dr Minoru Okada took out a patent as early as 1935 and performed successful experiments on a phase-comparison radio beacon system; this worked on the same principle as one of the most important present-day navigational aids, VOR (VHF omni-range). Dr Okada took out a total of 15 patents in the field of radionavigation between 1934 and 1940 [5].

The name of Professor Kinjiro Okabe is associated with work on the magnetron from the late 1920s onwards [7]. He succeeded in developing magnetrons giving appreciable continuous-wave power on 10 cm in 1936, while worthwhile pulse emission was obtained in late 1941, and complete centimetric radar sets were installed in April 1942 in the battleships *Ise* and *Hyuga* [4]. The manufacturer of the Japanese high power pulse magnetron was the Nippon Musen Company, although it is likely that some other firms, such as Toshiba and Nippon Denki, were also involved. Wilkinson [8] gives a value of 22 kW for the peak power output of the Nippon Musen magnetron type S-51 which operated on a nominal wavelength of 3 cm. A report by Dr S. Nakajima (9) quotes a peak power output of 130 kW for a 4·6 cm wavelength magnetron with a 'Mandarin-orange' type of anode.

Japanese Navy No. 32 10 cm radar being used in a coastline defence role.

The experiments of Professor Okabe at metric wavelengths were continued from 1936 onwards. By 1939, a small set giving 3 W output was successfully employed at Hankow. In 1940, a variety of sets with power outputs ranging from 3 W to 400 W were built. It must be noted that while technical experts in the Imperial Navy promoted the development of radar, there was a bias against it from senior operational staff; this was owing to an understandable philosophy of preserving 'radio silence' during engagements, since any emission of radio waves from ships could disclose their position and do away with the element of surprise [4]. After the battle of Cape Matapan, when the Italian Navy suffered because of lack of radar support, this policy was changed [4].

In 1937, the Navy's research laboratory had experimented with FM continuous-wave radar; ranges of up to 5 km were obtained from ships in a large fleet-parade held in Tokyo Bay. Because of dissatisfaction with the results, the experiments were not continued. As in the case of the radio direction finders two decades earlier, when foreign equipment was purchased and studied by the Navy before they designed and manufactured their own sets, so also with radar. Considerable assistance was received both from the Germans and from studying enemy sets which were captured; to be specific, in 1940 a Japanese Technical Commission spent several months in Germany and returned with reports of pulsed radars being built in Britain. This acted as a stimulus to officers in both the Army and the Navy to initiate development programmes. Again, in 1942, one Japanese army officer and two civilians, all engineers, went to Corregidor and Singapore to inspect captured British and American radars; at Corregidor one American SCR-268 was in complete working order [2]. Later, design specifications for a 10 cm airborne search set, the Rotterdam-Gerät (based on the British H2S), were received by radiotelegraph from Germany. This resulted in the type 51 set being constructed, which developed 6 kW peak output power but which, disappointingly, achieved a maximum range of only 20 km [8].

One can point to 1941 as the year when serious radar work was initiated by both the Army and the Navy. In understanding the development and the performance of Japanese radar, two facts must be borne in mind. First, there seems to have been a complete lack of co-ordination and co-operation between Army and Navy in the whole field of radar design, development, manufacture, operation and training. The word 'enmity' has been used to describe the situation [1]. Secondly, civilian scientists working on radar projects in the laboratory were not given an opportunity to observe the fruits of their work; there was no effective feedback from the battle-front to the designers.

There were two broad classifications of Japanese radar sets, Type A and Type B. The former were continuous-wave or Doppler systems, while the latter were pulsed radars. The Navy operated Type B sets only, while the Army used both types, and operated them side by side in the Air Defence System which guarded the Japanese homeland. Another broad classification was that of detectors and locators. Detectors referred to long-range early warning sets, whereas anti-aircraft and searchlight control sets were referred to as locators. Wilkinson [1, 2] discusses in some detail the system of radar nomenclature employed by the Japanese. One example only will be given here, the Tachi–Tase–Taki designation of the Army sets. After the development in 1941 of the first detector by the Sumitomo Communications Industries Company and the first locators by the Tokyo Shibaura Company, an Army Institute called the Tama Research Institute was formed to promote the expansion of radar research. In designating sets, the prefix 'Ta' for Tama Institute was used with one of three suffixes as follows:

Tachi, meaning land-based ('chi', from tsuchi, meaning earth)

Tase, meaning shipborne ('se', meaning water)*

Taki, meaning airborne ('ki', from kūki, meaning air)

The first Army pulse radar set to be built was the prototype of what was known as the Tachi-6 warning set, of which some 350 were built. Frequency was within the band 68 MHz to 80 MHz. Peak power output ranged between 10 kW and 50 kW. Ranges up to 300 km were obtained, targets being displayed on an A-scope. Range accuracy was ±7 km and azimuth accuracy was ±5°. The set corresponded to the American SCR-270 and was first manufactured in 1942. A Tachi-6 system comprised a transmitter which radiated in all directions or into a 90° sector, with from 3 to 6 receivers spaced about it at a distance of some hundred yards. Polarisation was horizontal. The receiving antennae were Yagi arrays which could be hand-rotated. In operation, one or more receivers were assigned a particular search sector. While one antenna tracked a flight of aircraft, the others could continue searching.

Before the Japanese attack on Pearl Harbour, an Air Defence System had been organised by the Army to protect the Japanese mainland and this comprised three districts with an information centre in each district. These centres were at Tokyo, Osaka and Fukuoka. The centres received information from a network of radars around the coast and from observers. The first radars to be used were Type-A systems, bistatic continuous-wave installations. These remained in operation throughout the war. It was the intention to replace them eventually with monostatic pulse radars except in certain mountainous regions. With a transmitter at one point and a receiver at another distant point, a fence was created. An operator would report to the Information Centre that an aircraft had passed through the beam or fence. On the display board in the Centre, each link of the Type-A Chain was denoted by a line which could be illuminated in red whenever an aircraft had passed through the beam. The frequencies of these Type-A sets lay in the band 40 MHz to 80 MHz. Typical transmitter power outputs were 3 W, 10 W and 100 W with distances between transmitters and corresponding receivers varying from about 40 miles to 150

---

* This derivation is not altogether clear: while 'se' may mean 'rapids' in a river, the common word for water is 'mizū'.

miles [8]. The longest Type-A line used was not in Japan itself, but between Taiwan (Formosa) and Shanghai, a distance of over 400 miles.

The Navy operated, completely independently of the Army, its own network of stations, all Type-B sets, although towards the end of hostilities they were contemplating the installation of some Type-A systems. The coverage of their system in many cases duplicated that of the Army, but they were principally concerned with protecting their own installations.

Towards the end of the Pacific war, when Japan was very much on the defensive, deficiencies became apparent in the working of their Air Defence System, particularly because of a lack of effective GCI and IFF equipments.

A small number of large civilian firms produced the radar equipment for both the Army and the Navy. The principal companies were the Nippon Musen Company, the Tokyo Shibaura Company and the Sumitomo Communications Industries. The Nippon Musen Company was affiliated to Telefunken of Germany and it employed some 16 500 people in four plants. Its research section at Mitaka became noted for vacuum tube development and, in particular, for a series of 10 cm and 3 cm high power magnetrons developed during the war.

Radar training schools for operators and maintenance personnel were run by both the Army and the Navy. A school for the Navy was opened first in March 1942, while an Army training school did not formally open until April 1944. The Navy's school, at Choga, opened in September 1944, and could accommodate 7000 students at any one time. The standard of training there was considered by the inspection team from the United States Air Force to be comparable to that attainable in the United States.

Towards the end of hostilities in 1945, the Army and the Navy each had available all the common types of land-based, shipborne and airborne radar sets. They also possessed radar counter measures equipment, both search receivers and jamming transmitters or 'disturbers'. The Army produced both FM and pulse radar altimeters, and was developing a hyperbolic navigation system somewhat similar to Loran. The Navy developed FM altimeters. Wilkinson [2, 8] classifies the radar equipment developed by the Japanese Army and Navy and indicates that 52 Army types were designed, while the Navy carried out design studies on some 62 types of sets. In Appendix I are listed the characteristics of the principal Japanese Navy sets available at the end of the war [4].

4.7 Beginning of radar in Russia

A common opinion regarding the manner in which radar was introduced into the USSR has been that Russia acquired its first equipments during the Second World War from Britain and from the United States through the 'Lend-Lease' programme, and that subsequently it copied some of the designs of its allies before independently developing its own systems. This view may be substantially correct as far as the ultimate post-war development of Soviet radar is concerned [1], but it does not take into account significant achievements by Russian scientists and engineers in the field of radar in the 1930s. The origins of radar science in Russia go back as far as 1932 and in a more supportive political, and indeed military, climate the achievements before 1941 might have been far more substantial.

The short account which follows is based largely on material available in John Erickson's paper [2].

The first significant set of tests, which were successful and which are well documented [3, 4], were carried out by LEFI (Leningradskii Elektrofizicheskii Institut), the Leningrad Electro-Physics Institute, in July and August 1934. The equipment used, designated RAPID, was a bistatic continuous-wave system; the transmitter gave an output of 150–200 W on a wavelength of 4·7 m, and with the carrier frequency modulated at 1000 Hz. The receiver was of the super-regenerative type and employed a small horizontal dipole antenna, which, judging by a photograph of the apparatus [5], was mounted on a short vertical rod that could be held in the operator's hand and rotated. The presence of an aircraft was indicated by a beat note in the receiver headphones. Tests were carried out on the 10th July 1934 with a separation between transmitter and receiver of 3 km, and on the following day with a separation of 11 km. Aircraft up to heights of 1000 m and within a radius of 3 km from the receiver were detected. Further tests were carried out on the 9th and 10th August 1934 to help establish the ultimate range of the equipment, and it was discovered that maximum ranges of 75 km were obtainable. The tests of July and August 1934 are well documented by both Oshchepkov [6] and Khoroshilov [7].

The results obtained on the 9th and 10th August 1934 are shown in Table 4.1. The tests on the 9th August were carried out in the Krasnoguardeisk region, while those on the 10th August were carried out in the region of Siverskiy station. Krasnoguardeisk (Gatchina) is 25 miles SSW of Leningrad, while Siverskiy is some 15 miles further south.

March 1932 may be taken as a starting point for the story of Russian radar, when a new sound-locator and searchlight system was being tested for Air Defence purposes. The subsequent developments followed a pattern which had its counterpart in other countries, and particularly in Britain and the United States; the discovery of inherent prohibitive limitations in the use of sound locators, the obtaining of some encouraging results with infra-red detection, but only in a limited number of situations, and finally the realisation that some method of radio-detection was the only satisfactory solution. In March 1932, the Military-Technical Administration of the Red Army was running a series of tests with a new sound-locator system. In these tests, an aircraft approached a target at night-time, with the target protected by sound-locator and searchlight. Even with ideal conditions of still air pertaining and no evasive manoeuvres being performed by the aircraft, the success rate for locating and illuminating the target was only just over 50%. The locator

Table 4.1

Place and date of test	No. of test	Time	Route	Aircraft Altitude m	Results	Remarks
9.10.1934, Krasnoguardeisk transmitter 50·4 km from aircraft. Aircraft type: 'R-6'	1	17.58	towards Leningrad	1200	good	On curved paths the distance of aircraft from receiver was 5 km–7 km
	2	18.02	curved path		good	
	3	18.04	towards Krasnoguardeisk	2500	good	
	4	18.08	curved path		weak	
	5	18.10	towards Leningrad	3300	good	
	6	18.13	curved path		good	
	7	18.15	towards Krasnoguardeisk	3900	good	
	8	18.19	curved path		weak	
	9	18.21	towards Leningrad	4500	good	
	10	18.25	curved path		good	Behind cloud
	11	18.29	towards Krasnoguardeisk	5000	good	
	12	18.32	curved path		unclear	
	13	18.36	towards Leningrad	4200	good	
	14	18.38	descending		good	
	15	18.41	descending		unclear	
	16	18.48	descending		weak	Direct signal received was weak and uncertain
	17	18.49	descending		good	
10.10.1934, Siverskiy transmitter 70·6 km from aircraft. Aircraft type: Fighter	1	19.00	various			When the receiver antenna was raised the signal improved
	2	20.00	various			

was put into service with the Air Defence Forces (Voiska Protivo-vozdushnoi oborony: PVO), but even with special training of operators (and the employment of blind people with a heightened sense of hearing), the system was not of much value.

The Main Artillery Administrative (Glavnoe artilkerisskoe upravlenie: GAU), realising the limitations of acoustical methods of detection, approached the All-Union Elektro-Technical Institute (Vsesoyuznyi Elektro-technicheskii Institut: VEI) for assistance. Here, Professor V. L. Granovskii had been developing an infra-red detector for use against aircraft. He produced in 1932 a 'thermal course and bearing indicator', which used a 150 cm searchlight mounting [8]. The heat, caught by the parabolic metal dish, was focused on to a heat-sensitive element, which, in turn, converted the heat into electrical energy whose variations were then amplified, producing an aural indication in a set of headphones. M. Lobanov was in charge of the

Russian 'RUS-2' 75 MHz medium-range early-warning set.

equipment. In the absence of cloud, a bomber could be detected up to ranges of 10 to 12 km. However, the apparatus proved of value when mine sweepers were detected from one of the Kronstadt forts (Gulf of Finland) at a range of 22 km. When the general infra-red programme was eventually terminated in 1935, special work for the naval forces was still maintained.

The GAU, considering a radio-wave solution, then made an approach to the Scientific-Research Experimental Institute of Red Army Signals for technical assistance, but were informed, and possibly understandably so by a fully communications oriented organisation, that the venture was not realistic. The GAU then approached prominent radio-physicists and in August 1933 held a conference with the personnel of the Central Radio Laboratory (Tsentral'naya Radiolaboratoriya: Ts RL) and, as a result, Ts RL agreed to conduct research into 'radio-technical means' of detecting aircraft, a formal agreement being signed in October 1933. Yu. K. Korovin, using apparatus developed by Professor N. D. Papaleski between 1930 and 1932 for geodetic measurement, demonstrated the feasibility of the concept, and in January 1934 a new agreement between GAU and Ts RL was signed whereby Ts RL would produce a radio-locator set for AA guns. The GAU also contracted with the Leningrad Institute, LEFI, for work on radio-detection. Engineer B. K. Shembel, directed by Academician A. A. Chernyshev, supervised the work.

The PVO, or Air Defence Command, independently of the GAU, became interested in the radio-detection of targets and, with a broader objective than the GAU one of gun-laying. The PVO had acquired Pavel Oshchepkov, an electrical engineer, who had graduated in 1931 and who had gained some experience working on modifications to the optical sights of AA guns. Oshchepkov was to play an important part in the development of Soviet radar. In June 1933, he drafted a paper on behalf of the PVO for submission to Defence Commissar K. E. Voroshilov seeking permission and the necessary funds to set up a research unit. A meeting followed between the PVO and both K. E. Voroshilov and M. N. Tukhachevskii, Head of Ordnance, which resulted in an initial budget of 250 000–300 000 roubles* and a high priority programme of research being approved, the first part of the programme to be carried out in 1934. In August 1933, Oshchepkov visited the Soviet Academy of Sciences in Leningrad where he met its president, A. P. Karpinskii, and other Academicians and returned to Moscow quite hopeful about the practicability of radio-detection. In January 1934, Oshchepkov met Academician A. F. Ioffe to make arrangements for a conference. The conference was held on the 16th January 1934 and it was attended by a large body of radio specialists and radio-physicists which included Academician Ioffe and B. K. Shembel, and by Khoroshilov, Oshchepkov and Zhukoborskii from the PVO command. The conference affirmed that the most promising method for detecting aircraft, at night-time, at great height or in conditions of poor visibility, was a radio-detection method, but it also recommended that the

* Sterling equivalent in 1933 was slightly in excess of £25 000–£30 000.

development of other methods of detection, including the use of infra-red radiation, should be continued. Although the conference specifically referred to centimetric and decimetric radio-waves, this qualification was struck out of the minutes by Ioffe.

On the 19th February 1934, the PVO signed an agreement with LEFI to carry out research into the measurement of electromagnetic energy reflected from various objects and materials. LEFI were authorised to build a transmitter and receiver. This led to the experimental apparatus 'RAPID' and the tests of July and August 1934 already referred to; a successful demonstration was given later in August in Moscow to senior officers and officials, and the PVO placed an order for 5 factory-produced experimental sets on the 26th October 1934.

Much had been achieved in a relatively short space of time, but with hindsight one can judge that the choice of a continuous-wave solution would ultimately prove unfruitful; on the other hand, it must be said that the aim of the PVO was to produce an electromagnetic fence or screen at heights up to 10 000 metres along the Soviet frontier [2], so that a continuous-wave system, as used by the Japanese, would have been adequate.

The 'RAPID' system was no solution to the fire-direction problem of the GAU. It should perhaps be mentioned that the Soviet radio industry, because its expertise was exclusively in the longer wave equipments and techniques, was of little assistance to any of the radar research being undertaken. In 1935, LEFI was absorbed into the Television Institute and re-named NII-9.* In 1936, NII-9 produced the 'BURYA' ('STORM'), a search and tracking radar [9], which operated on a wavelength of 18 cm and gave a continuous-wave output of 6–7 W. It had two parabolic antennae, one for transmission and the other for reception. The set was capable of detecting an aircraft at a range of 10 to 11 km with an angular accuracy of 3 to 4°. Neither range nor angular precision was good enough for anti-aircraft purposes. This prototype equipment was improved by increasing the transmitter power output and receiver sensitivity and by increasing the angular resolution; the latter was achieved by means of a double lobing equi-signal technique. 1936 was a year of intensive research, and while the GAU worked with NII-9, the PVO co-operated with Ioffe and the Leningrad FTI (Fiziko-tekhnicheskii institut) and with Rozhanskii and the Ukranian FTI. The PVO programme resulted at the end of 1936 in a prototype pulse radar operating on a wavelength of 4 m which achieved a range of 7 km but which, following improvements in transmitter power output, detected an R-5 aircraft flying at an altitude of 1500 m at a range of 17 km.

Opposition from the Red Army Signals command to the whole radar venture grew from 1935 onwards. However, in 1937 the new Army Commander, A. I. Sedyakin, ran a large-scale air-defence exercise in which sound locators and searchlights were used and in which they again showed

* NII-9 = нии-9 = научно-исследовательский институт № 9 = Scientific Research Institute No. 9.

their inadequacies. During the exercises, he was briefed on the work which had already been done on radio-detection and he had a discussion with Lobanov. It was Sedyakin's intention, apparently, to submit proposals to the Government to have the radio-detection programme speeded up. Before anything could be achieved, however, the great purge of 1937–1938 affected the high command of the Army and, in fact, totally weakened the complete army and its whole officer corps [2, 10, 11]. Indeed, on the 11th June 1937, Marshal Mikhail Tuckachevsky, Vice Commissar of Defence and Commander in Chief of the Red Army, together with seven of his top generals, was charged with spying for 'a foreign state' (Germany). Pavel Oshchepkov was arrested and spent the following 10 years in prison. The Signals Research organisation, the NIIIS KA, took over Oshchepkov's laboratory and indeed took over his production programme. Previous to this, in the midst of the purges, the antagonism of the Signals Command against radar development had been made manifest by their formal enquiry into the scientific work of NII-9. Now, their attitude was changed and they promoted the development of Oshchepkov's equipment which was titled 'REVEN' ('RHUBARB'). It went into production as 'RUS-1' and was tested in 1939. Some 45 of these sets were made and a number of them were installed in the Soviet Far East and in the Trans-Caucasus.

NIIIS KA in July 1938 started a series of tests with a pulse equipment which developed 1 kW peak power and which had fixed transmitting and receiving antennae. The set detected an aircraft at a range of 50 km and at heights up to 1500 m. In the following year, they co-operated with Ioffe's FTI group to produce a mobile set called the 'REDUT', which detected an aircraft at a range of 95 km at a height of 7500 m, and at a range of 65 km at a height of 3000 m. Tests were carried out near Sevastopol in October and November 1939 and, using antennae at a height of 160 m above sea level, aircraft at a range of 150 km could be detected. Ten 'REDUT' sets were manufactured. Throughout 1940, the Signals Command, the Leningrad FTI and factory engineers successfully worked on the problem of using a single antenna for transmission and reception which, among other advantages, resulted in an improvement of the bearing resolution of the equipment. 15 of these improved sets, which became known as 'RUS-2' were produced after 1940.

The NIIIS KA had assigned the production of an AA radar set to the Ukranian FTI in 1937. The 'ZENIT' set was produced in 1938 and operated on a wavelength of some 60 cm; during tests in October 1938 it could achieve a range of only 3 km against a bomber, but after lengthy modifications, it achieved, in mid-1940, a range of over 25 km with a range error of ±(500 m–800 m) and an angular accuracy of about 2°. The measurement of target co-ordinates required 38 seconds so that the equipment was completely unsuitable for automatic fire control. It was not until 1943 that the set, developed into a single antenna system and designated 'RUBIN', went into use.

Mention must be made of microwave valve development. NII-9, working under the direction of Professor Bonch-Bruevich and B. A. Vvedenskii, developed a klystron operating on a wavelength of 15–18 cm and producing continuous-wave power of 20–255 W. D. E. Malyarov and N. F. Alekseyev, both of whom worked under Professor Bonch-Bruevich, produced a series of cavity magnetrons between 1936 and 1937; demountable, water-cooled, and sealed-off, one-, two-, and four-cavity valves were manufactured, and the highest powered four-cavity demountable design was estimated to have produced a maximum power of 300 W at a wavelength of 9 cm [12].

NII-9 also achieved success with a pulse anti-aircraft range-finder called 'STRELETS', which gave a peak power of 16 kW at 80 cm and which could locate an R-5 aircraft at a range of 20 km with accuracies of 160 m in range and 3° in elevation angle. Then, at the end of 1939, NII-9 were also authorised to produce three prototype anti-aircraft sets, one designated 'BURYA-2' and two of a type designated 'BURYA-3'; the latter type proved the more successful, giving a range of 17·5 km with an azimuth accuracy of 1°. This particular evolution of gun-laying radars led to development work in 1940 on a set designated 'LUNA', which incorporated the azimuth measuring facilities of 'BURYA-3' and the range and elevation measuring techniques of 'STRELETS'.

Opposition to radar came again in the summer of 1940, this time from proponents of the acoustic method of detection and in particular from the staff of the Department of Acoustics in the Dzerzhinskii Artillery academy. A conference was called at which Academician N. D. Papaleksi, a radio physicist, and N. N. Andreyev, Professor of Acoustics, authoritatively and successfully intervened on behalf of the radio-detection method.

Germany invaded Russia on the 22nd June 1941. The State Defence Committee (Gosudarstvennyi Komitet Oborony, GOKO) on the 13th July 1941 issued instructions that a 'LUNA' equipment should be ready by the 5th August 1941 and that production of the model should start on the 20th August 1941. However, the swift German advance and the ensuing evacuation of industries to the east prevented this.

Eventually, in February 1942, the State Defence Committee again returned to the promotion of radar development. A combined industrial and research institute was set up, which included a considerable number of the staff of NII-9, and this institute produced two models of a gun-laying radar, designated 'SON-2A', by November 1942. Shortly afterwards, this set went into quantity production.

4.8 Beginning of radar in Holland

The Dutch developed a working radar before the Second World War and did so apparently without any knowledge of the work in progress in other

Model of Dutch (Signaal) 70 cm system developed 1939 by J. L. W. L. Von Weiler and M. Staal and brought to Britain, 1940.

countries. In 1934, Jhr. Ir.* J. L. W. C. Von Weiler, who worked for the Dutch Armed Forces in the Dutch Physics Laboratory near the Hague (and now part of TNO,† the Netherlands Organisation for Applied Scientific Research), was experimenting with what were then ultra-short waves (1·25 m wavelength) for use in telecommunications. He had two assistants, Ir Gratama and Ir Staal. With a point-to-point link, it was noticed at times that the signal at the receiving end fluctuated and the fluctuations seemed to be caused by the passage of birds flying through the path between the transmitter and receiver. Next, reflections which were confirmed to be coming from aircraft were noted.

These observations led to the development of a pulse radar working on 70 cm. A common antenna for transmitting and receiving was used. This

* Jhr. = Jonkheer, rank of nobility; Ir. = Ingenieur = engineer.
† TNO: Organisatie voor *t*oegepast-*n*atuurwetenschappelijk *o*nderzoek.

consisted of 4 rows of vertical dipoles, each row containing 8 antennae, placed in front of a wire mesh reflector. The prototype equipment was mounted on a truck. The antenna was turned by means of bicycle pedals and could be elevated by means of a handle. The power output attained was low and at a demonstration given to the military authorities, the range obtained on an aircraft was about 8 nautical miles. The purpose of the equipment seems to have been to use it in conjunction with a searchlight for anti-aircraft gun control. Possibly because there was an acoustic output from the receiver, the apparatus was called by the Dutch 'an electrical listening device' ('Electrisch Luistertoestel').

A contract was drawn up with Philips for the production of the device, production to start in 1939. Only four sets were ready by April 1940 and these were marked for interservice testing. When the Germans invaded Holland in May 1940, one of these four sets was used against enemy aircraft and performed well [1]. One set was cannibalised and its components were brought to Ijmuiden and shipped to England, while Von Weiler and Staal escaped from Scheveningen to England with a complete set of blueprints. The remaining devices and drawings were destroyed and did not fall into enemy hands [2].

Both Von Weiler and Staal, on their arrival in Britain, were sent to the Signal School at Portsmouth where 50 cm equipment was being built. They, with some other fellow-countrymen, were integrated into radar development in Britain for the duration of the war. One of the sets was assembled and installed in the Royal Dutch Navy destroyer *Isaac Sweers* which, although incomplete at the time of the invasion, had been towed to England and completed there. The radar was designated by the British 'type 289 of Dutch origin'. Von Weiler accompanied the destroyer on her maiden voyage to the Mediterranean on the 13th September 1941. The destroyer was later sunk in the Mediterranean during the war.

Both Von Weiler and Staal held naval commissions during the war. The former was made a Professor in the Delft University of Technology after the war, while the latter became Technical manager of Signaal, which is now a member of the Philips Group.

4.9 Beginning of radar in Hungary

The emergence of radar in Hungary is rather singular in that it occurred during the Second World War virtually in the midst of an advanced radar technology, but totally uninfluenced and unaided by it. The development took place without either German initiative or help, and then, when success was achieved in 1944, the development team set to work, at the beginning of March 1944, on a project to obtain radar echoes from the moon. Because of delays caused by upheavals at the end of the war, the equipment was not completed until January 1946 and then, on the 6th February 1946, the first successful lunar-echo experiment was carried out. This was just weeks after both the

American Signal Corps and the Russian successes, and must be regarded as a great achievement.

For an understanding of why Hungary should receive no technical support from Germany, it is necessary to refer to the political situation that pertained there during the Second World War [1]. At the outbreak of war, the Government, under the leadership of Prime Minister Count Pál Teleki, sought to manoeuvre politically between maintaining Hungary's alliance with Germany and maintaining friendly relations with the Western Powers. This stance could not be maintained. After Hungary's attack on Yugoslavia on the 11th April 1941 and its declaration of war on the Soviet Union on the 27th June 1941, Britain declared war on Hungary on the 7th December 1941. Nevertheless, the alliance with Germany was not a full and open one and, indeed, on the 19th March 1944 Germany invaded Hungary.

The key figure in the story is Dr Zoltán Bay, who was head of the radar development team in Hungary which became known as the 'Bay Group'. Before his involvement with radar in 1942, Dr Bay had, since 1938, been Professor of Atomic Physics at the Technical University of Budapest. He had also, since 1936, been director of the Tungsram Research Laboratory. In 1948, he went to the United States where he continued his research in the general area of particle physics and where he has held several senior academic appointments. Thus it was only under pressure of war, and then for a limited time, that Dr Bay became involved in radio science.

After Hungary entered the Second World War, and when enemy bombings became a possibility, the Ministry of Defence, on the suggestion of the Advisory Committee for Military Technology, decided that microwave experiments should be undertaken. The aim was two-fold: establishment of radio links and the detection and ranging of aircraft. The Institute of Military Technology knew of the existence of these two areas of microwave technology, but no more. When the Bay Group was set up, consisting of about 10 research workers and 30 technical support staff, Lieutenant-Colonel J. Jaky, head of the communication section of the Institute of Military Technology, was appointed co-ordinator between the military authorities and the Group. He endeavoured, without success, to obtain technical information from the Germans and to arrange a visit of Zoltán Bay to Berlin [2].

The United Incandescent Lamp and Electrical Co. Ltd ('Tungsram') at Újpest (close to Budapest) was commanded by the Hungarian Ministry of Defence in early 1942 to pursue microwave experiments. Tungsram was a natural choice. In the 1930s it was the third largest factory in Europe (after Philips and Osram) manufacturing incandescent lamps, radio valves and radio equipment, and 20% of radio valves in Europe were supplied by it. Tungsram possessed a large industrial research laboratory with about 25 people with academic qualifications supported by a technical staff of 75. Some half of these people were involved in research and development in the field of radio valves and radio circuitry. When radar work began in 1942, it was organised in such a way that Tungsram carried out the research work and developed the necessary transmitting and receiving valves while Standard Ltd. (a Hungarian subsidiary of ITT) manufactured the equipment in close collaboration with Tungsram. By the summer of 1944, some equipments were available for the Hungarian Army [3].

When the Group commenced their work, the first decision to be made was the choice of transmitting valve. They were aware of the properties of the klystron and the magnetron, but decided not to attempt their development and compromised by using wavelengths above 50 cm. By 1944 a valve (EC 108) had been developed which could generate 40 W in CW operation and a peak power of 10 kW in pulse operation. The research team was divided into subgroups, each of which concentrated on one of the following areas [2]:

Theory
Transmitter valves
Transmitter circuitry
The receiver
Pulse generation
Cathode-ray tube display circuits
Antennae and direction finding

At the beginning of 1944, an experimental model with a 3 m parabolic antenna was put together. More or less concurrently with this 50 cm equipment, a longer range radar operating on a 2·5 m wavelength was developed by Tungsram and Standard. A few of these latter sets were installed in the vicinity of Budapest in the summer of 1944.

In the beginning of March 1944, Dr Bay suggested that the techniques acquired in terrestrial radar be used in pure scientific work, and one such project would be the transmission of microwaves to the moon and the attempted reception of reflected signals. Work on the project, based on a wavelength of 50 cm, started in the summer of 1944. The radar laboratory was moved in June 1944 from Újpest to Nógrádveróce, and then back again to Újpest in September. After the German occupation of Hungary, all work on the project ceased. It started again after the siege of Budapest, but only briefly because then in April 1945 the Russian Army dismantled the Tungsram factory and transported everything to the Soviet Union. Finally, in 1945, the project was restarted, but now on a wavelength of 2·5 m, principally because parts for the 2·5 m radar were still available in Standard and at the Ministry of Defence.

The transit time of a radar signal to the moon and back is about 2·6 s. A pulse of 0·06 s duration was transmitted every 3 seconds. The estimated peak pulse power was 150 kW. The antenna was a system of 36 end-fed dipoles (6 by 6) with an aperiodic mesh reflector giving a gain of over 9 dB. The bandwidth of the receiver, after demodulation, was reduced to a mere 20 Hz to increase the signal/noise ratio as much as possible. Calculations showed that, notwithstanding this, the signal/noise ratio would be of the order of 0·1 and hence well below the limit of observability. It was at this juncture that the critical technique of the experiment was introduced [4, 5].

This was a method of integration, or cumulation, whereby the weak echoes received back from the moon for each firing of the transmitter were preserved and added over a lengthy period of time. If, in a given time, n pulses were sent out, then the amplitude of the received echoes would be multiplied by a factor of n, whereas the RMS fluctuations of noise during the same period would be multiplied by a factor of $\sqrt{n}$, thus resulting in an improvement in signal/noise ratio of $\sqrt{n}$. With a pulse transmitted every 3 seconds, the process of integration carried out for a period of 50 minutes (1000 pulses) would give an improvement in signal/noise of $\sqrt{1000}$, thus improving the original value of 0·1 to about 3. After considering the various possibilities available to him, Bay decided to use a bank of 10 water voltameters (coulometers) in which hydrogen formed at the positive electrodes* could be stored and measured. A rotating switch whose arm revolved once every 3 seconds connected each of the coulometers in succession to the output of the receiver. The amount of hydrogen liberated in each over a period was proportional to the integrated output of the receiver, so that 9 voltameters received noise only, while one received noise plus the sought for signal.

References and bibliography

Section 4.1 references

1   CLARK, Ronald W.: 'Tizard' (Methuen and Company Ltd., London, 1965), Chapter 5
2   AVIA/349: Public Record Office, Kew, London, UK
3   AVIA/137: Public Record Office, Kew, London, UK
4   'General notes on GL and SLC systems', communicated by C. S. Slow to Professor B. K. P. Scaife, Trinity College, Dublin
5   Transcript of lecture sent to writer
6   SAYER, A. P.: 'Army radar' (The War Office, London, 1950), p. 50
7   Reference 6, p. 71
8   'Shooting at aircraft', Report MR105, A. C. Cossor Ltd., London, 1943, copy received from L. H. Bedford
9   'Study of AF.1', Report HRD9, A. C. Cossor Ltd., London, 1942, copy received from L. H. Bedford
10 BEECHING, G. H.: 'Checking the angular accuracy of precision fire-control radar', *J. Instn Elect. Engrs,* 1946, **93**, Part IIIA, pp. 519–526
11 GARRATT, G. R. M.: 'One hundred years of submarine cables', (HMSO, London, 1950)
12 RATSEY, O. L.: 'As we were: fifty years of ASWE history 1896–1946' (Admiralty Surface Weapons Establishment, Portsdown, 1974). Copy available Naval Historical Library, Empress State Building, London, UK
13 ROSS, A. W.: 'Problems in shipborne radar', *J. Instn Elect. Engrs,* 1946, **93**, Part IIIA, pp. 236–244

* A KOH solution of 3% concentration was found to be the most suitable electrolyte.

14   O'DEA, W. T.: 'Science Museum Handbook of the collections illustrating electrical engineering. II Radio communication. Part I: History and development' (HMSO, London, 1934)

15   RAWLINSON, J. D. S.: 'Development of radar for the Royal Navy 1935–1944', *Naval Electrical Review,* 1975, July, pp. 51–57

16   COALES, J. F., CALPINE, H. C. and WATSON, D. S.: 'Naval fire-control radar', *J. Instn Elect. Engrs,* 1946, **93**, Part IIIA, pp. 349–379

17   FRIEDMAN, Norman: 'Naval radar' (Conway Maritime Press, Greenwich, 1981)

18   Air Ministry: 'Signals Volume IV: radar in raid reporting' (A.P. 1063, Air Ministry, London, 1950), p. 72, available AIR41/12, Public Record Office, Kew, London, UK

19   For a short description of an Australian produced radar set, reference can be made to: BOWEN, E. G.: 'a textbook of radar' (University Press, Cambridge, 1954), 2nd edn, p. 521

20   The radar work initiated by Sir Basil Frederick Jamieson-Schonland at the Bernard Price Institute, University of Witwatersrand, Johannesburg, is alluded to in: 'Standard Encyclopedia of Southern Africa Vol. 3', (Nasan Ltd., Cape Town, 1971), p. 460

21   MIDDLETON, W. E. Knowles: 'Radar development in Canada: The radio branch of the National Research Council of Canada 1939–1946' (Wilfrid Laurier University Press, Waterloo, Ontario, 1981)

22   EGGLESTON, Wilfrid: 'Scientists at war' (Oxford University Press, London, 1950)

23   SCOTT, Otto J.: 'The creative ordeal: the story of Raytheon' (Athenæum, New York, 1974)

24   Bowen Papers: EGBN 4/1. Archives, Churchill College, Cambridge, UK

25   Bowen Papers: EGBN 4/2. Archives, Churchill College, Cambridge, UK

26   GUERLAC, H. E.: 'Radar in World War II' (unpublished history of Division 14 of the National Defense Research Committee, 1947), Section BII. Microfilm copy available from: Photoduplication Service, Library of Congress, Washington DC, USA

27   An interview with Dr Edward G. Bowen. Conducted by Dr David K. Allison, Historian, Naval Research Laboratory. Part 5 of tape, available at Naval Research Laboratory, Washington DC, USA

Section 4.1 bibliography

WATSON-WATT, Robert: 'Three steps to victory' (Odhams Press Ltd., London, 1957)
    An autobiographical study, rich in anecdotal content, which furnishes a good overview of the development of British radar.

ROWE, A. P.: 'One story of radar' (The University Press, Cambridge, 1948)
    Relates the history of the Telecommunications Research Establishment (TRE). Of considerable value for dates and events.

SAYER, A. P.: 'Army radar' (The War Office, London, 1950)
    This book was compiled by authority of the Army Council and furnishes a comprehensive account of the development of Army radar systems and of the organisational structure within which this development was carried out. The evolution of Gun Laying, Search Light Control and Coastal Defence sets is treated in detail.

'The story of radiolocation' (an authoritative account prepared for the use of the Ministry of Information, London, 1945)
    Copy available in Science Museum Library, London, UK. A clear concise account of the development of British radar, with separate sections devoted to the Army, Navy and Air Force.

Air Ministry: 'The Second World War 1939–1945, Royal Air Force, Signals Volume IV, radar in raid reporting' (A. P. 1063, Air Ministry, London, 1950). Copy available: AIR 41/12, Public Record Office, Kew, London, UK

A detailed account of the evolution and operation of the Chain Home and related systems. The various theatres of war are dealt with and comprehensive references and appendices are provided. This volume could be regarded as essential reading for the student of British radar history*.

RATSEY, O. L.: 'As we were: fifty years of ASWE history 1896–1946' (Admiralty Surface Weapons Establishment, Portsdown, 1974). Copy available at Naval Historical Library, Empress State Building, London, UK

Includes an account of British naval radar development.

'The Institution of Electrical Engineers Radiolocation Convention', March 1946, *J. Instn. Elect. Engrs,* 1946, **93**, Part IIIA, pp. 5–1620

The selection of papers contained in this volume provides a comprehensive technical treatment of British radar technology to the end of the Second World War. The authors were involved in key areas of development.

'The Institution of Electrical Engineers Radiocommunication Convention', March 1947, *J. Instn. Elect. Engrs,* 1947, **94**, Part IIIA, pp. 1–1038

The papers deal, among other things, with various aspects of direction finding and with CW navigational aids and so supplement the information contained in the previous item.

GUERLAC, H. E.: 'Radar in World War II' (unpublished history of Division 14 of the National Defence Research Committee, 1947). Microfilm copy available from; Photoduplication Service, Library of Congress, Washington, DC, USA

Chapter 6, Section A provides quite a detailed account of the origins of British radar and includes useful references.

Air Ministry: 'The Second World War 1939–1945, Royal Air Force, Signals Volume VII, Radio Counter-Measures' (A. P. 3407, London (1950). Copy available: AIR 41/13, Public Record Office, Kew, London, UK

A detailed account of defensive and offensive radio and radar counter-measures employed by the British against German forces during the Second World War. Contains useful references and appendices.

SMITH, R. A. (Chief writer): 'Technical monographs on wartime research and development in MAP: ASV (the detection of surface vessels by airborne radar)'. Copy available: Bowen Papers (EGBN 2/4), Archives, Churchill College, Cambridge, UK

A history of the development of British air-to-surface-vessel radars, both metric and centimetric. Includes a chapter each on the theory of echoing from surface targets and on sea returns. There is a short chapter on the future outlook for radar and sea warfare.

'Proceedings of Royal Commission on Awards to Inventors', copy available at Science Museum Library, London. Forty-four days of hearings between 12th April 1951 and 14th December 1951, inclusive.

As the proceedings comprise the evidence given before the Commission of key participants in the British development of radar, they are a very fruitful and accurate source of information.

'A discussion on the effects of the two World Wars on the organisation and development of science in the United Kingdom', *Proc. Royal Soc.* 1975, **342**, No. 1631, pp. 439–591.

Ten contributers, including Professor R. V. Jones who organised the Discussion. The contribution of Professor Jones, which includes a valuable bibliography, is particularly relevant to the development of radar in Britain.

JOHNSON, Brian: 'The secret war' (British Broadcasting Corporation, London, 1978)

Book is based on the BBC television series 'The secret war'. A highly readable, well illustrated, popular account of the electronic warfare waged between Britain and Germany in the Second World War; a considerable portion of the book is devoted to radar.

HEZLET, Arthur (Vice-Admiral Sir): 'The electron and sea power' (Peter Davies, London, 1975)

---

* Signals Vol. V, 'Fighter control and interception', has now been declassified and is available: AIR 10/5485, Public Record Office, Kew, London, UK.

An historical study covering the period from the mid-nineteenth century to the present day; while it deals principally with Naval communications, there is an extensive coverage of radar matters.

PRICE, Alfred: 'Instruments of darkness (the history of electronic warfare)' (MacDonald and Jane's, London, 1977), 2nd edn
A well written account of the role played by radar and electronic warfare in the British–German conflict.

PRICE, Alfred: 'Aircraft versus submarine' (William Kimber, London, 1973)
While dealing with the broader subject of anti-submarine warfare, it provides useful background information on the role of air to surface vessel radar.

POSTAN, M. M., HAY, D., and SCOTT, J. D.: 'Design and development of weapons (studies in Government and industrial organisation)' (Her Majesty's Stationery Office and Longman's Group Ltd, London, 1964)
A volume in the history of the Second World War, United Kingdom Civil Series. Useful information given on radar development and on Government Research Establishments.

JONES, R. V.: 'Most secret war' (Hamish Hamilton, London, 1978)
A valuable history of scientific intelligence pertaining to British and German activities in the Second World War. Provides useful background material to the history of radar.

HINSLEY, F. H.: 'British Intelligence in the Second World War, Volume 1' (Her Majesty's Stationery Office, London, 1979)
This is an official history of British Intelligence during the war years. Contains a translation of the 'Oslo Report' and provides information on German Air Force navigational aids together with useful background material to the radar story.

HINSLEY, F. H.: 'British Intelligence in the Second World War, Volume 2' (Her Majesty's Stationery Office, London, 1981)
This is a continuation of the previous item and covers military operations between the summer of 1941 and the summer of 1943. Again, it is a useful background to the radar story and in particular it sets down, with references, the manner in which knowledge of the technical and operational characteristics of the principal German radars was acquired.

STREETLY, Martin: 'Confound and destroy' (MacDonald and Jane's, London, 1978)
Deals with the activities of 100 Group of Bomber Command of the Royal Air Force and provides a well documented and illustrated treatment of the use of electronic warfare in Europe from 1943 to 1945. Descriptions are provided of several British and German radar systems.

'Radar: a report on science at war' (His Majesty's Stationery Office, London, 1945). (This is a reprint. The original was published in the United States of America by the Government Printing Office; it was released by the Joint Board on Scientific Information Policy for Office of Scientific Research and Development, War Department, Navy Department.)
Some 50 pages with 14 short chapters and an appendix giving a technical description of a radar system. Of interest because it is the official (American) account of radar, released towards the end of hostilities, describing its role in the Second World War.

ROWLINSON, Frank: 'Contribution to Victory' (Metropolitan-Vickers Electrical Company, Limited, Manchester, 1947)
Gives an account of some of the special work carried out by Metropolitan-Vickers during the Second World War. There is a chapter on radar which contains interesting photographs of equipments.

DUMMELOW, John: '1899–1949' (Metropolitan-Vickers Electrical Company, Limited, Manchester, 1949)
An illustrated history of the Metropolitan-Vickers Company. Refers to the work of C. R. Burch, F. P. Burch and J. M. Dodds on continuously evacuated valves. These valves contributed significantly to the successful operation of the Chain Home transmitters.

BURCH, L. L. R: 'The Flowerdown link' (L. L. R. Burch, 1980)

A nostalgic history of the Royal Air Force Apprentice School Flowerdown. Mention is made of the fitting, from 1939 onwards, of AI, ASV and IFF equipments into various aircraft.

KINSEY, Gordon: 'Orfordness – secret site' (Terence Dalton Limited, Lavenham, Suffolk, 1981)

Includes an account by Arnold Wilkins of the first year of development of British radar at Orfordness.

KINSEY, Gordon: 'Bawdsey – birth of the beam' (Terence Dalton Limited, Lavenham, Suffolk, 1983)

Tells of the part played by Bawdsey in the development of British radar. This includes an account by Arnold Wilkins, which is a continuation of that given in the previous item.

NEALE, B. T: 'CH – the first operational radar', *GEC Journal of Research*, 1985, **3**, pp. 73–83

An excellent summary of the construction, operation and performance of the Chain Home system.

ROBINSON, Denis M: 'British microwave radar 1939–41, *Proc. Am. phil. Soc.*, 1983, **127**, pp. 26–31

An interesting anecdotal account of some of the people who pioneered British microwave development. The article highlights the importance of Rutherford's influence on British physicists in the years before the War.

Section 4.2 references

1 TRENKLE, Fritz: 'Die deutschen Funkmeßverfahren bis 1945' (Motorbuch Verlag, Stuttgart, 1979)

2 WILLISEN, H. K. V.: 'Die Geschichte der Funkmeßtechnik bis 1935', unfinished manuscript, c.1950, copy received from Herr F. Trenkle

3 Discussion with Herr F. Trenkle, September 1979

4 Report written in 1971 concerning Dr Runge's activities in Telefunken. Extract sent by Dr Runge to writer

5 RUNGE, Wilhelm T.: 'Rückstrahltechnik bei Telefunken in den dreißiger Jahren', 1981, copy of report sent by Dr Runge to writer

6 SIECHE, Erwin: 'German Naval Radar to 1945 Part 1', *Warship*, 1982; **21**, pp. 2–10

7 HOFFMAN-HEYDEN, Adolf-Eckard: 'Die Funkmeßgeräte der Deutschen Flakartillerie (1938–45)' (Verkehrs-Und Wirtschafts-Verlag GmbH, Dortmund, 1957). Quotation is taken from: NORTON, K. A. and OMBERG, A. C.: 'The maximum range of a radar set', *Proc. Inst. Radio Engrs*, 1947, **35**, pp. 4–24

Section 4.2 bibliography

In the following bibliography, the capitals DGON are an abbreviation for Deutsche Gesellschaft für Ortung und Navigation e.V., Am Wehrhahn 94/96, D4000 Düsseldorf 1, Postfach 2622, Federal Republic of Germany.

PLENDL, Hans: 'Impuls-Peilung', *Luftfahrtforschung,* 1936, **10**, pp. 367–370

Describes pulse technique as used in ionospheric research, and then treats of its use in high-frequency direction finding, where it helps to distinguish between the direct wave and the ionospheric reflected wave. The Telefunken ground system and the Lorenz airborne set are described.

DZIEWIOR, K.: 'Zusammenstellung der bekannt gewordenen Vorschläge zur Abstandsbestimmung und zur Rückstrahlpeilung auf funktechnischem Wege und ihre kritische Betrachtung', *Luftfahrtforschung,* 1939, **16**, pp. 326–338

A comprehensive treatment of devices for measuring distance, including radio-altimeters. Covers work carried out in countries other than Germany. The information in this paper would supplement that given in Section 3.9.3.

THEILE, Ulrick V.: 'Hohenmeßverfahren auf funkentechnischer Grundlage', *Luftfahrtforschung*, 1939, **16**, pp. 339–347

Deals with electrical methods for measuring altitude including the capacity altimeter, and radar-type continuous wave and pulse devices. A continuation of the material in the previous item, and is likewise supplementary to Section 3.9.3.

GROOS, O. H.: 'Die Erzeugung von Zwergwellen mit dem Magnetfeldröhrensender', *Hochfreq. Tech. Elektroakust*, 1938, **51**, pp. 37–43

Discusses the design of split-anode magnetrons for the generation of microwaves.

'Wellenberatung der Truppe durch die Funkberatungsstelle' (Erprobungsstelle der Luftwaffe Rechlin, 1942), available from DGON

Treats propagation in a qualitative fashion with a particular emphasis on HF.

'Einsatzfibel für Funkmeßgeräte' (Flugmeldedienst Heft 9, 1944), available from DGON

Deals with aspects of VHF propagation which affect the operational use of radar; the radio horizon, the effect of reflection from the earth on antenna vertical coverage patterns and the siting of radar sets.

BRANDT, Leo: 'Von der Rundfunkwelle zur Zentimeterwelle des Radar (Aus 20 Jahren Hochfrequenzentwicklung)' (Wissenschaftlicher Verein für Verkehrswesen e.V. in Essen, 1951). Copy available from Universitätsbibliothek und Technische Informationsbibliothek (TIB), Welfengarter 1B, 3000 Hanover 1, W. Germany

The text of a lecture given on the 8th February 1951. Gives an informative and broad overview of German wartime radar systems.

BLEY, Curt: 'Geheimnis Radar (Eine Geschichte der Wissenschaftlichen Kriegfuehrung)' (Rowohlt Verlag GmbH, Hamburg-Stuttgart, Baden-Baden, Berlin, 1949)

Some 39 pages of a general and operational history of German radar, but with considerable details of contemporary British achievements. Commences with an historical description of the technical background to radar and traces events until the end of the Second World War.

'Institutes of the Bevollmaechtighter Fuer Hochfrequenz-Forschung' (Combined Intelligence Objectives Sub-Committee, G-2 Division, SHAEF (Rear), HMSO, London, 1945). Copy available from DSIR National Lending Library for Science and Technology.

The contents deal with the investigation made in 1945 of the work carried out at some six Institutes during the war years. The material, which is contained in over 200 pages, is very varied and encompasses radio-navigational aids, devices and fundamental research.

'Fiat Review of German Science 1939–1946 Electronics, Incl. Fundamental Emission Phenomena Part I' (Senior authors: GOUBAU, George and ZENNECK, Jonathan), (Office of Military Government for Germany, Field Information Agencies Technical, Dieterich'sche Verlagsbuchhandlung, Inhaber W. Klemm, Wiesbaden, Germany, 1948)

Five chapters which deal with electron emission and electron flow, electronic devices, oscillators, modulation and scanning, rectification. Detailed treatment with comprehensive references.

'Fiat Review of German Science 1939–1946 Electronics Incl. Fundamental Emission Phenomena Part II' (Senior authors: GOUBAU, George and ZENNECK, Jonathan), (Office of Military Government for Germany, Field Information Agencies Technical, Dieterich'sche Verlagsbuchhandlung, Inhaber W. Klemm, Wiesbaden, Germany, 1948)

Nine chapters which deal with amplification, shot-effect in electron tubes and limits of amplification, non-emitting electromagnetic systems, antennae, circuit theory, wave propagation, direction finding, television (including image scanning), high frequency measuring technique. Detailed treatment with comprehensive references.

BRANDT, Leo (ed.): 'Funkortung in der Luftfahrt. Frankfurter Fachtagung 1953'; Teil I: Zur Geschichte der Funkortung. Deutsche und Britische Beiträge; Teil II: Auswahl Physikalischer Grundfragen der Funkortung; Teil III: Die Funkortung der Deutschen Flugsicherung; Teil IV: Funkortungs-erfahrungen deutscher und ausländischer Flieger; Teil V: Sonderfragen der Funkortung in Luftraum und im Weltall. (Verkehrs und Wirtschafts-Verlag GmbH, Dortmund, 1953), available from DGON

Proceedings of a conference on radar held in Frankfurt in 1953. In five parts; each part contains papers dealing with more or less a specific aspect of radar. The first part deals with the history and the development of radar. Part two deals with specific areas of radar technology. Part three is concerned with navigational aids and air traffic control. Part four deals with the operational aspects of aircraft navigational aids. The final part is concerned with the application of radar techniques to meteorology, ionospheric research and astronomy.

HOFFMANN-HEYDEN, A. E.: 'Die Funkmeßgeräte der Deutschen Flakartillerie 1938–45' (Verkehrs und Wirtschafts-Verlag GmbH, Dortmund, 1957), available from DGON

A detailed and excellent account of the requirements, conception and development of anti-aircraft defence or 'Flak' control radars. System principles, schematic diagrams and performance figures are given for each equipment, together with illustrative photographs. The text also provides a good indication of the status and evolution of German radar technology from 1938 to 1945.

'Berlin-Fibel' (1944), available from DGON

A basic but comprehensive handbook on the FuG224 (Berlin) set. Includes circuit, waveform and performance diagrams. This was a microwave airborne navigation set similar to the British H2S.

BRANDT, L. (ed.): 'Decknamenverzeichnis der deutschen Funkmeß und Ultrarot-Entwicklungen der Jahre 1944–1945', available from DGON

Special names or code names for various radar, radio-navigational and infra-red equipments. A brief description of each item is given, together with the names of manufacturers.

BRANDT, L. (ed.): 'Decknamenverzeichnis der deutschen Funkmeßgeräte-Entwicklungen nach dem Stand vom 1 März 1944', available from DGON

Special names or code names for various radar equipments and accessories. Brief description given for some items.

BRANDT, L. (ed.): 'Sitzungsprotokolle der Arbeitsgemeinschaft Rotterdam (März 1943–Sept. 1944)' (Düsseldorf, 1953), available from DGON

A British Stirling bomber with an H2S (9 cm, ground-mapping) radar on board was shot down near Rotterdam, Holland, on the 2nd February 1943. The radar contained a cavity magnetron. General Wolfang Martini, head of the German Air Force Signal Service, set up a special 'Rotterdam' Commission (Arbeitsgemeinschaft 'Rotterdam') to examine the new centimetric device and to devise necessary countermeasures. The minutes of 18 sessions of the Commission are recorded in this volume. These sessions were attended by prominent service and industrial personnel, and the records afford a valuable insight into German radar technology of the mid-war period.

'Würzburg-Fibel', Teil I, Teil II, (Herausgegeben vom Oberkommando der Kriegsmarine Amtsgruppe Technisches Nachrichtenwesen in Einvernehmen mit dem Reichsluftfahrtminis-terium Chef Nachrichtenverbindungswesen, Oktober 1943), available from DGON

A primer on the Wurzburg Set in two parts. The first part covers general principles of radar and techniques applicable to the Wurzburg in addition to explanatory descriptions of circuits. Part two contains further explanatory descriptions of circuits and describes additions to, and accessories for, the basic set.

TRENKLE, Fritz: 'Bordfunkgeräte der deutschen Luftwaffe 1935–1945', (Düsseldorf, 1958), available from DGON

A comprehensive and excellent treatment of Luftwaffe airborne equipment, including communication, radio-navigational, radar and radar-countermeasures sets. Includes two appendices with photographs and tables.

TRENKLE, Fritz: 'Deutsche Ortungs-und Navigationsanlagen (Land und See 1935–1945)' (Düsseldorf, 1966), available from DGON

In four parts. A comprehensive enumeration and description of German radio-navigational and radar equipments for the period 1935 to 1945. Deals mainly with direction finders, aircraft radio-navigational systems and radars, including secondary radars and also with missile guidance systems and infra-red and optical systems. A relevant bibliography included. A most important sourcebook. This work clearly highlights the high status reached by Germany in these areas during the war years.

VÖGELSANG, C. W.: 'Radar gestern und heute', *Köhlers Flieger-Kalender,* 1962, **14**, pp. 47–58, Wilhelm Köhler Verlag, Minden (Westf.)
A very general exposition of radar with some references to early German developments.

HAHN, Fritz: 'Deutsche Geheimwaffen 1939–1945' (Erich Hoffmann Verlag, Heidenheim, 1963)
Gives quite a detailed description of 322 German secret weapons, including radar devices.

BRANDT, L.: 'Zur Geschichte der Radartechnik in Deutschland und Grossbritannien' (Pubblicazioni dell' Istituto Internazionale delle Comunicazioni, Genova, 1967)
Based on a lecture given to the 'XV Convegno Internazionale Della Communicazioni' at Genoa from the 12–15th October 1967.
A very good overview of the development of radar in Germany and Great Britain. Covers the period from Hulsmeyer's experiments in 1904 to the post-war years of the 1950s and the building of radio-telescopes. A liberal number of selected photographs is included.

BONATZ, Heinz: 'Die Deutsche Marine-Funkaufklärung 1914–1945' (Wehr Und Wissen Verlagsgesellschaft MBH, Darmstadt, 1970)
Concerned with Naval Intelligence, and particularly Signal Intelligence, in the German Navy for the period 1914–1945. Is of value as background information, particularly with regard to the U-boat warfare and the battle of the Atlantic.

PRAUN, Albert: 'Vernachlässigte Faktoren in der Kriegsgeschichtsschreibung', *Wehrwissenschaftliche Rundschau,* 1970, **20**, pp. 137–145
The paper is not concerned with radar, but would be of value to the radar historian in that it addresses itself to broad relevant principles of military science.

ANDONE, Marius: 'Batalia Radarului', *Viata Militara,* 1974, No. 4, pp. 24–26
This Rumanian paper deals with certain intelligence aspects of German–British radar history, including the *Admiral Graf Spee* Seetakt radar, the Oslo report, the Brunneval raid and the use of 'Window'.

TRENKLE, Fritz: 'Die deutschen Funkmeβverfahren bis 1945' (Motorbuch Verlag, Stuttgart, 1979)
Includes a concise history of the beginning of German radar. This is followed by a clear exposition of ground, airborne and naval radar systems employed during the Second World War. Good appropriate photographs and a valuable bibliography.

TRENKLE, Fritz: 'Die deutschen Funk-Navigations und Funk-Führungsverfahren' (Motorbuch Verlag, Stuttgart, 1979)
An excellent treatment of German radio and radio-navigational aids and related systems. Good bibliography.

BURKEL, Helmut: 'Die Radartechnik bei AEG-Telefunken' (AEG-Telefunken, Ulm, 1979)
A popular account of radar from Hülsmeyer to the present day with the emphasis on the achievements of AEG-Telefunken. Over 70 pages.

TRENKLE, Fritz: 'Die deutschen Funklenkverfahren bis 1945' (AEG-Telefunken Aktiengesellschaft, Ulm, 1982)
A comprehensive account of German remote controlled devices and remote controlled techniques. Photographs, illustrations and bibliography.

'Exercise "Post Mortem". Report on an investigation of a portion of the German raid reporting and control system' (Air Ministry, London, 1945)
A report on an Exercise carried out between 25th June 1945 and 7th July 1945. The object of the Exercise, as stated in the report, was as follows:

The Exercise is being arranged to investigate the German methods of operating their defence organisation of raid reporting and control against our bomber attacks, and in particular to ascertain the effect of British radio countermeasures on this organisation and on German radar equipment. . . . A detailed study of German methods by British personnel will be of great value in planning the post-war raid report and control system of Fighter Command, . . .

The intention was to:

observe selected stations and control points of the ground organisation of the German air defence system in the province of SCHLESWIG-HOLSTEIN and in EASTERN DENMARK, manned by German personnel operating under raid conditions simulated by heavy bomber aircraft of Bomber Command, and radio countermeasure aircraft of No. 100 Group.

Apart from the importance of the results in assessing German equipment and air-defence methods, the report provides a good technical description of the following German radar systems: Giant Würzburg; Freya; Mammut; Wasserman; Jagdschloss; Elephant and Associated Russel.

REUTER, Frank: 'Funkmeß' (Westdeutscher Verlag, Opladen, 1971)
A comprehensive treatment of the history and development of German radar. Plentiful references. A source book for any in-depth study of the subject.

BEKKER, Cajus (Berenbrok, Hans Dieter): 'Augen durch Nacht und Nebel (Die Radar-Story)' (Gerhard Stalling Verlag, Oldenburg, 1964)
A good overview of the history of German radar, with considerable operational material provided.

FRIEDMAN, Norman: 'Naval radar' (Conway Maritime Press, Greenwich, 1981), pp. 205–206
Useful information on some nineteen types of German naval radar sets.

SIECHE, Erwin: 'German Naval Radar to 1945 Part 1', *Warship*, 1982, **21**, pp. 2–10
Provides an account of the origins of German radar and deals with the systems of coding or designation for German sets.

SIECHE, Erwin: 'German Naval Radar to 1945 Part 2', *Warship*, 1982, **22**, pp. 146–157
Deals with installations on battleships, cruisers, destroyers and motor torpedo boats.

Among miscellaneous material on German radar equipment which is available at the Science Museum, London, UK, the following four books are considered to be important sources of information:

(*a*) 'Ground Electronic Equipment of the G.A.F.'
(*b*) 'Ground Communication Equipment of the G.A.F.'
(*c*) 'Electronic Equipment of the German Navy'
(*d*) 'Airborne Electronic Equipment of the G.A.F.'

The following set of microfilms are available from: Chief of Circulation, The Albert F. Simpson Historical Research Centre, USAF, HOA, Maxwell AFB, AL 36112, USA.

Against each reel number a summary of contents is supplied. This summary is that provided by the Albert F. Simpson Historical Center. Generally, each microfilm covers far more subject material than is indicated by the summary.

Reel Number

A1007   Combined Intelligence Committee German radar Industry. C.I.O.S. Evaluation report No. 105 CIC report 75/113, 3rd July 1945 (File No. 119.0412–105)

A1246   Assistant Chief of Air Staff Intelligence.
Impact (Magazine)
How German radar works in locating our planes. January 1944, pp. 15–17 (File No. 142.036)

A1290   AC/AS, Intelligence.
British report on German radio and radar including report by Sir Robert W. Watt on radio and radar aids to defeat of the rocket, 1944 (File No. 142.53–1)

A2622–A2623 Proving Ground Command.
Collected intelligence reports by 1st proving ground electronics unit on German electronics equipment, 1944
(File No. 240.6541–1 thru 10)

A2891   Air Room, Intelligence Department, AAFSAT. German radar in the invasion of Europe; Air Room interview, 25th June 1944 (File No. 248.532–49)

A5070   United States Forces European Theater, 1st November 1945.
ETOUSA
Charts giving performance data on the different German Air Force and German Naval Radar sets. Prepared by RCM Division, Office, Chief Signal Office, 15th April 1944 (File No. 502.654A)
ETOUSA
Broad band Rehbock U5, a German apparatus used to adjust radar sets in the frequency range of 360-600 Mc/s. Intelligence report based on translation of a German document, 10th November 1945 (File No. 502.654B(u))
ETOUSA
German Radar for Photo Interpreters, December 1944 (File No. 502.654E(u))

A5187   SHEAF
Combined Intelligence Objectives Sub-Committee Report, July 1945.
Design of radar test equipment, Siemens Halske plant, Munich (File No. 506.620 V.59)
SHEAF
Combined Intelligence Objectives Sub-Committee Report. Institut Fur Physikal-ishche Forschung, Neu Drossenfeld (File No. 506.620 V.61)

A5188   British Air Ministry
C.I.O.S. Target: Electronic equipment aboard German Naval Units; reported by Lt. G. M. Robertson, USNR (File No. 506.620(u) V.91)

A5210   Headquarters United States Forces in Austria.
Interrogation summary: High frequency research projects of the Plendl Institute 13p. April 1946 (File No. 507.654)

A5220   Combined Operational Planning Committee
German Aircraft Reporting Service: Intelligence data, 1943–1945 (File No. 508.654–21)

A5252   British Air Ministry
The rise and fall of the German Air Force, 1933–1945: Development of airborne radar. Pp. 188–191 (File No. 512.042–248(u))

A5405   British Air Ministry
Radio and radar equipment in the Luftwaffe, XI–radar aids to flak. ADIK reports No. 396, 1945 (File No. 512. 619B–30(u))

A5415   British Air Ministry
C.S.D.I.C. Air (MF report A498): 'Some Notes on German Radar Sites, Factories and Training'. (See also A504 and A519; A524 and A531, 1945). (File Nos. 512.619C–7A; 512.619C–7B; 512.619C–7C)

A5422   British Air Ministry, A.D.I. (Science)
Anti-jamming procedures for flak control radar. Air Scientific Intelligence Technical Translation No. 3, 12th January 1945 (File No. 512.621J–3(u))

A5425   British Air Ministry
German documents on signals organisation and personnel, and radar equipment, held by General Martini, Director General of G.A.F. Signals and his Chief of Staff (untranslated), 1945 (File No. 512.625T)

A5455   British Air Ministry
The Luftwaffe; a 3-volume collection of material on the German Air Force gathered from various sources, 1945 (File No. 512.631U)

A5476   British Air Ministry
Jagdschloss (German ground radar)
Air Scientific Intelligence Interim Report, 6th October 1944 (File No. 512.654D(u))

Combined Intelligence Objectives Sub-Committee.
Various reports on Telefunken GmbH German radar and Radio Company, 1945 (File No. 512.654A)
British Air Ministry
Operational instructions and training principles for German fighter controllers on AN procedure; translation, May 1944 (File No. 512.654–1(u))
British Air Ministry
German ground radar characteristics. Air Scientific Intelligence Report No. 20, 14th December 1943 (File No. 512.654D–20(u))
British Air Ministry
German control of fighters by the Benito and Egon methods. Air Scientific Intelligence Report No. 25, 14th June 1944 (File No. 512.654D–25)

A5477   British Air Ministry, A.D.I. (Science)
'Mannheim'. German radar set for flak control. Air Scientific Intelligence Report No. 31. 25th January 1945 (File No. 512.654D–31(u))
British Air Ministry, A.D.I. (Science)
Characteristics of German ground radar
Air Scientific Intelligence Report No. 34, April 1945 (File No. 512.654D–34)
British Air Ministry
German aircraft reporting Vol. IV of Air Defence of the Reich (manuals), September 1944 (File No. 512.6541–4(u))

A5510   2nd Tactical Air Force
Radar reports; RAF Air Technical Reports, Series R, describing captured German radar sites in Germany, Denmark and Holland, 1945 (File No. 515.6504–4(u))

A5701   USSTAF
Post Hostilities Investigation – German Air Defenses, Vol. III, 1938–45: German Air Force Signals and Air Defense. Contains descriptions of German electronic equipment with photographs (File No. 519.601A–3(u))

A5704   USSTAF
Air staff post hostilities intelligence requirements on German Air Force, Sec. IVB, 1935–45: Characteristics of German radar equipment and operational employment in the aircraft warning and fighter control system (File No. 519.601B–4(u))

A5708   USSTAF
Air staff post hostilities intelligence requirements on German Air Force, Sec. XI, 1935–45: Radar communications in the German Air Force (File No. 519.601B–11(u))

A5723   USSTAF
Air Intelligence Summary, 23rd April 1944: a German tail-warning radar. Some date (*sic*) on the FUG 216, which shows aircraft approaching from the rear. Pp. 19–20 (File No. 519.607A–24(u))
USSTAF
Air Intelligence Summary, 21st May 1944: German airborne radar. A brief summary of the principal types according to their operational functions. Pp. 10–11 (File No. 519.607A–28(u))

A5724   USSTAF
Air Intelligence Summary, 29th October 1944: New German radar system. Long-range 'Jagdschloss' equipment may improve ground control set-up (File Nos. 519.607A–51(s); 519.607A 72(s); 519.607A–76(s))
USSTAF
Air Intelligence Summary, 7th January 1945: 'Klein Heidelberg' is a new German radar system. Pp. 21–22 (File No. 519.607A–61(u))

A5726   USSTAF
          Report of interrogation of General der Flieger Martini, Commanding General of
          the G.A.F. Signals. 10th October 1945 (File No. 519.619–16(u))

A5728   USSTAF
          The Signal Intelligence Service of the German Luftwaffe in the course of World
          War II, 1945 (File No. 519.6314–10)

A5729   USSTAF
          Minutes of Flak conference held in London, 1–11th June 1945 (File No. 646–2(u))

A5731   USSTAF
          Technical Intelligence Reports, 'R' series, pertaining to German radio and radar
          equipment and operating sites. June 1944–June 1945 (Filed by report number
          following dash). (See also 519.6543–1) (File No. 519.6505R(u))

A5735   USSTAF
          FuMG 404 – Jagdschloss, a German radar
          Electronics Intelligence Report Diamond (PT) No. 7, 15th April 1945 (File No.
          519.6541–7(u))
          USSTAF
          Wismar in the Würzburg. Electronics Intelligence Report EISS/SP/R.9, 31st
          October 1944 (File No. 519.6541–9(u))
          USSTAF
          Survey of German airborne radio and radar equipment. Electronics Intelligence
          Report Diamond (PR) No. 9, 22nd May 1945 (File No. 519.6541–9A(u))

A6092   M.A.A.F., Intelligence Section
          A report of the investigation of several flak defenses in North Italy by radio
          countermeasures personnel. Gives basic aspects of radar flak intelligence, 7th June
          1945 (File No. 622.654–3)

A6115   Mediterranean Allied Tactical Air Force collection of reports, correspondence,
          and messages concerning enemy radar. May–July 1944; April–August 1944 (File
          No. 626.654)

B1757   Military Intelligence Division, W.D.G.S. Military attache report on German Air
          Forces reactions to Bomber Command attacks. 14th March 1946, p. 17 (File No.
          170.2278–4D(u))

B1761   Military Intelligence Division, W.G.D.S. Technical analysis of German AI Radar
          (FuGe 202 – Lichtenstein)
          Military attache report, 10th January 1944, enclosing Royal Aircraft Establish-
          ment Technical Note Nr. Rad. 180, subject 'Report on FUG 202 Radio Equipment
          (Lichtenstein)' (File No. 170.2278D–1(u))

K1027-S   Karlsruhe Document Collection
          Employment of radar, 1938–1945 (German text) (File No. K113.3019–10)

K1029-R   Karlsruhe Document Collection
          Procurement and development of radar for the German Air Force, 1939–1945
          (German text) (File No. K113.8671–3)

K5003   British Air Ministry
          German centimetric wave technology at the end of the Second World War, by
          Professor Leo Brandt
          Text of a lecture given at a radar conference held at Frankfurt-am-Main, 1953
          (German translation VII/160) (File No. K512.621)

REINHARDT, W.: 'Radar-Bibliographie'
      (Deutsche Versuchanstalt für Luftfahrt E. V. Bericht Nr. 73, Westdeutscher Verlag, Köln
      und Opladen, 1959)

VAN DER HULST, Roelof: Private communication
  Miscellaneous set of documents relating to the development of German radar within the period 1933–1942; documents obtained from Frank Reuter. Summary forwarded by Mr Van der Hulst to writer.

Section 4.3 references

1 GUERLAC, H. E.; 'Radar in World War II' (unpublished history of Division 14 of the National Defense Research Committee, 1947), Section AIV
  Microfilm copy available from: Photoduplication Service, Library of Congress, Washington DC, USA

2 HOWETH, L. S.: 'History of Communications-Electronics in the United States Navy' (United States Government Printing Office, Washington, 1963), Chapter XXXVIII

3 ALLISON, David K.: 'The origin of radar at the Naval Research Laboratory: a case study of mission-oriented research and development' (Princeton University, Ph.D., 1980), Microfilm copy available from University Microfilms International, 300N Zeeb Road, Ann Arbor, MI 48106, USA

4 ALLISON, David K.: 'New eye for the Navy: the origin of radar at the Naval Research Laboratory', NRL Report 8466, Naval Research Laboratory, Washington DC, 1981; this is a revision of Reference 3.

5 GEBHARD, Louis A.: 'Evolution of naval radio-electronics and contributions of the Naval Research Laboratory', NRL Report 8300, Naval Research Laboratory, Washington DC, 1979

6 ENGLUND, Carl R., CRAWFORD, Arthur B., and MUMFORD, William M.: 'Some results of a study of ultra-short-wave transmission phenomena', *Proc. Inst. Radio Engrs*, 1933, **21**, pp. 464–492

7 (Quotation from Reference 6 above):

Field Fluctuations from Moving Bodies
It is well known that the motion of conducting bodies, such as human beings, in the neighbourhood of ultra-short-wave receivers produces readily observable variations in the radio field. This phenomenon extends to unsuspected distances at times. Thus, while surveying the field pattern in the field described above, we observed that an airplane flying about 1500 feet (458 metres) overhead and roughly along the line joining us with the transmitter, produced a very noticeable flutter, of about four cycles per second, in the low frequency detector meter. We then made a trip to the near-by Red Bank, N.J., airport, distant about 5½ miles (8·8 kilometres) and observed even more striking reradiation phenomena. Near-by planes gave field variations up to two decibels in amplitude, and an aircraft flying over the Holmdel laboratory and towards the landing field was detected just as the Holmdel operator announced 'airplane overhead'. These were all fabric wing planes. If the reradiation field to which such an airplane is exposed is of inverse distance amplitude type while the directly received ground fields are of more nearly inverse distance square type, as in Fig. 2, it is easy to see that at five miles an overhead airplane is exposed to a field intensity about ten times (20 decibels) that existing at the ground, and for ordinary airplane heights a high energy transformation loss in the reradiation process can occur and still give marked indications in the receiver meter. This airplane reradiation was mentioned at various subsequent times, sometimes when the airplane itself was invisible. A set of theoretical beat frequency versus distance curves are given in Fig. 7.

8 Transcript of a tape-recorded interview with Dr Robert Maurice Page, 26th and 27th October 1978
  Conducted by David K. Allison, Historian, Naval Research Laboratory. Available Historian's Office, NRL, Washington DC, USA

9 MESNY, René: 'Constantes De Temps, Durées D'Établissement, Décréments', *Onde Élect.*, 1934, **13**, pp. 237–243

10   TERMAN, Frederick Emmons: 'Radio engineering' (McGraw-Hill, New York, 1937), p. 151
11   PAGE, R. M.: 'The early history of radar', *Proc. Inst. Radio Engrs,* 1962, **50**, pp. 1232–1236
12   Letter from Robert M. Page to Professor B. K. P. Scaife, Trinity College, Dublin.
13   HENNEY, Keith: 'The Radio Engineering Handbook' (McGraw-Hill, New York, 1941), pp. 665–670
14   PAGE, Robert Maurice: 'The origin of radar' (Anchor Books, Doubleday and Company, Inc., New York, 1962)
15   FAGEN, M. D. (ed.): 'A history of engineering and science in the Bell System: National Service in war and peace (1925–1975)' (Bell Telephone Laboratories, Inc., New Jersey, 1978)
16   DAVIS, Harry M.: 'History of the Signal Corps Development of US Army Radar Equipment Part I', Fig. 14, Signal Corps Historical Section, Office of the Chief Signal Officer, Washington, (1944). Microfilm copy available from Photoduplication Service, Library of Congress, Washington DC, USA
17   Reference 16, p. 38
18   Reference 16, pp. 97, 98. Copy of specification reproduced at Appendix J
19   Reference 16, p. 71
20   AIR 2/2681: Public Record Office, Kew, London, UK
21   Reference 16, p. 85
22   DAVIS, Harry M.: 'The Signal Corps Development of US Army Radar Part II' (Office of the Chief Signal Officer, Washington, 1945), p. 25 *et seq.*
     Microfilm copy available from Photoduplication Service, Library of Congress, Washington DC, USA
     Radio Set SCR-268-T1 is historically important. Although the precursor of all subsequent Army radar development, only one model of it ever existed. It was the first radar set in the United States to employ double-lobe tracking (beam switching) and it was the last detector to combine radio and heat methods; from then on, thermal detection played no part in the development programmes of radar equipment. It was also the first US Army radar set to receive type classification. The SCR nomenclature was used for security reasons (SCR = Signal Corps Radio set, complete)
23   'The SCR-268 Radar', *Electronics,* 1945, **18**, pp. 100–109
24   RIDENOUR, Louis N.: 'Radar system engineering' (Radiation Laboratory Series, Vol. 1) (McGraw-Hill, New York, 1947)
25   TERRETT, Dulany: 'United States Army in World War II. The Signal Corps: The Emergency' (Office of the Chief of Military History, Washington DC, 1956)
26   Reference 1, section AV
27   Reference 4, p. 140. The footnote giving Dr Robert Page's comment is of interest
28   BUSH, Vannevar: 'Pieces of the action' (William Morrow and Company, Inc., New York, 1970)
29   Reference 1, section B-1, p. 13.
30   SMITH, Phillip H.: 'Electronic applications of the Smith chart' (McGraw-Hill, New York, 1969)
31   SOUTHWORTH, George C.: 'Survey and history of the progress of the microwave art', *Proc. Inst. Radio Engrs,* 1962, **50**, pp. 1199–1206

Section 4.3 bibliography

HIGHTOWER, John M.: 'The story of radar', Senate Miscellaneous Documents, 78th Congress, (Doc. No. 89), US Govt. printing Office, Washington DC, 1943
     A 19 page document which relates the story of the development of radar in the United States Navy. The style is journalistic, but useful details of times, people and events are given.
HOWETH, L. S. (Captain): 'History of communications-electronics in the United States Navy' (United States Navy Department, Bureau of Ships and Office of Naval History, USGPO, Washington DC, 1963), Chapter 38

A 26 page informative account of the development of radar in the United States with particular reference to the work of the Naval Research Laboratory. There is reference also to the work of the Army Signal Corps and to the National Defense Research Committee.

PAGE, R. M.: 'The early history of radar', *Proc. Inst. Radio Engrs.,* 1962, **50**, pp. 1232–1236

An authoritative and valuable account of the pathway of development of radar at the Naval Research Laboratory from the observations of Taylor and Young in 1922 until the installation of the first United States shipborne radars.

PAGE, R. M.: 'Radar – a retrospective in perspective', *Report of NRL Progress,* 1973, July, pp. 21–24

An overview of the development of radar in the United States and Great Britain, with particular emphasis on the work of the Naval Research Laboratory including a reference to post-war over-the-horizon radar.

GETTING, Ivan A.: 'SCR-584 radar and the Mark 56 Naval Gun Fire Control System', *IEEE on Aerospace and Electronic Systems,* 1975, **AES-11**, No. 5, pp. 921–936

The paper is based on an address delivered by Dr Getting after he received the Pioneer Award of the IEEE Aerospace and Electronic Systems Society in 1975. The material, which is concerned with the development of precision radar and anti-aircraft fire control sets, relates to work carried out under the auspices of the Radiation Laboratory from 1940 onwards.

'The SCR-268 radar', *Electronics,* 1945, **18**, pp. 100–109

A description of the operation of the various circuits of this Signal Corps fire-control radar which was developed before the war and which was declared obsolescent only in 1944.

BAXTER, James Phinney: 'Scientists against time' (Little Brown and Company, Boston, 1946)

This is the brief official history of the Office of Scientific Research and Development. While there is an extensive treatment provided of the role of radar in the war, other material, such as the proximity fuse and the atomic bomb, is also covered. The order from President Roosevelt in June 1940 establishing the National Defense Research Committee, and also the order of June 1941 establishing the Office of Scientific Research and Development, are reproduced in appendices

ALLISON, David Kite: 'New eye for the Navy: the origin of radar at the Naval Research Laboratory', NRL Report 8466), Naval Research Laboratory, Washington DC, 1981

An in-depth study of the development of radar at the Naval Research Laboratory. Provides extensive references and a valuable guide to archival material and secondary sources.

'Selected bibliography on radar' (Northwestern University Evanston Ill. Technological Institute Library, Evanston, Ill, 1946)

39 pages of useful references divided into the following categories: historical background; terms; in the war; industry; photography; scientific applications; scientists; transport applications; schools; prospects

HIGGINS, Thomas James: 'A classified bibliography of publications on the history and development of electrical engineering and electrophysics', University of Winsconsin Engineering Experiment Station Reprint No. 198, Engineering School University of Winsconsin. Reprinted from *The Bulletin of Bibliography,* **20** (Nos. 3 to 7, Sept–Dec 1950 and Jan–April 1952)

Some 25 pages of references in the fields of general electrical engineering and electrophysics. Although there is very little material specifically on radar, several items would be of interest when investigating its electronic and radio background

FAGEN, M. D. (ed.): 'A history of engineering and science in the Bell System: National Service in war and peace (1925–1975)' (Bell Telephone Laboratories Incorporated, New Jersey, 1978)

This, the second volume of a series, devotes much of its material to the important role Bell Laboratories have played in the development of radar technology in the United States. Other associated topics in defence technology are also covered.

PAGE, Robert Morris: 'The origin of radar' (Anchor Books, Doubleday and Company Incorporated, New York, 1962)

This was issued under the Science Study series and so is directed at the young student and the layman. It explains the principles of radar and outlines the action of the major circuits and devices used in pulse radar; in the process, several events in the development of United States radar, with which the author was connected, are referred to. A useful chronology of radar to the year 1940 is included.

TERRETT, Dulany: 'United States Army in World War II: The Technical Services. The Signal Corps: The Emergency' (Office of the Chief of Military History, United States Army, Washington DC, 1956)

This volume, except for its first introductory chapter, deals with the history of the Signal Corps from the end of the First World War until December 1941. It covers the birth of radar and its early development within the Signal Corps. There are excellent footnotes and references. An appendix gives a list of Signal Corps equipment, including radar sets, used during the Second World War

THOMPSON, George Raynor, HARRIS, Dixie R., OAKES, Pauline M. and TERRETT, Dulany: 'United States Army in World War II: The Technical Services. The Signal Corps: The Test' (Office of the Chief of Military History, Department of the Army, Washington DC, 1957)

This, the second of three volumes, deals with the period December 1941 to July 1943. There is considerable information on United States Army and other radar equipments, on radar development and on radar counter-measures. Valuable footnotes are provided in the text.

THOMPSON, George Raynar and HARRIS, Dixie R.: 'United States Army in World War II: The Technical Services. The Signal Corps: The Outcome' (Office of the Chief of Military History, United States Army, Washington DC, 1966)

This third, and last, volume in the history of the Signal Corps deals with the period mid-1943 to the beginning of 1946 when, in January of that year, echoes were obtained from the moon using a modified SCR-271 equipment. The use of radar, and of radio and radar counter-measures, during the latter part of the war is dealt with. Useful footnotes abound

DAVIS, Harry M.: 'History of the Signal Corps Development of US Army Radar Equipment, Part I' (Office of the Chief Signal Officer, New York, 1944), available in microfilm copy from Library of Congress, Photoduplication Service, 10 First Street, S.E., Washington DC 20540

Deals with Signal Corps experiments in acoustics and heat detection methods and outlines the early work, until 1937, on radar.

DAVIS, Harry M.: 'History of the Signal Corps Development of US Army Radar Equipment, Part II' (Office of the Chief Signal Officer, New York, 1945), available in microfilm copy from Library of Congress, Photoduplication Service, 10 First Street, S.E., Washington DC 20540

Deals with developments from 1937 onwards, but particularly with the design, testing and use of the SCR-268 series of equipments.

GOLDBERG, Alfred (ed.): 'A history of the United States Air Force' (D. Van Nostrand Company Incorporated, New Jersey, 1957: reprint edition Arno Press, New York, 1974)

While little reference is made to the role of radar in the Second World War and one chapter covers the Air Defence of America in the post-war period, the book provides a useful synopsis of the employment of the Air Force.

NORBERG, Arthur L.: 'The origins of the electronics industry on the Pacific Coast', *Proc. IEEE*, 1976, **64**, pp. 1314–1322

Outlines the growth of the electronics industry in California. Of relevance to radar because of the particular companies or institutes discussed; Eitel-McCullough, Litton Engineering Laboratories and Stanford University are among those selected for mention.

Massachusetts Institute of Technology: 'Radiation Laboratory Series' (McGraw-Hill Book Company, New York 1947–1953)

28 volumes including an Index Volume. The series, with over 15 000 pages and 53 principal authors, is a distillation of the work carried out by the Radiation Laboratory within the period 1940 to 1945, inclusive. The United States principally, but also Britain and Canada, contributed to the Research and Development work, which is reflected in these volumes. The

series was relevant in the post-war era not only in the realm of radar systems, but right across the whole domain of electronic engineering. The Index Volume provides a short history of the Radiation Laboratory, a description of its organisation and an account of the preparation of the Series.

STOUT, Wesley W.: 'The Great Detective' (Chrysler Corporation, Detroit, Michigan, 1946). A 99 page illustrated account of the Chrysler Corporation's effort in designing and manufacturing the antenna mechanism for the SCR-584 radar set, together with related information on the development of radar in the Second World War.

Section 4.4 references

1 AVIA 7/244: Public Record Office, Kew, London, UK
2 A. P. 1063: 'Signals Volume IV; radar in raid reporting' (Air Ministry, London, 1950), Chap. 7 and Appendix No. 57. Available under AIR 41/12, Public Record Office, Kew, London, UK
3 GUERLAC, H. E.: 'Radar in World War II' (unpublished history of Division 14 of the National Defense Research Committee, 1947), Chap. A-VIII, p. 16, microfilm copy available from Photoduplication Service, Library of Congress, Washington DC, USA
4 Reference 2, Appendix No. 57
5 BION, J.: 'Le Radar', *Revue marit.*, 1946, July, pp. 331–346
6 VASSEUR, A.: 'De la TSF à l'électronique' (Agence Parisienne De Distribution, Paris, 1975)
7 Personal notes of P. David. Copy given to Professor B. K. P. Scaife, Trinity College, Dublin
8 ENGLUND, Carl R., CRAWFORD, B. and MUMFORD, William M.: 'Some results of a study of ultra-short-wave transmission phenomena', *Proc. Inst. Radio Engrs*, 1933, **21**, pp. 464–492

Section 4.4 bibliography

GUTTON, Henri and BERLINE, Sylvain: 'Essais sur le Propagation des Ondes Électromagnétiques de 16cm de Longueur', *Comptes Rendus*, 1938, **207**, pp. 325–326
Deals with point-to-point transmissions at a wavelength of 16 cm and with the use of the SFR magnetron type M16. The results of tests over a particular path of 152 km between two mountains are discussed
GUTTON, Henri and BERLINE, Sylvain: 'Recherches sur les Magnétrons: Magnétrons SFR pour Ondes Ultra-Courtes', *Bull. de La Société Française Radioélectricien*, 1938, **12**, pp. 30–46
Published in both French and English. Describes the construction, theory of operation, and performance of SFR magnetrons. Paper relates to period 1934 to 1938.
MM. ÉLIE, GUTTON, H., HUGON et PONTE: 'Détection d'Obstacles à la Navigation Sans Visibilité', *Société Franc. Elect. Bull.*, 1939, **9**, pp. 345–353
Properties of very short radio waves discussed, including phenomena of reflection and diffraction. Means of detecting objects and of measuring bearing and range are described. Equipment designed around the SFR M16 magnetron is outlined. Results of tests are quoted.
BRENOT, P.: 'Réalisations d'Un Grand Centre de Recherches Industriel Pendant et Malgré l'Occupation', *Onde Élect.*, 1945, **20**, pp. 29–46
The main centres of industry and research in France just before the war and during the occupation are discussed. Achievements in the manufacture of various items during the period of occupation are dealt with. These include electronic valves of various types, multiplexers, radio-communication equipment, electron microscopes and electronic materials.
PONTE, M.: 'Sur des Apports Français à la Technique de la Détection Électromagnétique', *Annls Radioélect.*, 1946, **1**, pp. 171–180

Discusses the work of CSF and SFR from the early 1930s until 1942. In particular various centimetric equipments, which were used for the detection of objects, and research on the magnetron are treated.

BION, J.: 'Le Radar', *Revue marit.*, 1946, July, pp. 331–346

A general pre-history of radar, its basic principles and some early French history of radar are covered. The first of two papers.

BION, J.: 'Le Radar II, Historique du Radar', *Revue marit.*, 1946, August, pp. 457–471

The second of two papers. Concerned in a general way with the history of radar during the first half of the Second World War and includes useful references to French radar.

DAVID, P.: 'La Guerre des Ondes', *Revue marit.*, 1947, June, pp. 666–685

Discusses the various uses, and problems in usage, of radio waves during the Second World War. Deals with secrecy in communications, with navigation and guidance and with radar.

GIBOIN, E.: 'L'Évolution de la Détection Électromagnétique dans la Marine Nationale', *Onde élect.*, 1951, **31**, pp. 53–64

A discussion of radar policies and equipment in so far as they relate to the Navy, from the early 1930s until after the liberation of France. Details on equipment performance are given.

OUDET, L.: 'Le Radar et les risques d'abordage', *Revue marit.*, 1953, July, pp. 837–846

A discussion on the operational uses and limitations of radar at sea as an aid to safety.

GIBOIN, E.: 'La Marine et l'Industrie du Radar en France', *Revue marit.*, 1954, September, pp. 1158–1182

Dwells on various topics concerning radar systems. No particular chronological order is adopted, but items concerning the pre-war, war and post-war operations of French naval radars are briefly mentioned.

OGER, J.: 'Pré-Histoire du Radar', *Revue marit.*, 1955, **108**, pp. 433–469

Traces from 1904 to 1945, in chronological order, the important milestones in the development of radar. Events in France, Germany, Great Britain, and the United States are covered.

LECONTE, A.: 'Conception et Realisation des Radars des Marine et des Balises Radar', *Nouveautés tech. marit.*, 1957, pp. 217–227

Affords an early post-war period view of the uses of maritime radar. Deals with navigation, surveillance and beacons. Gives the principal characteristics of several commercial equipments.

GIBOIN, E.: 'Les équipments radio-électriques des navires de la marine nationale', *Nouveautés tech. marit.*, 1962, pp. 138–149

A useful paper in that it indicates the status of post-war French naval equipment. Discusses the electronic equipment used by the French navy with particular emphasis on shipborne communication and radar systems. Gives typical French naval ships' complement of radio-electric systems.

RIVOIRE, J.: 'Fuyante Électronique', *Revue de Defense National*, 1963, **19**, pp. 402–412

An overview history of the radio and electronic sciences including radar and television, and with a special French flavouring.

'Électronique et Armement', *Revue Historique de l'Armée*, 1964, **2**, pp. 179–202

The military uses of radio-electric science in a chronological sequence; includes radio communication from the early days, radar, television, infra-red and electronic counter-measures. Where applicable, French contributions are mentioned.

DEMANCHE, M.: 'Technique et Évolution du Radar' (Editions Étienne Chiron, Paris, 1948)

Useful historical material is contained within the text.

DAVID, Pierre: 'Le Radar' (Presses Universitaires de France, Paris, 1969)

One of the popular 'Que sais-je' volumes. An exposition of radar for the layman. Contains some historical material including a map of the 'Barrages Électromagnétiques' which protected Brest in 1939.

VASSEUR, Albert: 'De la TSF à l'Électronique (Histoire des Techniques Radioélectriques)' (Agence Parisienne de Distribution, Paris, 1975)

An excellent source particularly with regard to people and institutions. The history of radar in France is covered in quite considerable detail.

'France-Marine et radar (radio direction and ranging), avant et pendant la Seconde Guerre Mondiale' (Service Historique de la Marine, Chateau De Vincennes, 94300 Vincennes, 1978) Obtained from 'Chef du Service Historique de la Marine' at above address. Over 15 pages of a chronological survey of French radar development, together with a list of sources. Covers the period 1923 to 1942. This account was relied on considerably for the historical summary given above.

Section 4.5 references

1   ISTED, G. A.: 'Guglielmo Marconi and Communication Beyond the Horizon', *Point to Point Telecom.*, 1958, **2**, No. 2, pp. 5–17

2   ISTED, G. A.: 'The later experimental work of Guglielmo Marconi', *Marconi Rev.*, 1974, **37**, No. 192, pp. 3–17

3   TIBERIO, U.: 'Some historical data concerning the first Italian Naval radar', *IEEE Trans. Aerosp. Electron Syst.*, 1979, **5**, pp. 733–735

4   Letter from Professor G. Latmiral of the Istituto Universitario Navale, Naples, to Professor B. K. P. Scaife of Trinity College, Dublin

5   TIBERIO, U.: 'Misura Di Distanze Per Mezzo Di Onde Ultracorte (Radiotelemetria)' *Alta Freq.*, 1939, **8**, pp. 305–323

6   CASTIONI, L.: 'L'Italia aveva il radar: perché non lo usò?', *Storia Illustrata*, 1979, No. 258, pp. 46–59

7   CASTIONI, L.: 'Storia dei Radiotelemetri Italiani' (Istituto Storico E Di Cultura Dell'Arma Del Genio, Roma, 1974)

8   TIBERIO, U.: 'Ricordo Del Primo Radar Navale Italiano', *Riv. maritt.*, 1976, Dicembre

9   FRIEDMAN, Norman: 'Naval radar' (Conway Maritime Press Ltd, Greenwich, 1981)

10 Letter and draft copy of history of Italian Naval Radar sent by L. Castioni to Professor B. K. P. Scaife of Trinity College, Dublin

11 Letter from Professor Nello Carrara, President of the SMA Company of Florence, to Professor B. K. P. Scaife of Trinity College, Dublin

12 SAFAR drawing of the EC-3 bis transmitter, dated 17th February 1942: part of the 'Bozza 0122 SAFAR', which was supplied by L. Castioni to Professor B. K. P. Scaife of Trinity College, Dublin

13 Documents supplied by L. Castioni to Professor B. K. P. Scaife of Trinity College, Dublin

14 Patent No. 369208, signed 15th March 1939: 'Dispositive per Rivelare la presenza a distanza di corpi quale navi, aeroplani, anchi se questi corpo sono invisiblé cause l'oscurità della notte o del la nebbia e per un certo grado anche Vincendo la curvatura terrestre'. Copy sent by L. Castioni to Professor B. K. P. Scaife of Trinity College, Dublin

Section 4.5 bibliography

'Rivelatore di Ostacoli SFR D16 a microonde', *Alta Freq.*, 1936, **5**, pp. 314–319
   Discusses briefly the detection of obstacles by means of microwaves. Describes the use of the SFR retarded-field triode and includes a diagram and photograph of the transmitter and receiver antennae mounting of the SFR installation on board the *Normandie* (reference may be made to Sections 3.11 and 4.4)

TIBERIO, U.: 'Misura di Distanze per Mezzo di Onde Ultracorte (Radiotelemetria)', *Alta Freq.*, 1939, **8**, pp. 305–323
   Although this paper was published in 1939, it details the work carried out by Tiberio in 1935 on the feasibility of detecting objects by means of ultra-short radio waves. A basic radar-range equation is developed.

TAZZARI, Oreste: 'Radiotelemetria', *Riv. maritt.*, 1940, Novembre, pp. 177–187
The potential uses of 'radiotelemetria' in naval warfare are discussed. The techniques and equipment referred to are those of the Barkhausen–Kürz, or similar, type of the mid-1930s. Besides its possible use in detecting the presence of targets and in measuring their distance, its utilisation in night-time formations and in the launching of torpedo attacks is also dealt with.

TIBERIO, U.: 'Sulla Valutazione del Rapporto Tra Segnale E Disturbo Nei Ricevitori Con Indicatore Oscillografico', *Alta Freq.*, 1943, **12**, pp. 316–323
A theoretical study of the detection of radar signals in noise on a cathode-ray tube type A display. By means of Maxwell–Boltzman statistics, and using the parameters of a pulse radar, the author obtains an expression for the probability of detecting a target.

TAZZARI, Oreste: 'Radiotelemetria', *Riv. maritt.*, 1945; Gennaio, pp. 23–29
The first of three papers giving a discourse on radar. The discourse is divided into two parts. The first part is contained here and in the next item, while the second part is contained in the item following that. The paper deals with the basic concept of radar; time and hence distance measurement, and the measurement of angle.

TAZZARI, Oreste: 'Radiotelemetria', *Riv. maritt.*, 1945, Febraio-Marzo, pp. 18–24
A continuation of the previous item. Introduces the uses of the Doppler effect in radar and treats of propagation and range performance.

TAZZARI, Oreste: 'Radiotelemetria: Parte Seconda', *Riv. maritt.*, 1945, Aprile-Maggio, pp. 5–17
Part two of the article. First part was covered in the previous two items. Treats of the emergence of radar and its military uses in various countries. A wide range of topics and events are covered in a general fashion; much of this information would have been new to the general public in 1945.

TIBERIO, U.: 'Radiotelemetria', *Ricerca scient.*, 1945, Settembre, pp. 219–229
A general essay on radar. Deals, first, with pulse methods used in ionospheric research, then with events which determined the emergence of radar and, finally, with the limitations of radar and with its employment during the Second World War.

LATMIRAL, Gaetano: 'La Superreazione', *Alta Freq.*, 1946, **15**, pp. 148–166
A very interesting paper on the phenomenon of super-regeneration and its applications. The difficulties of providing a fully quantitative treatment are pointed out. A qualitative explanation of super-regeneration is given, together with applications, such as its use in radar jamming apparatus as was employed by the author.

TIBERIO, U.: 'Introduzione alla Radiotelemetria (Radar): Apparecchi E Nozioni Entrati Nell' Uso Corrente' (Editore Rivista Marittima, Roma, 1946)
A comprehensive textbook on radar principles. The introduction (13 pages) gives some historical facts particularly concerning the development of Italian radar.

TIBERIO, U.: 'Minimum detectable radar signal', *Proc. Inst. Radio Engrs*, 1948, p. 1261
An interesting letter which discusses a formula of A. V. Haeff published in the *Proc. Inst. Radio Engrs.* of November 1946. Indicates that Tiberio had during the war addressed himself to the same problem.

TIBERIO, U.: 'The reduced range in a radar subjected to an external noise generator', *Proc. Inst. Radio Engrs*, 1954, **42**, pp. 1791–1798
The effect of noise sources on radar performance and particularly on the 'visibility factor' of a radar is analysed. Of interest because it indicates Tiberio's thinking in some basic areas of radar theory.

TIBERIO, U.: 'Cenni Sull' Opera Della Marina Italiana Nel Campo Radiotechnico Durante la Guerra 1940–1945', *Riv. maritt.*, 1948
Deals broadly with radio-science and with the Italian Navy during the war, but, because of the author's background, is a valuable source for the early history of radar in Italy.

TIBERIO, U.: 'La Guerra Elettronica Ieri E Oggi': Un ricedisturbatore antiradar Italiano del '42', Lettere al Direttore, *Riv. maritt.*, Estratto del fascicolo di Novembre 1976
A letter from Tiberio dealing with the anti-radar device of Professor G. Latmiral which

operated between 170 MHz and 220 MHz and which was used in 1942 against the British radar at Malta.

TIBERIO, U.: 'Ricordo Del Primo Radar Navale Italiano', *Riv. maritt.*, Estratto dal fascicolo di Dicembre 1976

An account of the 'Gufo' (English: 'owl') radar, together with a short account of the DETE (Detector tedesco) German radar of the Freya type which was fitted on some vessels. A short but good bibliography. The details on the 'Gufo' and associated sets and on the FIVRE valves are excellent.

CASTIONI, Luigi Carilio: 'Storia Dei Radiotelemetri Italiani' (Istituto Storico E Di Cultura Dell' Arma Del Genio, Roma, 1974) (Estratto dal: 'Bolletino Dell' Istituto Storico E Di Cultura Dell' Arma Del Genio')

A detailed history of the development of Italian radar from 1933 onwards. Complete with photographs and with outline specifications of equipment. A very good source book.

TIBERIO, U.: 'Introduzione Alla Tecnica Radio E Radar' (Masson Italia Editori S.p.A., Milano, 1977)

The work is divided into three parts which are contained in two volumes.

Volume One has two parts: Apparati fondamentali and propagazione ed antenne.

Volume Two covers the third part: Radioassistenza alla navigazione e tecnica radar. The first part contains historical material on the development of Italian radar.

ARENA, Nino: 'Il Radar Volume Primo: La Guerra Sui Mari' (S.T.E.M.-MUCCHI S.p.A., Modena, 1976)

A short summary of radar development in Germany, Britain, Italy, the United States, and Japan is given. A detailed treatment of the operational use of radar in naval warfare. Includes a description of the battle of Cape Matapan. Excellent photographs of radar installations and of related topics.

ARENA, Nino: 'Il Radar Volume Secondo: La Guerra Aerea' (S.T.E.M.-MUCCHI S.p.A., Modena, 1977) (Official Air Force history)

The use of radar in air defence in the various countries and in the various areas of conflict during the Second World War. Excellent photographs. A list of Italian radar sets.

ARENA, Nino: 'Il Radar Volume Terzo: La Caccia Notturna' (S.T.E.M.-MUCCHI S.p.A., Modena, 1977) (Official Air Force history)

Descriptions of the British, German and Italian air-interception systems; details of airborne systems; an account of the United States organisation in Europe and in the Pacific. Excellent photographs.

CASTIONI, Luigi Carilio: 'L'Italia aveva il radar: perché non lo usò?', *Storia Illustrata,* 1979, May N.258, pp. 46–59, Milano

The history of the development of Italian radar. A very readable account and of value for dates, names and general background.

DI CAPUA, Robert: 'Guerra Elettronica: Cenni Sulla Evoluzione', *Riv. Militara*, 1974, **97**, pp. 39–41

A very brief and general account of electronic warfare from 1911 to the date of publication of the article.

LOMBARDINI, Pietro: 'Possibilità Di Radio Sondaggi Astronomici Con Onde Metriche', *Commentationes Pontificae Scientiarum,* 1944, **8**, pp. 13–19

Of interest because it discusses the possibility of using metric radar to obtain echoes from the moon; the transmitted power and received field strength requirements are considered.

Section 4.6 references

1   'A short survey of Japanese radar', Prepared by 2nd and 3rd Operations Analysis Sections, FEAF and Air Technical Intelligence Group, FEAF (ATIG Report No. 115), Headquarters, Army Air Forces, Washington 25, DC, 1945. Copy available on microfilm, Reel No. A7277, from Chief of Circulation, The Albert F. Simpson Historical Research Centre, USAF, HOA, Maxwell AFB, AL36112, USA

2   WILKINSON, Roger I: 'Short survey of Japanese radar – I', *Electl. Engng*, 1946, **65**, pp. 370–377, based on material in Reference 1
3   IZUMI, Shinya: 'Radio Rokeita No Hanashi (radio locator)' (Kagakusha, 1944), p. 123
4   Letter, with paper on Research and Development of D.F. and radar in the Imperial Japanese Navy, from Hideo Sekino, Director Shiryo Chosakai, Tokyo, to Professor B. K. P. Scaife of Trinity College, Dublin
5   Letter from Dr Minoru Okada to Professor B. K. P. Scaife of Trinity College, Dublin
6   Letter from Professor Higasi to Professor B. K. P. Scaife of Trinity College, Dublin
7   HARVEY, A. F.: 'High frequency thermionic tubes' (Chapman and Hall Ltd, London, 1944), Chapter 5
8   WILKINSON, Roger I.: 'Short survey of Japanese radar – II', *Electl. Engng.*, 1946, **65**, pp. 455–463, based on material in Reference 1
9   NAKAJIMA, S.: 'The study of the microwave magnetron' (Japanese Radio Co., 1947: copy sent by Dr Nakajima to Professor B. K. P. Scaife of Trinity College, Dublin)
     A very interesting 16 page report on work carried out between 1933 and 1945.

Section 4.6 bibliography

IZUMI, Shinya: 'Radio Rokeita No Hanashi (radio locator)' (Kagakusha, 1944)
    A general account, covering 146 pages, of the origins and development of radar. It covers in sequence the necessity for radar, the first descriptions of it in the press and the mass media, the principles of radar and an account of American, British and Japanese radars. It states that the first disclosure of radar to the public in Japan was on the 16th April 1943 (p. 121), while it contends that Japanese radar equipment was superior to British and American equipment, with sets located at important military establishments (p. 122).

MATSUI, Mineaki: 'Nihon Kaigon No Dempatanshingi Kenkyu No Gaiyo – I' ('Summary of radar research of Japanese Navy – I'), *Heiki to Gijutsu* (*Weapons and Technology*), 1975, September, pp. 21–33
    Provides a background to the emergence of radar in Japan and in other countries. The research and production processes in the Japanese Army are summarised. A table of the nomenclature of radars in the Japanese Navy is provided. The research and construction of radars used in anti-aircraft and surface-search roles are treated. 19 figures, including 11 photographs and 2 flow diagrams of equipments, are provided.

MATSUI, Mineaki: 'Nihon Kaigun No Dempatanshingi Kenkyu No Gaiyo – II" ('Summary of radar research of Japanese Navy – II'), *Heiki to Gijutsu* (*Weapons and Technology*), 1975, October, pp. 15–23
    A continuation of previous item. Summary of research and development of anti-aircraft radar. Facts about United States and British radars, which were revealed after the war. Early radar research in the United States; research and development of radar in Great Britain. The problems of technological supervision and the question of providing administration for an EM wave oriented technology.

'Gijutsu Shiryo (Technical Reports)', No. 82, 1978, subtitled 'Radar development of the Japanese Army during WWII: anti-air radar Tago Type 2, Tago Revised Type 4', issued by Boei Cho Gijitsu Kenkyu Hombu Gijitsubu Chosaka
    Contents, in 197 pages, cover: documents in Kawasaki* research Institute between October 1944 and May 1945: the discovery of radar phenomena; development of 20 cm radar; use of 3 m radar for warning and of 20 cm radar for attacking roles; the beginning of the Pacific war and the start of production of radars based on information obtained from British equipment; the production of the Würzburg set; hyperbolic navigation system; Tachi-28 system; anti-ship attacking radars; notes on design of radar sets; specific designs for transmitters, receivers and display units, and data on thermionic valves, including cathode-ray tubes.

---

* 10 miles from Tokyo.

AIR 40/2185. Public Record Office, Kew, London, UK
This file contains an extensive treatment of Japanese radars; there are 17 separate items including: a preliminary report on Japanese centimetric radar development (6th November 1945); a table summarising the characteristics of Japanese Army Type-B radars, issued by Military Intelligence Division, US War Department (July 1945); details of Gun Laying and searchlight control equipment, RAF Signal Intelligence Service (10th June 1945); report on Japanese radar together with 19 pages of specifications, in resumé form, of various Japanese sets, RAF Signal Intelligence Service (30th November 1944). In this report the formal nomenclature 'Mark, Model, Modification, Type' is used.

NAKAJIMA, S.: 'History of Japanese radar development to 1945' (IEE seminar on the history of radar development to 1945, 10–12 June 1985, Savoy Place, London; full text of lecture sent by Dr Nakajima to Professor B. K. P. Scaife of Trinity College, Dublin)
An overview of the development of radar in both the Japanese Navy and the Japanese Army. Of particular interest is the reference to the work on magnetrons and to research, started in 1942, on the building of a super high-power magnetron for the shooting down of aircraft.

FRIEDMAN, Norman: 'naval radar' (Conway Maritime Press, Greenwich, London, 1981)
On p. 207 some wartime naval radars are discussed.

The following set of microfilms are available from The Albert F. Simpson Historical Research Center, USAF, HOA, Maxwell AFB, AL3 6112, USA

Against each reel number, a summary of contents is supplied. This summary is that provided by the Albert F. Simpson Historical Research Center. Generally, each microfilm covers far more subject material than is indicated by the summary.

Reel No.

A1166 USSBS
Japanese ground radar operations and tactics. USSBS interrogation #491 of Lt. Col. Masuda Ija. (File No. 137.73–15)

A2891 Combat Intelligence Branch, D/O&T, AFF Center Air Intelligence Digest No. 1, 1 July 1945, containing reports on Japanese tactics and technical intelligence about the Japanese and their equipment and including reports on Allied Operations (File No. 248.533–10)
Director of Operations & Training, AFF Center Air Intelligence Digest No. 2, 16th July 1943, with data on Japanese tactics, targets, technical intelligence, aircraft, radar intelligence and photographic intelligence (File No. 248.533–11)
Director of Operations & Training, AFF Center Air Intelligence Digest No. 3, 1st August 1945, with data on Pacific operations. Japanese tactics, Japanese targets, Goering interrogation, radar and aircraft (File No. 248.533–12)

A2892 Directorate of Operations & Training, AFF Center Air Intelligence Digest No. 4, 15th August 1945, containing data on Manchuria, tactics Japanese, Japanese targets, interrogation, radar intelligence, and aircraft recognition (File No. 248.533–13)

A7102 AAF POA
Report of enemy radar activity in the Pacific Ocean area. 20th December 1944 (File No. 702.654A)

A7237 GHQ SWPA
Current statements on Japanese radar. November–December 1944 (File No. 710.654B)
GHQ SWPA
Japanese radar experiments. This document is the translation of 2 captured Japanese documents. The first is entitled 'Notes on Submarine Search First Phase Experiments', the second, 'Search Position and Dead Space Charts'. September 1944 (File No. 710.654C)

A7277 2nd and 3rd Operations Analysis Section FEAF. A short survey of Japanese radar, 20th November 1945 (File No. 720.310A)
2nd and 3rd Operations Analysis Section FEAF. Japanese radio and radar equipment

investigated at the Nagano Plant of the Nippon Musen Company, 19th December 1945 (File No. 710.310B)

FEAF

A method of estimating enemy radar coverage, employment of countermeasures and evasive tactics: memorandum report, 19th December 1944 (File No. 720.310–58)

A7526 V Bomber Command

Charts showing locations and approximate coverage of Japanese radar in Netherlands, East Indies, and Formosa areas. 10th March–1st April 1945 (File No. 732.6541–1)

A7777 XX Bomber Command

Monthly enemy radar report, January 1945 (File No. 761.654–1)

A8028 Southeast Asia Command

Japanese radar – Eastern fleet intelligence summary, November 1944 (File No. 805.654–1)

Southeast Asia Command

Japanese radar – East Indies intelligence summary, January 1945 (File No. 805.654–2)

A8120 Eastern Air Command

Weekly intelligence summary No. 23, 2nd February 1945: Japanese radar stations in South-East Asia. Sec. 5, pp. 5–11

(See also later issues) (File No. 820.607)

A8125 Eastern Air Command

Collected reports on Japanese radar and radio, February 1944–March 1945 (File No. 820.654)

A8340 Fourteenth Air Force

Weekly intelligence summary, 27th July–2nd August 1944. Report on enemy (Japanese) radar. Sec. 2, pp. 5–9

(see also later summaries) (File No. 862.607)

B0907 7th Fighter Wing

Scope Dope: published by Signal Section. June 1944–August 1945 (File No. WG-7-SU-BU(FTR)C)

B1764 Military Intelligence Division, War Department Translation of Japanese technician's notebook on radar. Translation No. 34, 15th July 1944 (File No. 170.2287B-1(C))

B2059 Boca Raton Army Air Field, Fla

Designation of Japanese navy radar, December 1944. Training literature extract No. 80 from JE1A reports Nos. 5402 and 5741 (File No. 280.87–213)

K7309 Japan Air Defense Force

History, Directorate of Engineering, January–February 1952. Revisions to the Japan radar program, 1941–1945 (File No. K719.930)

Section 4.7 references

1   See for instance: SKOLNIK, Merrill, I.: 'Introduction to radar systems' (McGraw-Hill, New York, 1962), p. 12 and FRIEDMAN, Norman: 'Naval radar' (Conway Maritime Press, Greenwich, London, 1981), p. 186

2   ERICKSON, John: 'Radio-location and the air defence problem: The design and development of Soviet Radar 1934–40', *Science Studies*, 1972, **2**, pp. 241–263

3   Ощепков, П. К.: 'Жизнь и Мечта' (Моск. рабочий, Москва, 1967) OSHCHEPKOV, P. K.: 'Life and dreams' (Moscow Worker, Moscow, 1967)

4   Шембель, Б. К.: 'У истоков радиолокации в СССР' (Советское радио, Москва, 1977) SHEMBEL, B. K.: 'To the sources of radiolocation in the USSR' (Soviet Radio, Moscow, 1977)

5   Reference 3, p. 97

6   Reference 3, p. 74 *et seq.*

7 Хорошилов, П. Е.: 'Это начиналось так . . .' (Воениздат, Москва, 1970)
   KHOROSHILOV, P. E.: 'It began thus . . .' (Military Publications, Moscow, 1970)
8 Лобанов, М. М.: 'Начало советской радиолокации' (Советское радио, Москва,
   1975), p. 16 *et seq.* LOBANOV, M. M.: 'The start of Soviet radiolocation' (Soviet Radio,
   Moscow, 1975)
9 Reference 4, p. 53 *et seq.*
10 RAUCH von, Georg: 'A history of Soviet Russia' (Praeger Publishers, New York, 1972), 6th
   edn., p. 357 *et seq.*
11 HYDE, H. Montgomery: 'Stalin' (Rupert Hart-Davis, London, 1971), p. 238 *et seq.*
12 Алексеев, Н. Ф. и Маляров, Д. Е.: 'Получение мощных колебаний магнетроном',
   1940, *Zh.tekh. Fiz.*, **10**, pp. 1297–1300. ALEKSEEV, N. F. and MALYAROV, D. E.:
   'Generation of high-power oscillations with a magnetron', *Zh.tekh.Fiz.*, 1940, **10**, pp.
   1297–1300

Section 4.7 bibliography

Баранов, А.: 'Пути развития радиоэлектронной борьбы во второй мировой войне'
Вестн. противовоздушной обороны, 1976, No. 7, pp. 87–91
BARANOV, A.: 'Means of development of radioloelectronic contest in the Second World War',
Anti-Aircraft Defence Handbook, 1976, No. 7, pp. 87–91
Deals with the period 1940–1945. Concerned with radio intelligence and with electronic
warfare involving British, German and United States forces in the European theatres of war.
Гранкин, В. Змиевский В.: 'Из истории радиоэлектронной борьбы' Воен.-ист.журн.,
1975, No. 3, pp. 82–88
GRANKIN, V. and ZMIEVSKI, V.: 'From the history of radioelectronic warfare', *Military
History Journal,* 1975, No. 3, pp. 82–88
Deals with the Russian use of radio-propaganda after the Revolution and with various uses of
electronic warfare by the Russians during the Second World War.
Клюкин И. И.: 'Первые русские гидроакустики' Судостроение, 1967, No. 5, pp. 71–74
KLUKIN, I. I.: 'First Russian hydroechosoundings', *Shipbuilding,* 1967, No. 5, pp. 71–74
Concerned principally with Russian work in underwater communication mainly in the period
from 1906 to the First World War; treats the work of R. G. Niremberg. There is some mention
of ultrasonic detection work done before the Second World War.
Кобзарев Ю. Б. 'Первые советские имлульсные радиолокаторы', Радиотехника, 1974, No. 5,
pp. 2–6
KOBZAREV, Yu. B.: 'First Soviet impulse locators', *Radiotekhnika,* 1974, No. 5, pp. 2–6
After a short discourse on early developments in radio science including ionospheric research,
the history of Russian radar development from 1934 until 1940 is covered. Includes
photographs of a RUS-1 type A display and of a RUS-2 antenna system.
Лобанов М. М.: 'Из прошлого радиолокации. Краткий очерк' (М., воениздат, 1969)
LOBANOV, M. M.: 'From the history of radio location' (Military Publications, Moscow, 1969)
212 pages. Starting with the infra-red detection experiments of 1932, it provides a
comprehensive account of Russian radar development. Photographs of equipment and
personnel are provided. Includes a table comparing the performance of American, British
and German equipment with the RUS-2 and RUS-2s systems.
Лобанов М. М.: 'К вопросу возникновения и развития отечественной радиолокации',
Воен.-ист.журн., 1962, No. 8, pp. 13–29
LOBANOV, M. M.: 'Concerning the origins and development of soviet radiolocation', *Military
History Journal,* 1962, No. 8, pp. 13–29
An informative summary of the development of radar in Russia from 1932 until the early
1940s. Contains photographs of the author and of Oshchepkov, Korovin, Shembel,
Bonch-Bruevich and Shestokov.
Лобанов М. М.: 'Начало советской радиолокации' (М. "Сов. радио", 1975)

LOBANOV, M. M.: 'The origins of Soviet radiolocation' (Soviet Radio, Moscow, 1975)
    286 pages. A revised and slightly enlarged edition of Lobanov's 1969 publication, as above.
    Includes references at the end of each chapter and also biographical sketches of the author
    and of fourteen others associated with Russian radar development.

Ощепков П. К.: 'Жизнь и мечта. Записки инженера-изобретателя, конструктора и
    ученого' (2-е изд. М., Моск.рабочий, 1967)

OSHCHEPKOV, P. K.: 'Life and dreams: notes of an inventor, scientist and designer' (Moscow
    Worker, Moscow, 1967)
    295 pages. Provides a valuable insight into the work and mind of Oshchepkov. A detailed
    account of the origins of the Russian radar programme and of the experiments of 1934.

Ощепков П. К.: '40 лет со дня начала испытаний первой советской радиолокационной
    станции "Рапид" 1934 г.', Из истории авиации и космонавтики, 1974, **22**, pp. 87–90

OSHCHEPKOV, P. K.: 'The Fortieth Anniversary of the first tests of the Soviet "Rapid"
    Radiolocation Station', *History of Aviation and Cosmonautics,* 1974, **22**, pp. 87–90
    An account, with references, of the background to the 'Rapid' tests in 1934.

Покровский Р.: 'Из исторци отечественной радиолокации', Воен.-ист.журн., 1976,
    No. 1, pp. 73–78

POKROVSKI, R.: 'The history of Soviet radiolocation', *Military History Journal,* 1976, No. 1,
    pp. 73–78
    Summarises the development of Russian radar between 1932 and 1944 and outlines the
    performance of the principal systems.

Покровский Р.: 'Юбилей отечественной радиолокации', Радио, 1974, No. 10, pp. 18–19

POKROVSKI, R.: 'The jubilee of native radiolocation', *Radio,* 1974, No. 10, pp. 18–19
    A short reference to the work of 1934 and to the development of the first Russian early
    warning sets, "RUS-1' and 'RUS-2'.

Стогов Д. С.: 'Первая советская', Радио, 1974, No. 10, pp. 18–19, 35

STOGOV, D. S.: 'The first Soviet Aircraft Detection System', *Radio,* 1974, No. 10, pp. 18–19, 35
    A brief article similar to previous item.

Стогов Д. С.: 'Рус-1 – радиоулавливатель самолетов', -Радиотехника, 1974, No. 11, pp.
    9–11

STOGOV, D. S.: 'RUS-1 – radio control of aircraft', *Radiotechnique,* 1974, No. 11, pp. 9–11
    Deals with the development, use and performance of 'RUS-1' radar.

Хорошилов П. Е.: 'Это начиналось так . . .' (М., Воениздат, 1970, 67 с.
    с нлл. – История радиолокации в СССР)

KHOROSHILOV, P. E.: 'It began thus . . .' (Military Publications, Moscow, 1970)
    66 pages. A reference to B. L. Rosing, who in 1907 patented a method of transmission of
    images and who suggested the use of a cathode-ray tube for display; also to work on
    magnetrons in 1934. Book deals with the background to the trials of 1934 and with the tests
    themselves.

Чемерис М., Шошков Е.: 'Русские изобретатели гидроакустических средств', -Воен.
    -ист.журн., 1967, No. 3, pp. 103–108

CHEMERIS, M. and SHOSHKOV, E.: 'Russian inventors of radioaccoustic methods', *Military
    History Journal,* 1967, No. 3, pp. 103–108
    Concerned with material covered in Klukin, 1967, as above but principally with the people
    involved.

Шембель Б. К.: 'У истоков радиолокации в СССР' (М., "Сов.радио", 1977)

SHEMBEL, B. K.: 'The sources of radiolocation in the USSR' (Soviet Radio, Moscow, 1977)
    Deals in chronological order with the people, institutions and equipment of the early Soviet
    radar. Short bibliography provided.

WITT, V.: 'Die Entwicklung den sowjetischem Funkmesstecknik', *Militärtechnik,* 1970, **4**, pp.
    174–177
    A useful overview paper, with illustrations of early Russian radar.

Section 4.8 references

1 'Innovation and tradition' A concise history of Hollandse Signaalapparaten, B.V. Hengelo, The Netherlands 1922–1974 (Vander Loeff/Drukkers, Enschede, 1974)
2 Letter from A. Verbraech of TNO to Professor B. K. P. Scaife of Trinity College, Dublin

Section 4.9 references

1 PANLÉNYI, Ervin (ed.): 'A history of Hungary', (Translation) (Collets', London and Wellingborough, 1975)
2 BAY, Zoltán: 'Visszaemlékezés A Magyar Holdvisszhang Kísérletekre', 1976, *Fizikai Szemle*, **26**, pp. 41–53
3 Letter from Zoltán Bay to Professor B. K. P. Scaife of Trinity College, Dublin
4 BAY, Zoltán: 'Reflections of Microwaves from the moon', *Hungarica Acta Physica*, 1946, **1**, pp. 1–22
5 VAJDA, P., and WHITE, J. A.: '30th Anniversary of Zoltán Bay's Pioneer Lunar Radar Investigations and Modern Radar Astronomy", *Acta Physica Academiae Scientiarum Hungaricae*, 1976, **40**, pp. 65–70

Chapter 5

The British Story

5.1 Introduction

The origins of radar and its subsequent development in Britain are clearly outlined and well documented in the literature[1].

In the minutes of the first meeting of the Committee for the Scientific Survey of Air Defence, held at the Air Ministry on Monday 28th January 1935, is recorded the following [2]:

1. At the request of the Chairman, D.S.R.* outlined the intended functions of the Committee, the terms of reference of which are

   'to consider how far recent advances in scientific and technical knowledge can be used to strengthen the present methods of defence against hostile aircraft'.

Further on in the same document, Minute 3(A)(f) reads:

The possibility of detecting short wave electromagnetic radiation reflected from the metal surfaces of an aircraft, using a ground source, was discussed.
D.S.R. said that Mr Watson Watt had prepared a memorandum on the uses of short wave electromagnetic radiation for defence purposes, which would be circulated to the Committee. Mr Watson Watt considered that there was some hope of detection by these means. *It was decided* that further consideration should be given to this possibility after Mr Watson Watt's memorandum had been circulated.

Some time in January 1935, H. E. Wimperis, who was a friend of R. A. Watson-Watt, the then Superintendent of the Radio Department of the National Physical Laboratory, Slough, consulted him on the possibility of using electromagnetic radiation to damage aircraft, or incapacitate air crew. Watson-Watt discussed the matter with Arnold F. Wilkins, who was then a

---

* Director of Scientific Research, Mr H. E. Wimperis.

Scientific Officer at Slough. Wilkins made some calculations which ruled out the possibility of a 'death ray', but he then suggested to Watson-Watt that radio waves might be used for the detection of aircraft [3].

The two studies, one which ruled against the feasibility of disabling radiation and the other which proposed a technique of detection, were incorporated in two memoranda drafted by Watson-Watt which are reproduced in Appendix D.

Wilkins was undoubtedly assisted in his thinking at this time [4] by his recollections of the flutter effect perceived by Post Office engineers in 1931 in their 60 MHz experimental VHF link between Colney Heath and Dollis Hill. This variation in received signal occurred when aircraft from Hatfield aerodrome flew in the vicinity [5].

A letter from A. P. Rowe to Tizard, dated the 4th February 1935, is worth reproducing in full, as it captures the atmosphere of the time [6].

Dear Mr Tizard,

A copy of a secret memorandum prepared by Mr Watson Watt on the possible uses of electro magnetic radiation for air defence, is enclosed herewith. I have checked the arithmetic and the formidable figures for damaging effect appear to be correct on the assumptions made. It was agreed at our first meeting that detection is a much more promising, though doubtless still difficult, line of research.

2. With regard to the meeting provisionally arranged for 14th February, the Commander-in-Chief, Air Defence of Great Britain, is on leave until 18th February, and I am trying to arrange a meeting at Uxbridge for the 18th or 19th February. I will telephone you before making definite arrangements.

3. I now have the Draft Minutes of our first meeting, and should much appreciate an opportunity of discussing them with you. I am lecturing at the Imperial College on Wednesday, 13th February, from 2.30 to 4.30 p.m., and perhaps some other time on this date would be convenient to you.

Yours sincerely,
A. P. Rowe

H. T. Tizard Esq., F.R.S.,
Rector of the Imperial College,
Imperial College of Science and Technology,
South Kensington, S.W.7.

Before continuing with the historical narrative, the next important episode of which is the Daventry Experiment of 26th February 1935, it is proper to pause and consider two points.

5.2 Other inventors of radar

The first point is the possibility that someone may claim that radar-type experiments (other than those of Butement and Pollard in 1931) were carried out in Britain prior to the developments of 1935. This is certainly possible, but there is no evidence that they affected or influenced in any way the main stream of development. It is opportune to quote from comments made by Rowe on 2nd October 1947 when reviewing the draft of a history on radar [7].

> Watson-Watt has described how he saw Dr Tucker* years before radar started, and suggested that radio should be tried. Moreover, Sir Edward Appleton has shown me a photograph of what is, in fact, an RDF reflection obtained by the Post Office people, I believe, from an aircraft. Appleton could produce the photograph. It is not quite so simple to say that scientists were not asked to consider the problem.
>
> I remember a senior man at T.R.E., on hearing of radar, saying that it was not new, and that echoes had been obtained on television tubes for two years. When asked why nothing was done about it he said that it was not on the programme.

5.3 Air defence before the Second World War

The second point is that the Committee for the Scientific Survey of Air Defence, which was a sub-committee of the Committee of Imperial Defence, were by no means operating in isolation. The timely development of Britain's Chain Home System depended at least as much upon the drive and vision of Tizard as upon the efforts of Watson-Watt and his co-workers. Yet Tizard and all of the key people of the radar saga, whether political, scientific, military or administrative, were influenced by a policy which was primarily concerned with the protection of Britain against aerial attack and which can be traced to pre-1914 days [8].

At the outbreak of the First World War the Royal Navy assumed responsibility for the air defence of Britain. Winston Churchill, First Lord of the Admiralty, in a statement on 5th September 1914 propounded four distinct lines of defence, namely [9]

(a) The attack on the enemy's aircraft as close as possible to their point of departure as well as bombing the bases themselves by a special squadron of aeroplanes in Belgium;

(b) an intercepting force of aeroplanes on the East coast of Great Britain in close communication with the overseas' squadron;

---

* Dr W. S. Tucker was Superintendent of the Air Defence Establishment, Biggin Hill, and the alleged discussion took place at a demonstration of sound locators.

(c) the concentration of the gun defence at vulnerable points of naval and military importance rather than for the protection of towns;

(d) the passive defence of London and other larger towns by darkening the localities.

In 1914, also, a general scheme of anti-aircraft defence had been drawn up by the staff of the Home Defence Directorate at the War Office. The scheme assumed that the whole of Great Britain was open to air attack and aimed at giving protection to magazines and explosives factories in eight designated areas throughout the island from Scotland to Portsmouth. It included a detailed plan for the defence of London.

In January 1916, after Zeppelin air raids on the Midlands, the Government took action on the reorganisation of the warning system. With the exception of NW Scotland, the whole island was divided into 8 Control Areas (referred to as 'Warning Controls'), each with a warning controller who acted as the representative of the Commander-in-Chief of the Home Forces. The headquarters of each controller was located at a main telephone centre in the Control Area. Each Control Area was subdivided into warning districts of extent approximately 30 miles by 35 miles based on a one half-hour transit of a Zeppelin raider.

The national telephone grid was the basis of the whole reporting and control system. The controllers had many sources of information. Initial warning of raids was often given hours in advance by British wireless direction-finding stations who picked up radio signals from the Zeppelins as they became airborne. Once raiders crossed the coast, information could be sent back from observation posts which were set up all around the country. In each Control Area Operations Room a transparent map was used showing warning districts which were lit behind by appropriate coloured lights. When, for instance, a particular district was threatened, the controller pressed a switch and the district in the map turned green. When attack was imminent another switch turned the district red. These colour signals were relayed automatically to the Telephone Trunks manager at the GPO who then issued the appropriate warnings.

The air defence organisation in Britain reached its apex in the First World War with the formation of LADA (London Air Defence Area) control under the command of Major-General E. B. Ashmore. By early 1918 radio telephone receivers were installed in some of the fighter aircraft and the whole LADA system was working effectively. It is worth quoting Major-General Ashmore [10]: describing the operation of the LADA arrangement, he wrote

This central control consisted essentially of a large squared map fixed on a table, round which sat ten operators (plotters), provided with headphones; each being connected to two or three of the sub-controls. During operations, all the lines were kept through direct; there was no ringing up throughout the system.

When aircraft flew over the country, their position was reported every half-minute or so to the sub-control, where the course was plotted with counters on a large-scale map. These positions were immediately read off by a 'teller' in the sub-control to the plotter in the central control, where the course was again marked out with counters. An ingenious system of coloured counters, removed at intervals, prevented the map from becoming congested during a prolonged raid.

I sat overlooking the map from a raised gallery; in effect, I could follow the course of all aircraft flying over the country, as the counters crept across the map. The system worked very rapidly. From the time when an observer at one of the stations in the country saw a machine over him, to the time when the counter representing it appeared on my map, was not, as a rule, more than half a minute.

In front of me a row of switches enabled me to cut into the plotter's line, and talk to any of my subordinate Commanders at the sub-controls.

The central control, in addition to receiving information from outside, constantly passed it out to the sub-controls concerned; so that the Commander, say, of an anti-aircraft brigade, would know, from moment to moment, where and when hostile aircraft would approach his line of guns.

By my side, in the gallery, sat the Air Force Commander (Brigadier-General T. C. R. Higgins) with direct command lines to his squadrons, and a special line to a long-range wireless transmitter near Biggin Hill. This transmitter was used for giving orders to leaders of defending formations in the air, during day time, in accordance with the movements of the enemy, as shown on the control map.

When the war ended in 1918, there was a quite complex air-raid warning organisation in existence and in particular the London Air Defence Area was well defended. It comprised, apart from an observer and reporting organisation, barrage balloon aprons, gun and searchlight stations, night-aircraft patrols against aircraft and night aircraft patrols against airships. No reasonable effort was spared to ensure that installations and the public were well protected against air bombardment. In fairness to those who later endeavoured to perfect sound location of aircraft, it must be recorded that sound locators were of considerable use. In particular, a concave sound mirror perfected by the Munitions Inventions Department and set into the face of a cliff gave adequate warning of the approach of German bombers, sometimes detecting them 20 miles from the coast.

Our final observation on this period is that Britain appears to have had a far more sophisticated air defence organisation than either Germany or France. General Ashmore's thinking seems to have been a prime influence and, indeed, we find his philosophy being re-echoed and expanded in 1938 in a book 'Views on air defence' [11].

If we move forward 10 years to 1928 we find [12] concern for the detection of bombers and the concept of identification of friend or foe in the problem of distinguishing above cloud and out of sight one's own fighters from the attacking aircraft. Special sound techniques are advocated including sirens, resonant vibrating or singing wires, devices attached to the aircraft exhaust and the provision of specially trained listeners!

Then in 1933 we find the plans for an acoustical mirror system for the Thames Estuary. Let us look at this scheme in some detail. On 19th December 1933 a conference was held [13] in the Air Ministry concerning experiments to be carried out for developing an acoustical mirror warning system. The conference was chaired by Group Captain W. L. Welsh and among those present was Colonel A. P. Sayer, R. E., and A. P. Rowe. The chairman mentioned that protracted experiments had been carried out by the ADGB (Air Defence of Great Britain) and the ADEE (Air Defence Experimental Establishment) to prove the utility of an acoustical mirror system for giving an early warning of the approach of aircraft. The experiments had culminated in the ADGB exercises of 1933. As a result of the exercises ADGB had submitted a plan for the employment of a number of 200 ft strip mirrors and 30 ft bowl mirrors which would give a continuous warning screen around the south-east of England from the Wash to St Alban's Head.

The conference then dealt with the location of mirrors, their maintenance, the purchase and hire of the necessary land, the civil works required, personnel, the scope of further experiments, the signal services and the organisation and administration. The proposed scheme was for the following sites and apparatus (see Appendix G):

Site	Shire	Mirror type
Frinton	Essex	200 ft
Clacton	Essex	30 ft
Tillingham	Essex	30 ft
Asplin's Head	Essex	30 ft
Grain	Kent	30 ft
Warden	Kent	30 ft
Swalecliffe	Kent	30 ft
Reculver	Kent	200 ft and 30 ft

On 20th May 1935 a letter was sent from the office of the Air Chief Marshal, Air Officer Commanding-in-Chief, Air Defence of Great Britain to the Headquarters Fighting Area, Royal Air Force, Uxbridge, Middlesex, to the effect that the Air Ministry anticipated that the whole of the acoustical mirror warning system would be available for use early in 1936. Then on 23rd September 1935 the Air Commodore, Director of Organisation, Air Ministry,

London WC2, wrote to the Air Officer Commanding-in-Chief, Air Defence of Great Britain:

> Sir,
>
> I am directed to refer to Air Ministry letter of above reference dated 15th Aug., and to say that it has been decided, in view of the possible development of alternative methods of detection, to suspend until the end of March 1936, all work on construction of the acoustical mirror system in the Thames Estuary.
>
> 2.–It is realised that this decision will render it impossible in any event to complete the scheme until some time in 1937. If it is decided to adopt an alternative method of detection, it is hoped that some of the sites already acquired for mirrors may be found suitable.

5.4 The Daventry experiment

Events moved very rapidly after the first meeting of the Committee for the Scientific Study of Air Defence on 28th January 1935, and the purusal by the Committee of Watson-Watt's memorandum. Wimperis was prepared to accept the validity of the calculations, but Air Marshal Sir Hugh Dowding pressed for an immediate demonstration before Treasury approval was sought for research.

Again, Wilkins' experience in propagation work proved an asset. Watson-Watt's initial response was to operate the Slough ionospheric transmitter at 6 MHz but to so modify it that it would operate on short pulses and with a considerably increased peak pulse power beyond its then maximum of 1 kW. However, Wilkins considered this impossible in the time available and recollected that the BBC Empire short-wave station at Daventry (call sign GSA) operated on 49·8 metres and beamed its 10 kW of power in a southerly direction. The Daventry antenna consisted of an array of horizontal dipoles which produced an azimuth beamwidth of some 60° with a main vertical lobe at an elevation of 10°.

Wilkins, in calculating the effect of re-radiation from an aircraft, as postulated in the second memorandum, had considered a monoplane bomber with a typical wing span of about 75 ft or 25 metres and had assumed this effectively equivalent to a horizontal half-wave dipole operating at 6 MHz. This now was the frequency of the Daventry transmitter. Thus the plan for the demonstration was to position a receiver at a suitable distance away from the Daventry transmitter and in its main beam, and to fly a Heyford bomber up and down the beam noting any fluctuations in the received signal caused by interference between the direct signal and the signal reflected from the aircraft.

The equipment was set up on 25th February inside a van in a field near Weedon, Northamptonshire, and the experiment was successfully carried out the following morning between 09.45 and 10.00 hrs. Wilkins was in charge of the demonstration and Watson-Watt and Rowe both witnessed its successful conclusion. In the minutes of the third meeting of the Committee for the Scientific Study of Air Defence held at the Air Ministry on Monday 4th March 1935, the following is noted [14]:

> D.S.R. reported that, using the 50 metre Daventry beam, Mr Watson Watt had successfully demonstrated the reception of secondary electro-magnetic radiations from a Heyford aircraft; the aircraft had a metal structure but the wings were fabric covered. With regard to the possibility of preventing these secondary radiations, D.S.R. said that the use of wireless equipment on aircraft had led to the practice of bonding metal components, hence aircraft were metallically continuous; if, however, some degree of insulation were used as a defence measure, a shorter wavelength than 50 metres might need to be used.
>
> D.S.R. informed the Committee that he had visited Orfordness with Mr Watson Watt; the site was suitable for experiments and he saw no reason why work should not commence almost immediately.

Let us pause to review the equipment and the technique of the Daventry experiment.

The method employed was an adaptation of that used at Slough (which had been introduced by Dr Hollingworth and perfected by Wilkins) to measure the angle of incidence at the ground of downcoming waves from the ionosphere [15] (see Fig. 5.1).

A,B are two parallel horizontal antennae. Downcoming waves are incident at an angle θ to the vertical. Normally no ground wave would be present.

If we consider that there is but one downcoming wave, incident at angle θ, and its associated ground reflected wave (DA and FEA in the case of antenna A) then the amplitudes of the EMFs produced in the antennae are equal. However, the EMF in A will lead that in B by a phase-angle

$$\left[\frac{2\pi d}{\lambda} \sin \theta\right]$$

As indicated, each antenna is connected to a receiver. If identical antenna and receiver systems are assumed, then when the amplified outputs of the two receivers are connected to appropriate deflecting plates of the cathode-ray tube measurements carried out on the resultant elliptic trace give directly the value of

$$\frac{2\pi d \sin \theta}{\lambda}$$

and hence of θ.

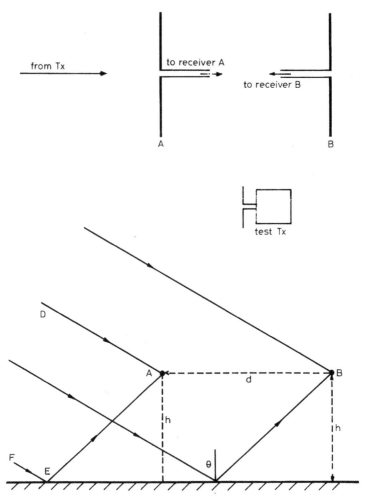

Fig. 5.1   *Method of measuring angle of downcoming waves at Slough*

Before a series of tests, it was necessary to equalise or 'line up' the receivers and also to ensure equality of the two antennae. The procedure for equalising the receivers was to common the two transmission lines and common the two receiver inputs and then adjust the receiver amplifiers for equal output as indicated by a 45° straight line trace on the cathode-ray tube. The equality of the antennae could be checked by placing a test transmitter equidistant from A and B as shown.

In the Daventry experiment the normal condition was the reception of a direct ground or space wave from the transmitter while the hoped-for phenomenon was a downward sky wave reflected from the aircraft when it flew through the beam. Antenna systems A and B were horizontal half-wave dipoles set up on poles about 15 ft above the ground, spaced about 50 ft apart and pointing directly at the transmitter. The transmission lines consisted of twisted lighting flex.

The equipment used was that employed in vertical ionospheric sounding and referred to as a 'pulse polarisation analyser'. At Slough a variety of techniques had been developed for high frequency direction finding, for the analysis of atmospherics, for measuring the angle of incidence and the polarisation of distant skywave high-frequency signals and for measuring the height and behaviour of the ionospheric layers. All of these were centred about the use of the cathode-ray tube and much common circuitry was employed.

The normal arrangement of the radio polarimeter is shown in Fig. 5.2*a*, while the special layout for the Daventry experiment is given in Fig. 5.2*b*. A detailed circuit diagram [16] is shown in Fig. 5.3.

A few comments on the radio polarimeter circuitry may be relevant:

(*a*) The local (or beat) oscillator of the frequency changer was common to both receivers so as to preserve the phase differences present at the signal inputs.

(*b*) The system was designed to handle pulse envelopes of 200 μs. The time differences between the arrival of the ground ray and the ionospheric ray was 1·5 ms or more for the F-layer, and 0·7 ms or more for the E-layer.

(*c*) The separation of transmitter and receivers was commonly of the order of 120 m so that circuits were liable to overload with a consequent flow of grid current, hence it was necessary that circuits have low time constants for quick recovery after reception of ground ray.

(*d*) The single-stage output IF amplifier was inserted after the two-stage IF amplifier proper for several reasons. The latter was a standard unit that had been used previously, but the extra stage allowed for the comparative insensitivity of the cathode-ray oscillograph when used for photography and also provided a push–pull output to prevent inter-receiver coupling.

For the Daventry experiment the connection to the horizontal deflecting plates of the cathode-ray tube from receiver B was removed. The signal voltage at the output stage of receiver B was paralleled with that from receiver A by inserting a lead between the grids of the output triode stages. The insertion of the phase changer introduced a loss in system B. Adjustment of the IF gain controls of the two receivers enabled the amplitudes of the two signal voltages at the output grids to be equalised, while adjustment of the phase changer enabled the phases of the two signals to be in opposition. Hence, when these two adjustments were correctly carried out, there was no deflection on the cathode-ray tube and so a negligible response of the complete system to the ground wave. If a 6 MHz signal should now arrive from any other angle, say reflected downwards from an aircraft, a vertical deflection would appear on the tube screen.

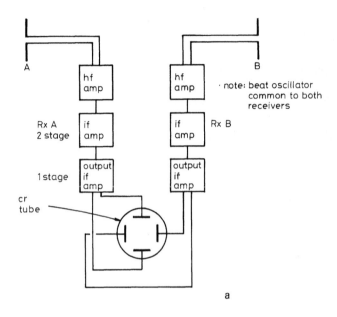

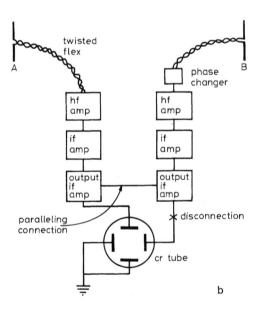

Fig. 5.2   *Polarimeter*
a Schematic
b Layout used in Daventry experiment

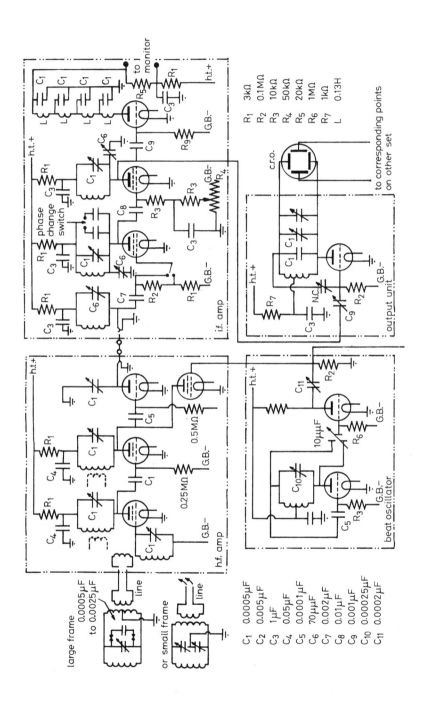

Fig. 5.3   *Circuit diagram of radio polarimeter*

The phase changer used was based on a goniometer-type phase shifting transformer comprising two primary field coils wound at right angles and a secondary search coil which could be pivoted through 360°. The primary coils were fed by currents which were 90° out of phase. In the circuit used at Daventry the signal from the antenna was applied to the grids of two tetrodes connected in push–pull and the field coils were inserted in the anode circuits, one coil detuned above resonance and the other detuned below resonance, so as to produce a rotating circular field in the volume swept by the search coil.

5.5 Development of the first radar system

The success of the demonstration meant a Treasury approval for development of the scheme with sanction to spend up to £12 300 in the first year, and this excluded the cost of aircraft flying time. It also meant that within almost a week Wilkins and Bainbridge-Bell were working at Slough on the project as outlined in Watson-Watt's memorandum, the former modifying an iono-spheric pulse transmitter. The memorandum had outlined a pulse radar working at 6 MHz with pulse widths considerably shorter than 200 µs and with a pulse repetition frequency somewhere between 50 and 1000. Sometime in mid-March Dr E. G. Bowen joined them and assisted Bainbridge-Bell on the transmitter development. The two principal objectives here were a shortening of pulse widths to about 15 µs and an increase in peak pulse power to about 50 kW [17].

On 13th May 1935 the team moved from the National Physical Laboratory at Slough to Orfordness in Suffolk. It consisted of Bainbridge-Bell and Wilkins, Dr Bowen (Scientific Officers), J. E. Airey (Assistant II) and G. A. Willis (Assistant III) [18]. Wilkins was occupied with receiver development and antenna problems, while Bowen and Bainbridge-Bell still worked on transmitters. The period in Slough had not been wasted. Army surplus stores had been searched for junk items and great assistance was obtained from the Naval Signal School at Portsmouth in obtaining naval-type silica valves [19].

For the first year the permanent radar team never exceeded 6 persons. Headed by Watson-Watt, who supervised Orfordness on a visiting basis, they were employed by the National Physical Laboratory but were paid by the Air Ministry. This point is mentioned because an obvious question to be asked is why Watson-Watt, as Superintendent of the Radio Research Station of the National Physical Laboratory at Slough, could not have used Slough and its organisation at least for the early development work. The fact is that he could not count on support from the National Physical Laboratory [20] and he was in effect compelled to break away from his career in order to help develop radar within a new organisation. (Perhaps his real greatness in the history of British radar lies in his administrative achievements and in the fact that he left a secure position and gambled his future on an as yet unborn technology.)

Orfordness, which was an airfield well equipped with unused buildings and power supplies, was ideally situated on the coast. It was also very close to the Aircraft and Armament Experimental Establishment at Martlesham. When the team moved in there on 13th May their transmitter was not yet operational, but was made so within three days. Within another five days the receiver was installed. The antennae in both cases were half-wave dipoles supported by 70 ft masts. On 31st May 1935, using the combined transmitter–receiver system, echoes from the ionosphere were observed. This leads us to an interesting point. The evolution of the working frequency from 6 MHz as used in these early experiments to the 22·7–29·7 MHz band for the Chain Home stations will be treated later, but at this time a reason for the choice of 6 MHz, apart from it matching the half-wave dipole model for a typical bomber's wing, was that transmissions would be propagated abroad by sky wave and mistaken for ionospheric research experiments.

A paper was published [21] by Watson-Watt, Wilkins and Bowen dealing with work carried out at Orfordness between May 1935 and May 1936 and which purported to establish stratified layers of ionisation in the troposphere and stratosphere (8·5 km to 13·5 km) which could return waves of frequency 6 MHz to 12 MHz at vertical incidence. It is worth quoting from this paper:

> Since a wide range of work in progress at the Radio Research Station, Slough, made it very undesirable to introduce a new high-power emitter there, the Air Ministry very kindly offered facilities at their Orfordness Research Laboratory, where the least possible disturbance to other radio work was ensured by the isolation of the station.
>
> A pulse emitter capable of giving large instantaneous outputs and giving well-shaped, substantially flat-topped, pulses of 20 μ sec duration, was installed, and was equipped with a directive aerial system favouring the vertical direction in emission. Some 300m away a corresponding directive receiving system was erected and connected to a receiver of sufficient band width to accept, with only moderate distortion, these very short pulses.

At this juncture it behoves us to provide as balanced a view of radar development as possible before discussing in any detail the actual equipments, and in the latter respect we must be selective, giving most emphasis to the Chain Home system. This can best be achieved by setting out events in chronological order and in tabular form. Before doing so, it is in order to record the first formal beginnings of Navy radar and of Army radar.

5.5.1 Navy radar
On 11th March 1935 a meeting of the Committee for the Scientific Study of Air Defence took place with Admiralty representatives only present, including the Signal School's Chief Scientist G. Shearing. The purpose of the meeting was to discuss means for detecting and locating aircraft through cloud and also Naval

AA gunnery and prediction methods. Strangely, no mention was made at this meeting of Watson-Watt's memorandum or of Air Ministry work being undertaken. In July 1935 a practical liaison was established between the Signal School and the Air Ministry.

5.5.2 Army radar

Colonel Worledge, the Signals Member of the RE and S (Royal Engineer and Signals) Board, visited the Radio Research Board at Slough on 4th July 1935. He had discussions with Watson-Watt and, as a result, it was arranged to include £30 000 in the draft estimates for 1936/7 for initial work under the RE and S Board.

5.6 CH (Chain Home) system

The sole function of the Chain Home stations was that of early warning. When we later describe in some detail the transmitter and receiver equipment used and their antenna arrays, it should be borne in mind that, while standardisation was more or less eventually achieved, many variations in detail occurred particularly between 1936 and 1940. In addition to the CH proper, which operated between approximately 22 and 50 MHz, there was associated with it the Mobile Radio Unit, AMES type 9, operating between 40 and 50 MHz and using much smaller masts, and the CHL (Chain Home Low), AMES type 2, stations operating at 200 MHz [22].

The Chain Home transmitter aerial system floodlighted an area in front of the station with a typical horizontal beamwidth of about 60°. A typical polar diagram is shown in Fig. 5.4. Direction finding methods employed in the receiving antenna array enabled rough information on bearing to be obtained; bearing errors as great as 12° were common. Sites were, of course, calibrated and correction curves were employed. Height finding was limited to angles of elevation between $1\frac{1}{2}°$ and 16°. The inherent deficiencies of poor bearing resolution, poor low-angle coverage and poor height resolution were later made good by other systems, but it must be stressed that the CH possessed a particular advantage in its simple technique or use of floodlighting. A conventional surveillance radar using a rotating pencil beam of radiation and corresponding rotating PPI trace may readily miss (or its operator may miss) seeing a target during one or more sweeps. This was not so with floodlighting, a fact which was appreciated by the operators and indeed by Watson-Watt himself.

It should be borne in mind that as early as 1935 in Orfordness and Bawdsey most of the later developments, including centimetric radar, were considered and discussed, but the chief priority until the outbreak of war was the Chain Home system. Indeed, what struck the writer forcibly while perusing pre-war files was the sense of urgency in getting on with the task, coupled with a sense

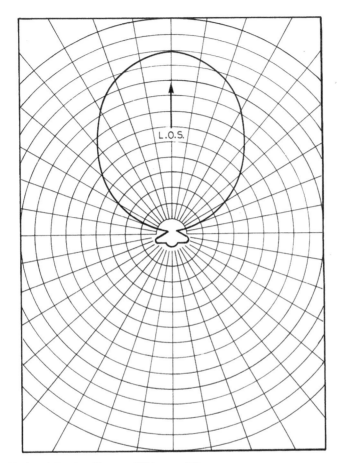

Fig. 5.4   *Horizontal polar diagram CH transmitting array*

of working to an almost definite deadline. Table 5.1 highlights what are considered to be key events in the evolution of the CH system.

5.6.1 Frequencies

The initial choice of frequency was 6 MHz, corresponding to a wavelength of 50 m, but this proved unworkable by the end of July 1935 because of interference from commercial broadcasting stations (it was suspected at Bawdsey that this interference was deliberate [24]). The wavelength was halved and another metre was added to avoid interference from stations operating on 24 m. This new wavelength of 26 m (11·5 MHz) was employed at Orfordness in the Autumn of 1935 and was used until the Spring of 1936, by which time the radar group were operating at Bawdsey. Again, jamming was experienced and it was necessary once more to change the wavelength. It was halved to 13 m and this remained the standard throughout the war.

Table 5.1

Date	Event	Remarks
28 Jan 1935	First meeting of the Committee for the Scientific Study of Air Defence (CSAD)	
26 Feb 1935	Daventry Experiment	
13 May 1935	Radar Team moved from Slough to Orfordness	
31 May 1935	Echoes from ionosphere observed using prototype transmitter and receiver	
5 June 1935	75 ft masts erected at Orfordness	
15 June 1935	Visit to Orfordness of CSSAD	Vickers Valentia aircraft flew as target. Watson-Watt, only, claimed to have seen echoes (at 17 miles). Bad thunderstorm. Radar team were not given sufficient time to prepare for demonstration. Bowen's words are worth recording [23]. Referring to 20th June 1935, he said: 'It is perhaps worthy of note in an historical survey of this kind that the first echoes were seen on that date by Watson-Watt, Wilkins and Bowen, the aircraft concerned being a Scapa flying boat flying at 17 miles'.

Date	Event	Remarks
16 July 1935	Further demonstration given to Rowe	Bristol type 120 aircraft picked up at 38 miles. 200 ft masts used
24 July 1935	Westland Wallace aircraft tracked out to 34 miles. Small formation of aircraft observed and number of aircraft in formation correctly 'counted'	
July 1935	Colebrook brought in to design special receiver	Receiver completed September 1935
Aug 1935	Early Warning Chain planned	
9 Sept 1935	Memorandum from Watson-Watt to the CSSAD on the state of RDF research with a formal proposal for an Early Warning Chain	A key document. Summarises outlook, and progress achieved at that time. Copy in Appendix E. Transmitters at every 20 miles along coast with receivers only at alternate stations envisaged
16 Sept 1935	Fifth meeting of Air Defence Sub-Committee of the Committee of Imperial Defence recommended the establishment of a chain of stations. It was estimated that some 20 stations would be required to provide detection and tracking of enemy aircraft at ranges between 65 miles and 5 miles from coast	This recommendation must, in retrospect, be considered remarkable because as yet no prototype RDF station existed. This decision involved the selection of sites, buildings and masts for transmitting and receiving stations. Proposed programme was held up mainly because of delays in antenna construction

Date	Event	Remarks
Nov 1935	Proposal by Air Staff for 7 stations to be available in August 1936: the first 3 to be operational in June 1936 (transmitter station at Orfordness; transmitter and receiver station at Bawdsey and transmitter station at Great Bromley)	By beginning of 1937 little progress made because of staff shortages
Early 1936	Siting of following stations carried out by Wilkins. (a) Bawdsey, Suffolk; (b) Great Bromley, Essex; (c) Canewdon, Essex; (d) Dunkirk, Kent; (e) Dover, Kent	
May 1936	Radar team moved to Bawdsey Manor	Remained there until 2 September 1939 just before outbreak of war
1 Aug 1936	Watson-Watt transferred to Air Ministry and became Superintendent Bawdsey Research Station	Watson-Watt considered his position and the proposed organisation too restrictive
Sept 1936	Air exercises with Royal Air Force. Scheduled 31 Aug–11 Sept. 13 MHz frequency	One transmitter tower, only at Bawdsey, ready. Exercise a disaster. Tizard was very annoyed. Only after recalibration of equipment on Sept 22 were worthwhile results obtained

Date	Event	Remarks
15 Jan 1937	Projected requirements were: 1 Station transmit and receive by March '37. Presumably Bawdsey 2 Stations transmit and receive by July '37. Bawdsey, Canewdon and Dover 5 Stations; 3 transmit and receive, 2 transmit by Dec '37. Latter to be Great Bromley and Dunkirk 20 Stations by Dec 1938	Ref. AVIA 10/47. Public Record Office, Kew, London, UK
19 Jan 1937	Intermediate programme for transmitters suggested. Production of 5 stations. Construction of parts by a number of contractors and assembly on site by Air Ministry. Transmitter design to be based on Rugby SW No. 9 set with certain modifications	Ref. AVIA 10/47. Public Record Office, Kew, London, UK
Feb 1937	Squadron leader (later Air Marshal Sir Raymund G.) Hart opened a training school for RAF personnel at Bawdsey	RAF personnel such as Squadron Leader Hart were posted to Bawdsey at the instigation of Air Marshal Dowding
March 1937	A radar chain station, which had been built within the confines of Bawdsey Research Station, was handed over totally to Royal Air Force personnel	This was the prototype of the CH Stations. It was designated Air Ministry Experimental Station Type 1 (AMES 1)

Date	Event	Remarks
April 1937	RDF exercise took place from 19 April to 1 May. Approximately ten aircraft each day took part. Frequency used was 22 MHz. Direction-finding and height-finding facilities were provided	Exercise was quite successful. Aircraft were plotted out to 80 miles with good range accuracy. Air Chief Marshal Dowding was somewhat reserved in his report of 7 June, concluding that RDF indicated generally the approach of aircraft to the country
July 1937	Dover station operational	
July 1937	Orders placed for the first commercially produced transmitters and receivers, the former with Metropolitan-Vickers Electrical Company Ltd, and the latter with A. C. Cossor Ltd	
Aug 1937	Canewdon station operational	
July 1938	Directorate of Communications Development (D.C.D.) formed. Watson-Watt Director	
July 1938	Great Bromley operational. Dunkirk completed but not calibrated	
Aug 1938	Royal Air Force Home Defence exercises. Five stations were operating and these worked on 22·64 MHz. The civilian scientific staff took part in operating the equipment at all stations, but particularly at Bawdsey	Difficulties in plotting and in Filter Room procedures became apparent. The expected inadequacy of radar cover at low elevation angles was confirmed

Date	Event	Remarks
Sept 1938	Dover, Dunkirk, Canewdon, Great Bromley and Bawdsey went over to a 24 hour watch. These were supplemented by mobile sets located at Drone Hill (to cover the Forth–Clyde area), West Beckham (to cover the Wash area) and Ravenscar (to cover the Tyne area)	Time of Munich crisis. The mobile sets were housed in wooden huts and used 70 ft masts. The stations were known as 'Advance' or ACH stations
Oct 1938	Filter Room opened at Bentley Priory, the Command Headquarters of Fighter Command	
April 1939	On Good Friday, the 9th April, the chain of stations extending from Ventnor, Isle of Wight, to the Firth of Tay went over to 24 hour watch	
Aug 1939	Main Home Defence exercise began 8th August and lasted for three days. The exercise involved the defence of a line from the Humber to the English Channel	The Chain and Filter Room worked well. Air Marshal Dowding was satisfied with the standard of the whole system. Identification Friend or Foe (IFF) set was fitted into some of the bombers
3 Sept 1939	At outbreak of war 18 of the CH stations were operational and connected to the Filter Room. Two other stations, Netherbutton and School Hill in Scotland, were functioning but were not connected to the Filter Room	Stations were operating on a single frequency in the band 22·0 to 27·0 MHz. This meant that only one system of antenna arrays was necessary. See Appendix G for location of stations

While a wavelength of 13 m, or rather one between 10 and 13 m, was standard, an alternative wavelength between. 6 and 7 m was also employed. In fact, the availability of four preselected frequencies as a protection against interference and jamming was inherent in the earliest thinking on the Chain Home stations. In one receiver that was used in the Chain, the type R3020, we find the facility of switching to four separate frequencies, each in one of the following bands [25]:

20–30 MHz
22–36 MHz
28–42 MHz
32–46 MHz

Even in the early days experimentation was not confined to the frequency in current use. In September 1935 H. Dewhurst employed a frequency of 22 MHz, while experimenting with a mobile type of station. During August and September 1935 tests were carried out to assess the effects of increasing frequency. 37·5 MHz was first tried and then, using valves obtained from the Navy Signal School, it was possible to operate on 75 MHz. Employment of the latter frequency enabled aircraft to be plotted out to over 60 miles. However, problems with a replacement set of transmitter valves led to the abandonment of this particular development.

As was mentioned earlier, the benefits of using much higher frequencies and, indeed, the employment of beam techniques were considered from the very outset of development in 1935. No useful progress in this area was then possible because of the overriding demand for an assured and immediate workable system and because of the limitations of staff and components.

The evolution to higher frequencies made a definite step forward with the use of 200 MHz by E. G. Bowen in airborne sets in 1937. The techniques Bowen used were adapted by Butement at Bawdsey in late 1938 in work which led to the development of 1½ metre 'beamed radar' in Coastal Defence (CD), Chain Home Low (CHL) and Ground Controlled Interception (GCI) equipments [26]. 600 MHz, or 50 cm, sets were employed by the Navy in equipments such as the type 281 radar, from 1939 onwards and we find this frequency used also in the AMES type 11, a stand-by for both CHL and GCI, and in the height-finding equipment AMES type 20. The valves used in the 600 MHz transmitters were evolved from the 'micropup' type [27]. The most common type was the NT99 (CV92), with indirectly heated oxide coated cathode and using tuned lengths of transmission line in the cathode and grid circuits.

An Air Ministry equipment of some interest which operated at 600 MHz but which was developed in the latter days of the war was the AMES type 16. It was designed as a Fighter Direction Station to assist air operations over France prior to the invasion of Europe, giving information on aircraft to ranges of 200 miles. It employed a wire netting paraboloid reflector 30 ft in diameter which effected a beamwidth of approximately 10° in both azimuth and elevation. The required vertical coverage was achieved by oscillating a dipole antenna in the focal plane of the paraboloid. This set was unlikely to be jammed by the Germans because their Würzburg operated on 600 MHz.

The development of the cavity magnetron in 1940 opened the way for a series of 10 cm and 3 cm equipments. The first design work on a 10 cm equipment was in relation to a gun-laying set which became the GL MkIII, whereas the first 10 cm set to see active service was the Naval type 271 radar which became available in March 1941.

With regard to the Chain Home itself, 10 cm equipment was employed from 1942 onwards in existing CH and CHL sites and in new sites for the detection of very low flying aircraft and surface vessels. These were given the blanket title of CHEL (Chain Home Extra Low). The equipments were modelled on the Naval type 271 set or the more high power type 277, and the first experimental system was installed at Ventnor in February 1942.

5.6.2 Polarisation

Horizontal polarisation was used in the Chain Home system. It is appropriate to recount some of the points discussed in Section 2.6 and to mention the factors which determined this choice of horizontal polarisation, not only for Chain Home but for virtually all the ground-radar systems.

The initial influence in the choice was the conception of the wing span of a typical approaching bomber as a resonant half-wave antenna. Progressing from this, one finds that the principal line of most targets, be they aircraft or ships, lies in the horizontal in any case. When one considers height finding using decametric or metric equipment, one finds that total reflection with π radians change of phase by the ground is desirable and this is achieved at all angles of elevation of interest by using horizontal polarisation.

The use of horizontal antennae meant that open wire feeders, which also possessed complete electrical symmetry with respect to earth, could be used. These open wire balanced feeders were admirably suited to the high voltages present with the high power transmitters and to the use of push–pull output stages in the transmitters.

In the very early days, before the design of the Chain Home type of station was finalised, vertical polarisation was in fact experimented with, but fortunately was discarded. Later work with CHL (200 MHz) and AMES type 11 (600 MHz) equipment using vertical polarization showed that back-scatter from the sea was considerable, particularly during stormy weather.

Another disadvantage in using vertically polarised waves is that their behaviour on reflection from the earth's surface is more complex. At low angles of elevation, reflection occurs with phase reversal, as in the case of horizontal polarisation, but as the angle of elevation is increased, the Brewster angle for the particular reflecting surface is reached where the coefficient of reflection becomes zero. Beyond this angle, reflection occurs again but with no change of phase.

A potential advantage in using vertical polarisation arises from the variation in reflection coefficient with angle whereby destructive interference is not so complete, with the gaps between lobes being rather indefinite so that a very natural form of 'gap filling' occurs. This was one reason for using vertical polarisation for both IFF (Identification Friend or Foe) and the Gee hyperbolic navigation system.

5.6.3 Radiogoniometer

The radiogoniometer principle as used in the Bellini–Tosi method of direction finding is well known [28]. Two large loop antennae, each of a single turn, are used. These are at right angles (see Fig. 5.5). In series with each antenna is a

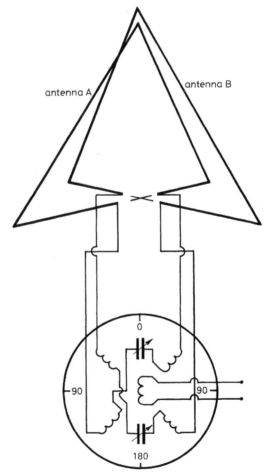

Fig. 5.5   *Bellini–Tosi antenna system*

small field coil. The two field coils are mounted with their planes at right angles and a search coil rotates within them. The field coils effectively reproduce, in miniature, the main field in which the two antenna loops are situated. The search coil turning within the pair of field coils is exactly equivalent to a single turn rotating loop directly sampling the received electromagnetic field, and direction finding is achieved by observing the position of the search coil when the currents induced in it are zero or a minimum.

An outline of the operation of the goniometer as is relevant to the Home Chain will now be given. Consider the loop antenna LM, as shown in Fig. 5.6a. The EMF E_r, in volts, induced in a single turn loop antenna of dimensions much less than a wavelength, is given by the expression

$$E_r = 2EA(\pi \cos \theta)/\lambda \tag{5.1}$$

where E, in volts/metre, is the strength of the electric field, λ is the wavelength

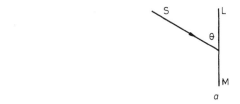

a

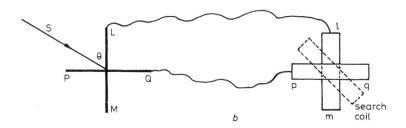

b

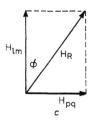

c

Fig. 5.6 *Principle of operation of goniometer*

in metres, θ is the angle which the incoming wave makes with plane of the antenna, and A in square metres is the area of the loop. To enhance the sensitivity of the system, the only factor which can be improved is A, all the other parameters being fixed. Within the constraint of its dimensions being much less than λ, A can be increased. A large loop would present difficulties in rotation, so the stratagem of using large fixed loops and associated small field coils is very effective.

An elementary explanation of the principle of the goniometer can be given by reference to Fig. 5.6b, where we have

$$E_{LM} = kA \cos \theta, \text{ where } k \text{ is a constant} \tag{5.2}$$
$$E_{PQ} = kA \cos (90° - \theta) = k A \sin \theta \tag{5.3}$$

from whence, with identical electrical constants of fixed coils and loops, the currents in the coils are

$$I_{lm} = \frac{kA}{R} \cos \theta = K \cos \theta \tag{5.4}$$

$$I_{pq} = \frac{kA}{R} \sin \theta = K \sin \theta \tag{5.5}$$

where R and K are constants. Knowledge of θ is what is sought.

The magnetic field produced by the two field coils will be proportional to the current in each loop, and if we assume that this field is uniform over the area of the search coil (because of its small size), then

$$\frac{H_{pq}}{H_{lm}} = \frac{I_{pq}}{I_{lm}} = \frac{K \sin \theta}{K \cos \theta} = \tan \theta \tag{5.6}$$

If the search coil is turned until the induced EMF in it is zero, then it will rest in the vertical plane containing the resultant magnetic field H_R, as indicated in Fig. 5.6c. Here

$$\tan \phi = \frac{H_{pq}}{H_{lm}}, \text{ so } \phi = \theta$$

In practice, electrical constants might not be balanced as has been assumed. R. A. Smith and C. Holt Smith provide an informative treatment of the effects of amplitude and phase variations, resulting from circuit instabilities, in crossed-dipole direction-finding systems, as used in the Chain Home stations [29].

5.6.4 Direction finding
In the Chain Home stations the transmitting antenna array floodlit a considerable area in front of the station, as indicated already in Fig. 5.4, and beamed its radiation at maximum intensity in what was referred to as the line of shoot of the station. The receiving array was basically a pair of horizontal

centre fed half-wave dipoles, as illustrated in Fig. 5.7a. The dipoles were set at right angles to each other with one dipole, LM, or the x-dipole, perpendicular to the line of shoot.

Suppose a reflected signal arrived from D at an angle α to the line of shoot. Voltages would be set up in both dipoles and a comparison of voltage amplitudes, effected by the goniometer method already described, would then determine α. However, there would be an inherent ambiguity in this reading as the signal could as easily have come from D'. To resolve this, a reflector was placed behind the x-dipole. It could be made operative by closing a relay R which was controlled by an operator at the receiver console. When R was closed, the amplitude of an echo received from direction D increased, whereas the amplitude of one received from direction D' decreased.

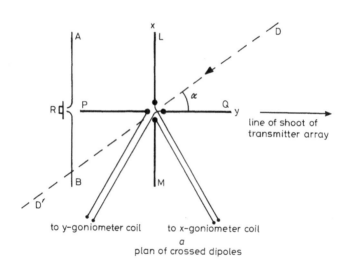

to y-goniometer coil   to x-goniometer coil
a
plan of crossed dipoles

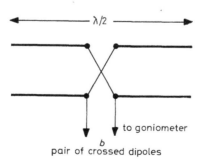

to goniometer
b
pair of crossed dipoles

Fig. 5.7   *Direction finding on Chain Home*
a Plan of crossed dipoles
b Pair of crossed dipoles

In practice, the main or 'A' D/F system on the receiver antenna consisted of two pairs of crossed dipoles at a mean height of about 215 ft and stacked as indicated in Fig. 5.7*b*. The errors in measured bearing, caused by the feeder system or by the terrain, varied with azimuth. The necessary calibration of a particular station was achieved by flight checking using balloons, autogyros or fixed wing aircraft.

5.6.5 Height finding

Let B, in Fig. 5.8, denote the position of an aircraft above the earth's surface, ACE, and at an angle α to the horizontal AD.

Suppose that the aircraft's height, h, which equals BC, is approximately given by [BD + DE]. Denote the range AC by R. It can readily be shown that

$$BD \text{ (ft)} = 107R\alpha \qquad (5.7)$$

where R is measured in nautical miles and α is measured in degrees.

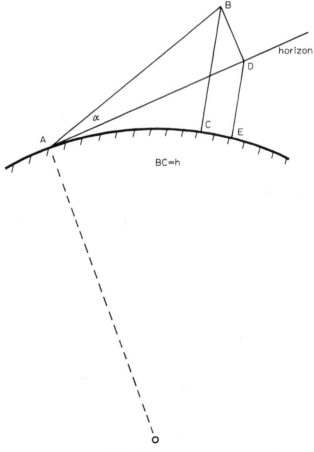

Fig. 5.8   *Calculation of an aircraft's height*

If we assume AD $\simeq R$, and the diameter of the earth to be 6686 nautical miles, then it can be shown that

$$\text{DE (ft)} = 0\cdot88R^2 \tag{5.8}$$

where R is again measured in nautical miles.

Thus we obtain an expression for the height, h, in feet

$$h = 107R\alpha + 0.88R^2 \tag{5.9}$$

This formula can be used to derive height–range–elevation curves or tables. Range was accurately known, so it was necessary for height finding to determine α.

5.6.5.1 Measurement of elevation angle: The first method of determining elevation employed was that of using spaced horizontal parallel dipoles, as perfected by Wilkins in 1933, for measuring the angle of downcoming ionospheric reflected signals. This was first tried out, and with success, in September 1935 at Orfordness. The system adopted in the Chain Home stations, and which was first used in the April 1937 exercises, was the comparison of the signal strengths received in two sets of dipole antennae positioned at different heights on the receiving tower mast.* The ratio of the signal strengths received enabled the angle of elevation to be measured. In the calculations the ground was considered to be perfectly flat and to have a reflection coefficient of -1.

In general terms we can say that the electric field strength E_ϕ obtained from an antenna of vertical polar diagram $f(\phi)$ at a height h above such a plane is given by

$$E_\phi = f(\phi)\sin\left[\frac{2\pi h}{\lambda}\sin\phi\right] \tag{5.10}$$

For each antenna being considered, the polar pattern is assumed to be symmetrical with $f(\theta) = f(-\theta)$ in both magnitude and phase. Then, if two similar antennae are used at heights h_1 and h_2, the ratio of the amplitudes of the signals received is

$$\frac{\sin\left[\dfrac{2\pi h_1}{\lambda}\sin\phi\right]}{\sin\left[\dfrac{2\pi h_2}{\lambda}\sin\phi\right]} \tag{5.11}$$

If h_1 and h_2 are known, ϕ is determined.

---

* A study of both the horizontal and vertical techniques for measuring the elevation angles of downcoming radio waves had been done by A. F. Wilkins in 1933. Precedence for evolving both methods goes to Harald Friis of Bell Laboratories [30].

Let us look at the geometrical optics underlying the amplitude ratio formula used above (see Fig. 5.9a). A represents a horizontal antenna at height h above the ground. A' represents its image.

Let the direct ray of the radio-frequency signal which arrives at ground level, at 0, be represented by $E \cos \omega t$. The direct ray at A, at height h, is then represented by

$$E \cos \left(\omega t - \frac{2\pi h}{\lambda} \sin \phi \right) \qquad (5.12)$$

$$= E \left[\cos \omega t \cos \left(\frac{2\pi h}{\lambda} \sin \phi \right) + \sin \omega t \sin \left(\frac{2\pi h}{\lambda} \sin \phi \right) \right]$$

The indirect ray at height h is

$$-E \cos \left(\omega t + \frac{2\pi h}{\lambda} \sin \phi \right) \qquad (5.13)$$

$$= -E \left[\cos \omega t \cos \left(\frac{2\pi h}{\lambda} \sin \phi \right) - \sin \omega t \sin \left(\frac{2\pi h}{\lambda} \sin \phi \right) \right]$$

Hence the resultant signal at height h is

$$2E \sin \omega t \left[\sin \left(\frac{2\pi h}{\lambda} \sin \phi \right) \right] \qquad (5.14)$$

$\sin [2\pi h/\lambda \sin \phi]$, and in particular the (h/λ) ratio, determine the lobing pattern in space. As an illustration, Fig. 5.9b shows the pattern for the case $h = \lambda$. Minima are at $0°$, $30°$ and $90°$. The positive and negative signs indicate that a phase reversal takes place in passing through a zero in the pattern.

The receiving antenna array in the East coast system consisted of three assemblies at mean heights of typically 215 ft, 90 ft and 45 ft. At 215 ft there were two pairs of crossed dipoles with reflectors, called the 'A' D/F system. At 95 ft there was a single height antenna with reflector and at 45 ft another pair of crossed dipoles with reflectors, called the 'B' D/F system. What was referred to as 'A' height was taken using the 215 ft antenna and the 95 ft antenna, whereas 'B' height was taken between the 95 ft antenna and the 45 ft antenna. The significance of the two separate systems will be indicated shortly. Fig. 5.10 depicts the position of the receiving antennae on both the East coast and West coast types of mast [31].

The actual measurement of the ratio of the signals received in a chosen pair of antennae was achieved by the operator pressing a height button which resulted in the top antenna feeding one field coil of the same goniometer which was used for direction finding, and the lower antenna feeding the other field coil. By a process similar to that used in determining azimuth, the echo signal was 'D/F'd out' and the position of the pointer then gave the ratio of the signals

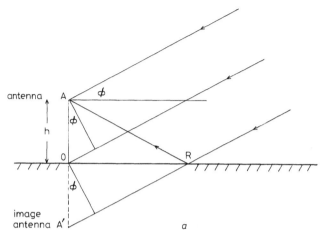

Fig. 5.9a   *Measurement of elevation angle*

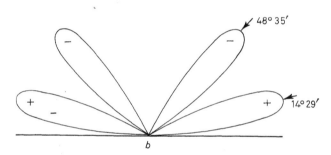

Fig 5.9b   *Antenna lobing angles*

received in the two antennae. The angle of the goniometer pointer was known as θ_H (theta height). A graph was available which related θ_H to the angle of elevation. The method of constructing this graph for a typical station, with antenna heights of 215 ft, 90 ft and 45 ft and operating at 22·69 MHz, is explained in Appendix F. In this Appendix height curves (i.e. θ_H versus α) for an ideal site are worked out. In practice the actual height curves could differ widely from the calculated ones and those used at a station were those obtained with the aid of flight calibration. Aircraft such as the Avro Anson were employed on calibration flights and latterly during the war years these were equipped with special calibration gear.

The necessity for two height systems, the A and the B, becomes clear from an examination of the θ_H/α curves. The A curve turns over at 8°. For a θ_H reading of 180° the target could be either at 6° or at 12°. This ambiguity is resolved by the B system, which gives unambiguous readings to elevation angles of about 24°. The A system curve is quite steep at lower angles and thus provides good sensitivity, whereas the B curve is very flat or insensitive at low

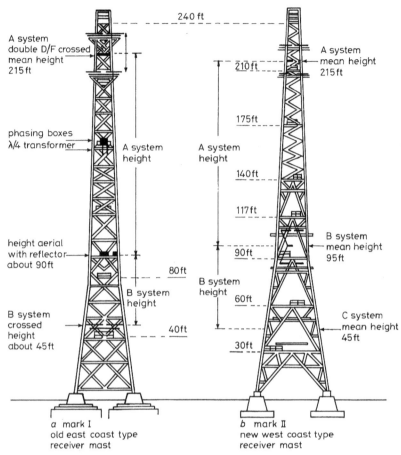

Fig. 5.10 *Chain Home receiver antenna arrays*
a Mk I, old East coast type receiver mast
b Mk II, new West coast type receiver mast

angles. The two systems were used judiciously by the operator in an ambiguity resolving procedure, with the A system being mostly employed for the lower angles of elevation and the B system for the higher angles of elevation.

α could have been read from the height curve and thus a calculation of height made from α and range. In practice, height was ascertained by the use of a manual calculator such as Calculator type 2 [31], which was calibrated for the station in which it was used, or by the use of the Electrical Calculator type Q [32]. This latter calculator, known as 'the fruit machine', was first put into operation at Poling, in Sussex, in June 1940. It was completely electromechanical in design, consisting of relays and motor-driven uniselector switches. The essential principle of its operation was that a θ_H versus α height curve could be wired into a switch whose 15 rows of 50 contacts represented a graph of 50 units

(θ_H axis) by 15 units (α axis). An operator using an actual θ_H versus α graph would, for a particular value of θ_H obtained from the goniometer, simply move a pointer horizontally until he intercepted the curve and then read off the corresponding value of α. The uniselector switch was equivalent to graph paper 50 units high and 15 units wide, and the wiring together of appropriate contacts on the switch bank was equivalent to a θ_H versus α curve. Similarly, height–range curves for various values of α from $\frac{1}{8}$ degree to $15\frac{5}{16}$ degree were wired in, the curves being obtained from the formula obtained as previously

$$h = 107R\alpha + 0\cdot88R^2 \tag{5.9}$$

Any corrections necessary after the recalibration of a station were easily done with a soldering iron by a linesman. When the operator pressed the appropriate keys on the control console, the position (map grid reference), height and number of aircraft in the formation were then displayed in the form of illuminated numerals. The number of aircraft in the formation, or 'raid strength', was estimated by the operator (from the size and appearance of the CR tube 'blip') and so the raid strength display was part of the electrical calculator proper. The information displayed was telephoned by a 'teller' to the Filter Room. Relays were provided in the calculator which enabled the grid reference and the raid strength, but not the height, to be stored.

5.7 Transmitters

The radar transmitters with which we are concerned here are the traditional pulse-modulated type where we envisage a sinusoidal carrier being multiplied in the time domain by a series of rectangular pulses. Essentially they did not differ from amplitude-modulated transmitters operating in the same frequency bands. In Chapter 2 we dealt with the spectrum of a pulse-modulated sinusoidal carrier wave and discussed, for rectangular pulse-modulation, the bandwidth requirements of a radar receiver. Here, the prime consideration will be to review the methods of pulse modulation which were employed in those early decametric and metric transmitters.

In a five-page memorandum [33], signed on 6th February 1937 by E. J. C. Dixon, a Senior Technical Officer at Bawdsey research station, the subject of 'RDF1 Transmitters – Intermediate Programme' was discussed. Included in the document is the following:

> On 19.1.'37 Mr Larnder suggested an intermediate programme, so far as transmitters were concerned, of producing 5 stations by Method 3.1*.

* Alternative methods of production had been discussed in a previous memorandum of 15.1. 1937 titled 'Technical Development RDF1'.

This method envisages construction of parts by a number of contractors and assembly on site by Air Ministry. It was agreed by the Superintendent, Bawdsey Research Station, after conference that the Post Office design of short wave transmitter with some omissions and modifications would be suitable and that a production programme on the basis of 3 stations, viz. Bawdsey, Canewdon and Dover by July 1937, and additional stations, viz. Great Bromley and Dunkirk by December 1937, should be drawn up.

Design and production Programme

1.0: Take P.O. design for 'Rugby S.W. No. 9' with following modifications.

1.1: Omit frequency control stages, i.e. include only 500 Watt (as master oscillator) stage, 3 kW amplifier stage 20 kW amplifier stage.
Output switch for transmission lines with provision for connection alternatively to 3 kW or 20 kW stage.

1.2: Arrange, at the outset, for four wavelengths between 12 and 13 metres, but design includes possibility of operation on any wavelength in the short wave band down to 12 metres. Provision is made for rapid wave-change to any one of 4 pre-selected wavelengths.

1.8: Omit, at the outset, the C.E.* screen grid valves in the final stage and replace by a triode stage with balance† condensers using Standard Telephone Valves type S.S. 1971 without water jackets. Design for support to be made at Bawdsey Research Station.

1.9: Omit water-cooling parts in final stage, but provide air-cooling in lieu.

2.0: Obtain prints of drawings (layout and details) and copies of specifications from the Post office Engineering Department, Radio Branch, in sufficient quantities to supply local needs and to obtain quotations from suitable manufacturers.

It is already arranged that 3 copies of prints and 1 copy of specification should be provided by the Post Office.

2.1: Modify prints where necessary.

2.2: Modify specifications where necessary and get copies typed.

9.0: Appropriate estimate – It is estimated that the total cost of the frameworks, parts and accessories to be provided by contract will be of the order of £3,000 per transmitter or a total cost of £15,000 maximum.

---

* Continuously evacuated.

† Balancing condenser formed part of a bridge neutralising circuit which was necessary with radio-frequency triode amplifiers. The triode valve was in fact superseded by the screen-grid tetrode because, with the latter, higher gains per stage without neutralisation were possible.

Tentative estimates of delivery dates are placed between 4 months and 6 months from date of order. It will almost certainly not be possible to get a better date than 4 months for the delivery of a completely fitted transmitter even when the responsibility is divided as well as possible between several competent manufacturers. To this must be added 1 month for getting out details and contracts and 2 months at least for wiring and testing. A better date than the end of July cannot therefore be anticipated for the first of these transmitters.

The story of the Chain Home and related transmitters has really three parts. The first part is concerned with the experimental work of the small team which included, first, Dr E. G. Bowen and then later Dr J. H. Mitchell and Bainbridge-Bell. They worked at Orfordness and later at Bawdsey on equipment similar to that used at Slough for ionospheric work. The second part of the transmitter story relates to the period of Dixon's memorandum which was quoted from above. It is concerned with those transmitters which were developed and manufactured at Bawdsey, transmitters such as the TF3 which helped to supply the embryonic Chain until such time as the Metropolitan-Vickers transmitters became available. The third part concerns the Chain Home transmitters proper (types T3026 and T3026A) which were developed and manufactured by the Metropolitan-Vickers Electrical Company in response to a specification issued by Bawdsey in March 1938. Dr J. M. Dodds of Metropolitan-Vickers was then the leader of the Radio Section and was the driving force in producing highly successful transmitters. Behind the success of these transmitters lies another story, that of the development of the Metropolitan-Vickers high-power demountable valves. Because of the extra reliability afforded by the use of demountable valves in the drive and output stages of the transmitters, these valves must be considered to have been a key factor in the overall success of the radar chain. To F. P. Burch and to his brother C. R. Burch, of Metropolitan-Vickers, goes the principal credit for the creation of these valves.

We will shortly review transmitters which are representative of each of these three phases of development.

For material of relevance in British radar transmitter design, one can go back to 11th February 1924, when E. V. Appleton, J. F. Herd and Watson-Watt applied for a patent (British Patent No. 235 254) for what has become known as the 'squegging oscillator'. The aim of the circuit was to provide a undirectional linear time-base for cathode-ray oscilloscopes. The patent specification states:

> With suitable values of circuit constants, the system oscillates in trains of oscillations. During each train the grid rapidly becomes negatively charged, and at the end of each train the grid charge leaks at a uniform rate through the saturated diode until the next train of oscillations occurs.

The time of this uniform leak is determined principally by the capacity of the grid condenser and by the electron emission in the saturated diode.

A limiting case of the squegging oscillator is the one-pulse blocking oscillator [34]. The transmitter T3056, which was of the CHL/GCI (Chain Home Low/Ground Controlled Interception) type and which operated on 2 metres wavelength, illustrates the use of these two circuits. This equipment, whose development at Bawdsey is primarily associated with Butement, employed in the initial stage of its modulator unit a blocking oscillator, while the pulses of oscillations developed in the oscillator unit were terminated by a squegging action.

On the subject of self-modulation circuits or squegging oscillators, it is pertinent to quote briefly from a 1932 report from the nursery of British radar development. In the report of the Radio Research Board for the year 1931 [35], we find

> Various methods have been described for the generation and radiation of short radio-frequency pulses necessary for the group-retardation method. In the work done for the Board, a simple but satisfactory type of emitter has been devised by utilising a principle that had already been employed for providing a linear and undirectional time-base for the cathode-ray oscillograph. Briefly, it can be recapitulated that, if the grid-leak resistance of a continuous-wave generator be raised to a high value and

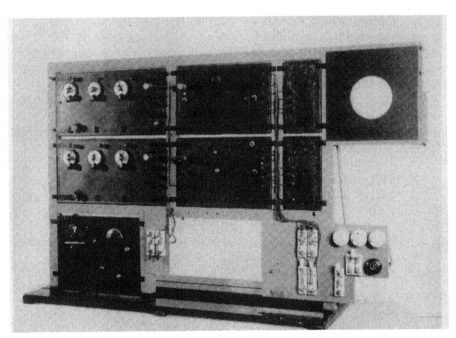

A view of receiver used in Daventry Test, 1934.

TF3 Transmitter, Bawdsey, 1936/37.

shunted by a condenser, with suitable numerical values of resistance and capacity, the oscillator can be caused to produce short pulses of radio frequency alternating with relatively long and uniform periods of quiescence. The action depends on there being a difference between the mean negative grid potentials for starting and stopping the generation of oscillations.

The modulation circuits of the Chain Home transmitters provided pulse lengths variable from 5 to 45 μs at a recurrence frequency of 50, 25 or 12·5 Hz. Grid modulation was used and the circuitry for achieving it would, by the end of the war, be regarded as a rather complex solution to a simple problem [36]. Grid modulators were employed in all the early transmitters that will be considered here. They come first in a classification of modulators which also includes hard valve modulators, condenser discharge modulators and inductance storage modulators. They required nothing special in the modulating

Chain Home 350 ft high steel masts: transmitting arrays were strung between masts.

Antennae of early Royal Navy radar: type 281 surveillance and gunnery ranging 90 MHz set.

valve and the power consumption in the modulating circuits was low. Whereas complex techniques later became necessary to modulate or pulse high power valves, and in particular the magnetron, operating at much higher frequencies with short pulse lengths and high pulse recurrence rates, grid modulation was adequate in the early transmitters. Its main disadvantages were that full HT voltage was applied to the modulated valves continuously with consequent risk of flash-over, and that, relatively speaking, the pulse shape was poor. At shorter wavelengths the use of valves with much smaller inter-electrode clearances and with oxide-coated cathodes made anode modulation essential and this trend was continued with the arrival of the cavity magnetron which could not be modulated in any other way.

The principle of grid modulation can be illustrated by reference to a simple schematic circuit (see Fig. 5.11a). V_1 represents a triode in a tuned-anode tuned-grid oscillator configuration. V_2 is normally conducting so that A is negative with respect to earth. The grid of V_1 is connected to A so that V_1 will be biased beyond cut-off. If a negative pulse is applied to B, V_2 will cease to conduct and then A will rise to earth potential and with it the grid of V_1, which will then oscillate. At the end of the negative pulse V_2 conducts again and V_1 stops oscillating. This technique, as we shall see, was used in the first 26 m transmitter built at Orfordness in 1935 and also in the Bawdsey-designed TF3 which, as has been mentioned, was used in some of the early Chain stations.

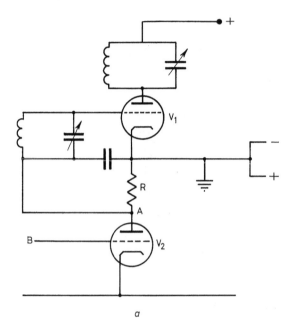

a

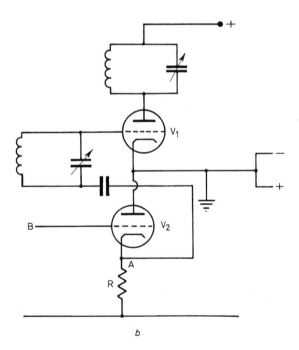

b

Fig. 5.11   *Schematic: grid modulation*

A variation on this simple technique, which was an improvement in so far as it allowed V_2 to be switched off between pulses, was used in the CHL/GCI transmitters. A schematic illustrating the method is shown in Fig. 5.11*b*. R biases V_2 almost to cut-off so that A is normally well below earth potential. A positive pulse applied to B allows V_2 to fully conduct and hence point A to increase positively towards earth potential, thus allowing V_1 to conduct and its circuit to oscillate.

Brief comment must be made on two items which were employed in virtually all radar transmitters, namely thyratrons (gas-filled triodes) and pulse transformers. Thyratrons, used as switches in modulator circuits, were originally of the mercury vapour type developed from pre-war industrial designs, but towards the end of the war more suitable hydrogen thyratrons were coming into use. The low mobility of mercury ions results in rather slow ionisation and de-ionisation times, whereas the low atomic weight of hydrogen promotes short de-ionisation times. In the grid modulation circuits with which we are here concerned, where the thyratrons functioned in an auxiliary role and not as the discharge 'switch' of a modulator handling large currents at a high recurrence frequency, there were no particular design problems. It is necessary only to allude to one principal limiting parameter of the mercury vapour thyratron, namely its de-ionisation time.

Hull [37] gave, for a typical grid-anode structure of discharge tube, an empirical equation for de-ionisation time, t,

$$t = \frac{0 \cdot 0012 p \, I^{0 \cdot 7}}{e_g^{3/2} x} \text{ s}$$

where p is the gas pressure in dynes/cm^2 (baryes), I is the anode current to be extinguished, in amperes, x is the distance between anode and grid in centimetres, and e_g is the negative grid potential relative to surrounding space.

Suppose that plasma near grid is at a potential of $+10$ V and that we choose typical values

$x = 2$ cm
$p = 5$ baryes
$I = 3$ A

grid potential $= -6$ V

Then

$$t = \frac{(0 \cdot 0012) \, (5) \, (3)^{0 \cdot 7}}{(6 + 10)^{3/2} \, (2)} = 101 \ \mu s$$

By the end of the war pulse transformers handling pulse power of 1 MW and of 1 μs duration were common. The main attributes of a pulse transformer are that it can step up the voltage and reverse the polarity of a pulse. It is of use also in matching the impedance of a load to that of a generator. When an

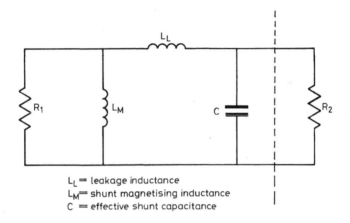

L_L = leakage inductance
L_M = shunt magnetising inductance
C = effective shunt capacitance

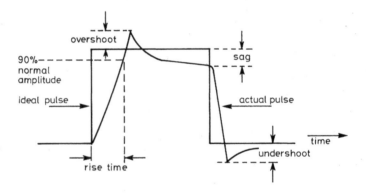

Fig. 5.12   *Pulse transformer schematic with sketch of a distorted pulse*

idealised rectangular pulse is transmitted, distortion such as rise time and overshoot will inevitably occur. An approximate equivalent circuit of a transformer (referred to the primary) and a sketch of a distorted pulse are shown in Fig. 5.12.

One can say that rise time and overshoot are mainly due to L_L and C. Design aims at achieving low L_L and C and high L_M. These are spatially conflicting requirements, and reduction of physical size means more attention to insulation is necessary. Core material of high resistivity, which increases skin depth penetration, formed into very thin laminations is used.

A post-war overview of the design and use of pulse transformers is available from W. S. Melville, who was a member of the British Thomson-Houston Co. Ltd. [38].

In the very first transmitter produced at Orfordness, a crucial item was a pulse transformer. This transformer effectively solved the problem of lowering the pulse widths of the order of 100 µs obtained with the ionospheric transmitters to the required width of the order of 10 µs. To E. G. Bowen must go the credit for appreciating the importance of a suitable pulse transformer and for providing one [39], and, of course, to him also goes the credit for the first working transmitter. As has been mentioned already, this was operational on a 50 m wavelength towards the end of May 1935 and in the autumn of 1935, because of interference, there was a change of wavelength to 26 m. A circuit diagram of this latter transmitter is shown in Fig. 5.13 [40].

The success of this transmitter was dependent on the NT46 Naval silica triode valves, which had been supplied by Hughes of the Admiralty Signal School, and on the pulse transformer. The valves, whose envelope was about 12 inches high and 5 inches across, had four long 'stalks', each 8 inches long, coming out from the envelope (see Fig. 5.14). The stalks carried the anode, grid and filament leads and to this configuration was attributed much of the high voltage, high power handling qualities of the valves. The filament rating was 20 V, 20 A and for pulsed operation an HT voltage of 12 kV could be applied while the filaments were overrun by a 10% increase in voltage. The transmitter operated on a pulse width of 20 µs and the only parameter one cannot be precise about is the peak pulse power, but it seems to have been in excess of 50 kW.

The NT46 valves were in a push–pull oscillator circuit. The balanced output impedance, a nominal 600 Ω, fed a simple half-wave dipole antenna using a y-match.* Modulation was ultimately achieved by the NT46R triode V_3 whose anode was fed from an HT supply maintained at earth potential. When V_3 ceased to conduct, its anode potential approached earth potential, and with it the grids of the two oscillator valves. The 'premodulation' circuit consisted of the thyratron V_1 and the triode V_2. A 50 Hz voltage derived from a 230 V, 50 Hz mains supply was fed to the grid of the thyratron V_1. A similar 50 Hz voltage leading it by some 90° was fed to the anode. When the grid potential exceeded the breakdown voltage the thyratron fired and remained conducting until the anode potential dropped below the quenching value. The time between strike and extinguish times was approximately 6 ms, and the net result was a positive pulse of voltage developed across R_2 in the anode circuit of V_1. This fed the oscillatory circuit L_1, C_6, VR_4 where damped oscillations occurred. The first positive half-cycle was sent as a positive 20 µs pulse to the grid of V_2, which was normally biased back just to cut-off. There then resulted a pulse of current through V_2 which passed through the primary of the pulse transformer and appeared as a sharp 20 µs pulse on the grid of V_3, thus cutting it off and allowing its anode to rise to earth potential.

---

* The two feed wires of the 600 Ω open line are tapped onto the dipole, each about 0·06 wavelength from either side of the centre of the dipole. Since 0·12 wavelength is greater than the distance between the feeder wires, the latter must be fanned out into a y-shape.

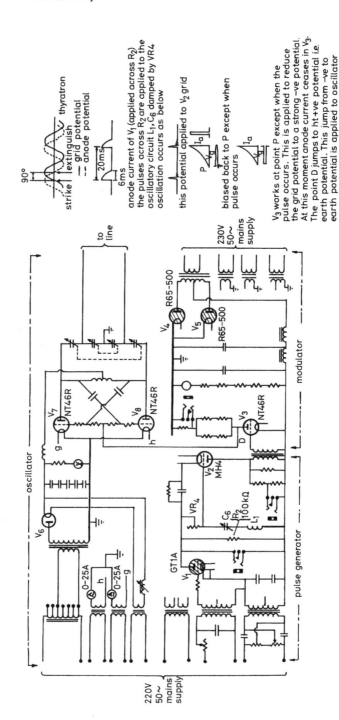

Fig. 5.13   Schematic: Orfordness 26 m transmitter

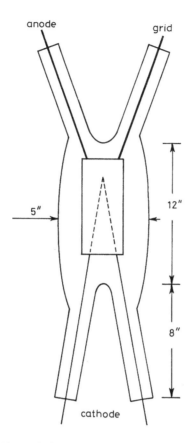

Fig. 5.14 *Dimensions NT46 triode*

Two comments may be made regarding this transmitter. The first would apply to any system of grid modulation, and that is that the potentials of both the grid and the anode of the oscillator valves rise exponentially, resulting in, relatively speaking, poor pulse shape. The second is that, while a single oscillator stage configuration was acceptable for a simple antenna and moderate power operation, for effective power generation some form of power amplification was desirable.

5.7.1 TF3 transmitter
This leads us to a consideration of the TF3 transmitter (see Figs. 5.15a and b) [40].
   Two NT46R valves (NT57s were also used), operating as a push–pull oscillator, drove two NT57 valves in a neutralised push–pull amplifier configuration.

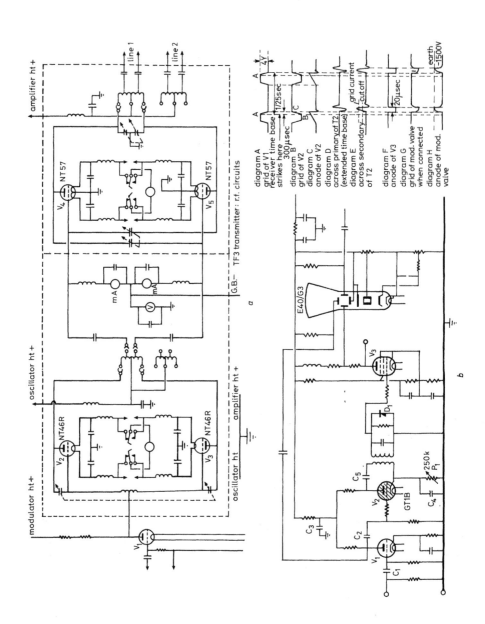

Fig. 5.15   *TF3 transmitter*

The NT57, which was another silica valve, developed by the Admiralty Signal School, required a filament supply of 15.2 V, 48 A and for pulse operation the filament voltage was again overrun by 10% with an HT voltage of approximately 15 kV.

In this transmitter the NT41A valve, V_1, of Fig. 5.15a, when cut off by a negative pulse, allowed the grids of the NT46R valves to be relieved of their negative bias. V_4 and V_5 amplified the pulses of oscillation from V_2 and V_3. Valve neutralising capacitors can be seen between anodes and grids of the push–pull pair. With NT60 valves having pure tungsten filaments in the amplifier, a peak pulse power of about 100 kW was claimed.

The premodulation circuits and accompanying waveform diagrams are shown in Fig. 5.15b. A locking 4 V positive-going trapezoidal pulse from the receiver, occurring 25 times per second, was fed to the grid of V_1. V_1 amplified and inverted it and passed it on to the grid of the GT1B thyratron V_2. The trailing edge slope of the locking pulse was some 300 μs in duration and the receiver time base fired at the start of the trailing edge. The 250 kΩ potentiometer P_1 in the cathode circuit of V_2 enabled control of its bias and hence of its firing point to be obtained (thus an optional time delay between the transmitter firing and the start of the receiver time base was possible). When the thyratron V_2 was not conducting, the capacitor C_5 charged up. When V_2 conducted, C_5 discharged through it providing a swing of oscillation or an impulse across the primary of T_2 (see diagram D).

The form of the impulse in the secondary is shown in diagram E. This was then shaped by bias and grid current limiting in V_3 into the required 20 μs pulse, while the negative-going portion of the oscillation disappeared due to the action of the diode D_1. The 20 μs pulse appeared as a negative pulse on the anode of V_3 and this was transferred to the grid of the modulation triode, thus biasing it off.

5.7.2 Chain Home transmitter T3026

The original specification [41] submitted in 1937 by Metropolitan-Vickers was quite brief and it is set out below:

1 The transmitter to be designed to send pulses of radio-frequency energy continuously, with a peak power of at least 200 kW.
2 The frequency of this energy would be in the range 20 to 55 MHz.
3 Adjustment to any one of four pre-selected frequencies in this range must be possible within 15 seconds, by which time the peak power must be at least 80% of normal.
4 The transmitter must be always within 0·05% of the stipulated value.
5 The pulse repetition rate to be either 25 or 50 per second, as required.
6 Provision must be made so that the pulse repetition could be locked to a 50 cycle wave, but be capable of adjustment in phase to any angle from 0 to 360° within 1·0°.

7 The pulse must not vary its point of incidence relative to zero phase angle of the wave by more than plus or minus 2·0 µs.

8 The radio-frequency voltage of each pulse must rise to 90% of maximum within 1 µs, must continue at maximum for a period which must be adjustable from 5 to 35 µs.

9 At the end of this period it must fall to 0·01 of its maximum within 2 µs.

10 The power radiated during quiescent periods must not exceed a few microwatts.

Since the transmitting valves, the SW5 double tetrode, the type 43 tetrode and eventually the type 45 tetrode, which was a continuously evacuated demountable type, were a key factor in the success of the transmitters, a brief mention of their background will be given.

Metropolitan-Vickers' first involvement with demountable valves arose from an interest in high-frequency induction heating. Initially, they successfully used parts of a disused broadcast transmitter (the Metrovick station 2ZY). They then built a larger furnace using a sealed water-cooled Western Electric 10 kW triode, but with this they experienced parasitic oscillations and general valve failure. Prompted by the effort and expense of circuit and valve replacement, the company produced a demountable oil diffusion-pumped triode. Many people saw the valve operating in 1930/31 and interest was shown in particular by the Radio Section of the Post Office who tested the valve and accepted it. An association with them then began which lasted until well after the war years.

Initially the Post Office's interest was in obtaining a demountable valve for the 500 kW VLF (GBR) station. This would have replaced the existing 54 10 kW water-cooled sealed valves. A less ambitious requirement was for a demountable screen-grid valve capable of handling about 50 kW for their short-wave telephone link service. It was J. H. Dodds who completed the successful development at Metropolitan-Vickers of a 60 kW screen-grid valve for the Post Office and it was he who, as leader of their Radio Section, undertook to provide special valves and transmitters for the Chain Home stations. The one was a propitious forerunner to the other.

5.7.3 Transmitter circuits

These comments on the transmitter circuits are based on the material in the paper by Dodds and Ludlow of Metropolitan-Vickers [42].

The original Chain Home transmitter was as schematically shown in Fig. 5.16. The SW5 double tetrode* tuned amplifier was the centre piece of the transmitter, its grids modulated by a pulse which varied from 4 to 45 µs at one of the chosen repetition frequencies of 50, 25, 16⅔, 12½ or 10 Hz. An Edison ES751 with tuned circuits in a thermostatically controlled enclosure acted as master oscillator. It fed the MV75 (75 W) tetrodes which, in turn, drove the

---

* Mounted in a single envelope.

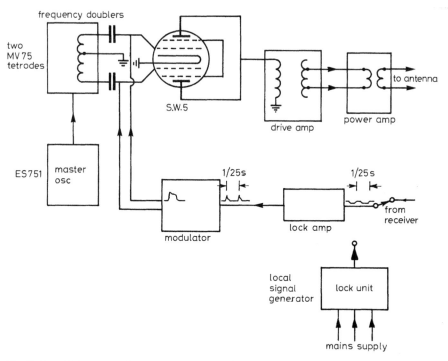

Fig. 5.16   *Schematic of Chain Home transmitter*

SW5. After the SW5 came a drive amplifier, a single type 43 tetrode. Its anode circuits were symmetrical and taps on it fed the control grids of the two type 47 valves which formed the final power amplifier. Four preset frequencies were provided for, so that, when tuning was required, four sets of tuned circuits had to be provided for; however, good mechanical design enabled one variable condenser only to be used for each bank of four circuits. It took less than 9 seconds to change from one frequency to another.

Before the Chain Home receivers had been fully developed, it was feared that the master oscillator and frequency doubler stages, which ran continuously during quiescent periods, might affect the receivers. Hence the final design of transmitter had the SW5 valve operating as a self-oscillator in a form of Hartley configuration. The RF circuits of this latter type of transmitter are shown in Fig. 5.17 [42]. Two frequencies only were used in this transmitter. The same circuit was employed for oscillator tuning and for drive-amplifier grid circuit tuning for the sake of mechanical simplicity. The driver stage and the power output stage were both set out in a push–pull arrangement.

18 kV on the driver stage valve and 36 kV on the output valve resulted in a peak power of 350 kW at 20 MHz. Modifications which allowed 36 kV on the driver valve increased this power to an average figure of 750 kW. By

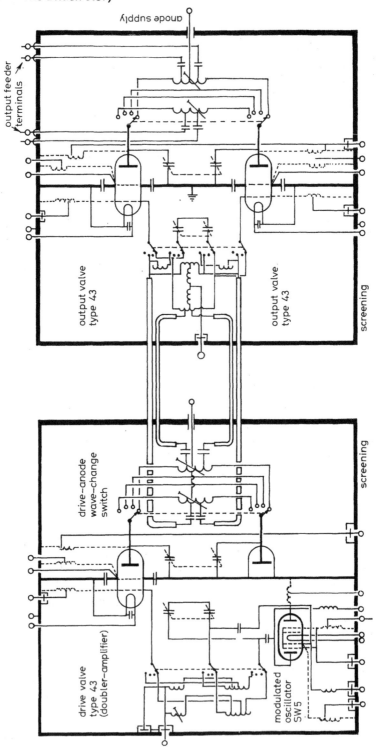

Fig. 5.17   *RF circuits Chain Home transmitter*

overrunning the filaments and increasing the HT to 40 kV, over 1 MW was obtained in tests.

The modulation circuits, which are shown in Fig. 5.18 [42], are more complex than those already described and might be considered to consist of three distinct parts.

The commencement of the modulation chain was either a 25 Hz, 4 V, square-wave locking signal from the receiver or the same form of signal created locally by a 420 V three-phase supply which was passed through a transformer with a selsyn type phase control resulting in a constant voltage of variable phase. This signal was passed to the grid of the triode V_{21} through a large resistance, so that a square wave was produced at its anode because of grid resistance stopper action and anode bottoming. The square wave was fed to the thyratron V_{22} and its filter network, producing pulses the same as those which were obtained directly over a remote line from the receiver, namely square wave with a front of the order of 100 μs.

Next, this square wave was sent into the 'Lock Amplifier' where valves V_1, V_2 and V_3 provided a positive pulse to trip the modulator proper. The cathode circuit of V_2, allowed the potential at the cathode to be varied and hence the valve's point of conduction, thus permitting the transmitter to fire and lock at any required point between 30 and 80 μs from the start of the receiver waveform. The modulator power unit had its positive pole earthed. A 1·0 μF capacitor C_3 was charged through V_{13} during the quiescent period of the receiver. When the thyratron V_7 was tripped, thyratron V_8 conducted and C_3 discharged through it and through the pulse-forming network in the cathode circuit of V_8. This pulse was sent directly to the control grid and screen grid of the SW5 oscillator valve.

The thyratrons V_8, V_9 and V_{10} formed the tail of the modulating pulse and determined the pulse width, the actual variable control being in the grid circuit of V_{12}. (In the earlier design when SW5 acted as an amplifier V_{10} and V_{11} formed a 'quench' pulse which was fed on to one of the control grids of SW5, resulting in full power in anti-phase being produced at the output stage and thus doing away with the 'tail' on the pulse.)

In practice, the modulator system proved completely stable and measurements carried out showed that variations in timings for the transmission of pulses were less than 0·025 μs. To assist the reliability and servicability of a station, the whole transmitter, including the water cooling system, was duplicated.

5.7.4 Transmitter antenna arrays

The original East coast transmitter antenna arrays were mounted on 360 ft steel towers, as indicated in Fig. 5.19, with platforms at 350 ft, 200 ft and 50 ft. The West coast transmitter antennae were curtain arrays of dipoles and reflectors mounted between guyed towers of about 325 ft height. Configurations differed in detail from station to station.

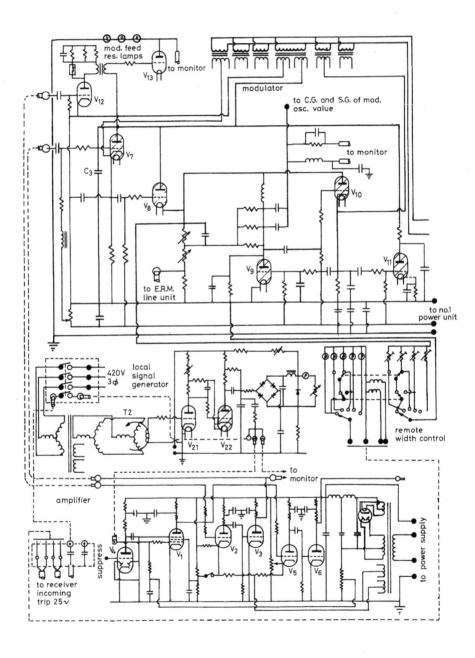

Fig. 5.18　*Modulation circuits Chain Home transmitter*

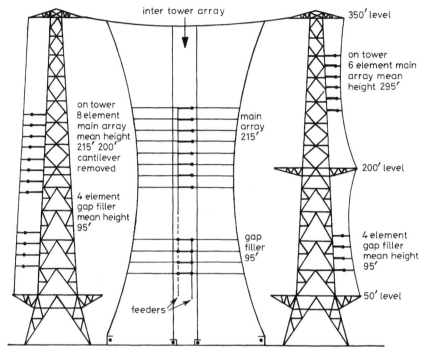

Fig. 5.19 *Chain Home transmitter antenna arrays: various configurations*

In the development days of the East coast chain, more simple systems were used or considered. Of interest in this regard is a memorandum from Bawdsey composed in December 1936 [33] and dealing with the development of RDF1. In it is written:

> The merits of the tilted wire antenna for T sites seem to be worth consideration. A horizontal antenna with wire length 4λ per side and tilt 65°, λ above the earth, gives a good 'floodlight' radiation comparable to the tier of horizontal dipoles.

A top stack of antennae at a typical effective height* of 290 ft with an alternative or complementary gap filling array at a typical effective height of 95 ft constituted a Chain Home transmitting antenna. It can be readily shown [31] that a vertical stack of six half-wave dipoles (Fig. 5.20), each separated by a half-wavelength, will in free space have a first null or gap in its vertical pattern $19\frac{1}{2}°$ above the horizontal. If this stack is placed at an effective height, h, of 290 ft above the ground, and if the station is working on a wavelength of

---

* A stack of vertical dipoles can, from the point of view of phase, be represented by a single dipole placed at the centre of the stack [31]. The height of this equivalent dipole above ground is known as the 'effective height' of the stack.

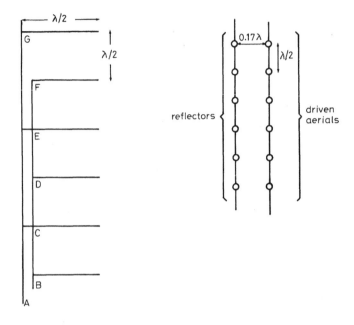

Fig. 5.20   *Stack of 6 dipoles*

12 metres, then, owing to reflections from the ground, the first gap above the horizon in the vertical pattern will occur at (see formula K37):

$$\alpha = \frac{94\lambda \,(\text{metres})}{h \,(\text{ft})}$$

$$\simeq 4°$$

with a corresponding lobe maximum occurring at 2°.

Likewise, with a stack placed at an effective height of 95 ft, and operating on the same wavelength, the first gap will occur at $\alpha \simeq 12°$, with a corresponding first maximum at 6°. Thus the vertical pattern of one array sufficiently complements that of the other to provide quite substantial gap filling.

Fig. 5.20 also shows the driven antennae of a stack spaced $0\cdot17\lambda$ away from tuned reflectors. This was an optimum distance determined by experiment. The driven antennae were made $0\cdot47\lambda$ in length and the reflectors $0\cdot49\lambda$ in length. When a curtain array was used, the reflectors consisted of closely spaced parallel horizontal wires at an optimal experimentally determined distance of $(5/8)\lambda$.

In order to determine the complete coverage of a Chain Home station, it was necessary to combine (that is, multiply) the polar diagrams of the receiving antenna array and of the transmitting antenna array. A gap in either meant, of course, a gap in the combination. Reference may be made to the 'Radar Supervisor's Manual' [31] for an illustration of this.

5.7.5 Performance diagrams

It is appropriate at this point to refer to performance diagrams with special reference to Chain Home stations. A performance diagram informs a radar operator at a glance at what ranges and heights he should pick up a specified target. Today, one generally refers to a performance diagram as a vertical coverage diagram, calculated for a 1 square metre target, say, and for a given probability of detection. Performance diagrams were loosely and inaccurately referred to by Chain Home staff as vertical polar diagrams (VPDs).

Reference may be made to Fig. 5.23 of Section 5.9 which, although it is concerned with GCI equipments, illustrates the layout of a performance diagram. The manner in which a performance diagram was constructed for a Chain Home station will now be outlined [31]. The co-operation of an aircraft, such as a Blenheim, which flew out from the station at a predetermined height (say 10 000 ft) and on a predetermined bearing, was necessary.

The signal/noise ratio as observed on the A-scope was denoted by z. As the aircraft flew away from the station, ranges and corresponding times were logged. The times when the various values of z were read from the A-scope were also noted so that a table of z versus range and hence versus elevation angle, α, could be made out. The noise at the output of any particular Chain Home station receiver could be regarded as having more or less a constant value. With an A-scope display, the limit of detectability of a signal was assumed to occur when $z = 1$. (At any point within the coverage of a station and for a particular type of aircraft flying away from the station, the value of z depended on, among other things, the power output of the transmitter, the sensitivity of the receiver and on the closeness of the flight path to the line of shoot of the station and to the centre of a vertical lobe of the combined transmitter and receiver antenna arrays' polar pattern.) Suppose A is the amplitude of the signal (E field) received by an aircraft one nautical mile from a station. At d nautical miles, on the same bearing, the amplitude would be A/d. The amplitude of the signal scattered back from the target at this range may be denoted by cA/d^2, where c is a constant. Hence

$$z \propto cA/d^2 \tag{5.16}$$

and

$$zd^2 \propto cA \tag{5.17}$$

Therefore

$$zd^2 = \text{constant} \tag{5.18}$$

A table of the following form was constructed:

Range (nautical miles)	z	α (degrees)	zd^2
23	25	4·59	13,225
27	22	3·87	16,038
30	10	3·45	9,000
34	2	3·00	2,212

For each angle of elevation, α, zd^2 will be constant. Hence, for any α, if one puts $z = 1$, one obtains the maximum range of the station. In the example above at $\alpha = 3$ degrees, the maximum range is $\sqrt{2212} = 47$ nautical miles.

5.8 Receivers

Just as the name of Metropolitan-Vickers Ltd. is associated with the manufacture of the Chain Home transmitters, so is the firm of A. C. Cossor Ltd. (now Cossor Electronics) linked with the production of the Chain Home receivers. The first receivers used in the 1935 experiment were, as was mentioned earlier, a development of the pulse-polarisation analyser receivers which had been used for ionospheric research at the Radio Research Station, Slough. This had been modified by Wilkins, principally by increasing the IF bandwidth, and then assistance had been received from colleagues at Slough to build a receiver which served as a prototype for the first commercially made Cossor set. It was most likely because of Cossor's expertise in the development and utilisation of the cathode-ray tube that Leslie H. Bedford of Cossor was initiated into the facts of radar and that Cossor were asked to develop and manufacture the Chain Home receivers. No doubt the fact that in 1936 Cossor were marketing two television receivers also influenced the choice. Leslie Bedford directed a special laboratory which was located in a house beside the Cossor factory at Highbury Grove, North London. In this laboratory, a small team, which included O. S. Puckle, produced the first commercially built Chain Home receiver [43].

The specification which was submitted to Metropolitan Vickers for the RDF transmitter has been quoted from in part. It is also of interest to look at parts of a similar specification for an RDF receiver [41].

SPECIFICATIONS B.R.S.* NO. 10002/a
SPECIFICATION OF REQUIREMENTS FOR R.D.F. RECEIVER
AT A FIXED STATION

    1.0:   General

           The equipment required is a high frequency receiver designed for use in conjunction with a high power transmitter which may be situated within 200 yards of the receiver.

* B.R.S.: Bawdsey Research Station.

The input to the receiver will be through a goniometer, of specified electrical characteristics, connected by pairs of screened transmission lines to an aerial system.

The goniometer and high frequency switch arranged to select combinations of a number of pairs of transmission lines will be supplied by Bawdsey Research Station. The receiver shall be designed, however, to accommodate these items internally and conveniently to the left hand of the operator.

The receiver output will be in the form of direct current pulses which shall be employed to operate a cathode ray tube device.

1.1: Flexibility of Design

The design shall preferably be in the form of a standard panel and rack construction, but in any event a certain degree of flexibility shall be incorporated in order to provide for the possible modification of the receiver after a period of service.

Flexibilities which are particularly required are as follows:

(*a*) Space for the introduction of single stage high frequency amplification before the goniometer;

(*b*) Provision for adjustment of H.F. and I.F. stages in point of gain and phase;

(*c*) Flexibility of arrangement of operator's desk apart from the cathode ray tube;

(*d*) Provision for controlling the beating oscillator frequency from that of the transmitter.

2.0: Electrical Characteristics

2.01: Type of Signal

The receiver shall accept a pulse of continuous wave radiation emitted by the transmitter and reflected from a distant object. The pulse shape is defined as follows:

'The pulse of continuous wave emission shall be repeated 25 or 50 times per second, as desired. Each pulse is to rise to 0·9 of its maximum amplitude within 1 microsecond, to continue at its maximum amplitude for a period of from 5–25 microseconds as determined by adjustment and thereafter to fall to 0·01 of its maximum amplitude within 2 microseconds'.

2.02: Sensitivity

With an input of 50 microvolts R.M.S. across the transmission lines to the goniometer field coils at a frequency of 30 megacycles per second, the D.C. output shall be such as to give a deflection of not less than 25 mm on the cathode ray tube. The deflection shall decrease in a linear manner with input from 25 mm to 0·5 mm. In this measurement the goniometer shall be set for maximum transfer of energy from the field coils to the receiver. Unless otherwise stated, all performance figures in this specification are to be obtained at the maximum linear sensitivity specified in this paragraph and at all frequencies between 20 and 30 megacycles per second.

2.03: Noise Level

With the first tuned circuit of the receiver short-circuited, the output due to noise generation in the receiver shall not exceed 0·5 mm defelection on the cathode ray tube.

2·04: Frequency Range

The receiver shall operate on any frequency in the range 20–45 megacycles, and shall normally be capable of immediate adjustment by means of a single control to any one of four pre-selected frequencies within this range. The division of the overall range into four frequency bands should preferably be such that any frequency between 24 and 41 megacycles can be obtained on either of two ranges. The tuning condensers shall be made accessible and shall be capable of rapid adjustment to within 5 degrees on a 180 degrees working movement and thereafter of fine adjustment about this setting.

2.05: Rapidity of Wave Change

The change-over from one to another of the four working frequencies must be capable of being made by a single operator in 10 seconds.

2.08: Paralysis

The receiver must be capable of accepting, without damage, a pulse continuous wave signal from the local transmitter not exceeding 20 volts radio frequency peak across the lines to the goniometer field coils, and shall recover its normal sensitivity in not more than 10 microseconds plus the length of the transmitted pulse. The length of the transmitted pulse is to be taken as the time between the instant when the transmitted pulse reaches 0·9 of its maximum amplitude and the instant when it has fallen to 0·01 of its maximum amplitude. The receiver shall incorporate a ground ray suppressor device which shall take the form of a separate pulse generator adapted to control the gain of the H.F. amplifier.

4.03: Insulating Materials

Insulators shall be of such material that their surface resisitivity shall not deteriorate over a period of months when working in sea-board atmosphere.

4.04: Valve Life

Valves or other thermionic apparatus shall be designed for long life which, under conditions of normal rating, shall not be less than 1,000 hours. This provision shall not apply to valves type 954 and 6L7.

4.05: Time for Changing Valves
The time taken to change a valve or other thermionic device at the end of its life shall not exceed five minutes.

20 sets made to this specification, at a cost of £1000 per set, were ordered from A. C. Cossor. On the 28th October 1937, Cossor were asked to build a further 19 sets; this figure of 19 was later increased to 40 to allow 1 standby set for each station. All these sets were built at the Cossor 'cabinet' factory at Hackney, London.

On the 29th June 1938, a conference on 'RDF receivers – final design' was held at the Air Ministry [41]. The attendance at the conference included Watson-Watt (D.C.D.),* Group Captain H. Leedham (D.D.C.D.)* and Robbins, Puckle and Bedford of A. C. Cossor. The notes of decisions taken included the following:

No prototype will be produced. The final specification is to be in accordance with the decision of this conference and is to apply to the 39 (19 + 20) sets. Slight modifications only will be permissible after approval of the 1st model.

The frequency is to be specified as $50 \pm \frac{1}{4}$ c/s instead of 47–53 c/s. Clauses 2.09 and 2.10.2 refer.

Goniometer: Working details to be supplied to Messrs Cossor to enable the contractor to build a 1st model for winding investigation at B.R.S. This work to be expedited.

The contractor will endeavour to deliver as follows: 1st set 5 months from 29.6.'38. Remainder 2 per week with a first interval of 2 to 3 weeks.

Fig. 5.21 shows a block schematic, dated 2nd June 1939, of a Cossor type RF5 receiver. The diagram has been extracted from the receiver handbook [44]. The handbook gives no clue to the actual use of the receiver, the nearest it approaches to doing so being the following:

The apparatus consists essentially of a radio receiver and a cathode ray tube with its associated time base . . . The output from the receiver portion of the equipment is arranged to appear as a vertical deflection superimposed on the horizontal trace . . .

The nomenclature T,H,L,C, etc., used in the schematic was applied to various panel-mounted units of the equipment. The complete receiver, by present-day standards, was a very spacious item. The units were mounted in a bay, consisting of four vertical iron racks closed at the rear by a metal curtain, which, when raised, cut off all HT supplies to the units. The receiver operated within the frequency range 20 MHz to 45 MHz in four bands. The first RF circuit panel was a 'pre-selector' stage, providing tuning but no amplification.

* D.C.D. Director Communication Development (Air Ministry) and D.D.C.D. Deputy Director Communication Development (Air Ministry).

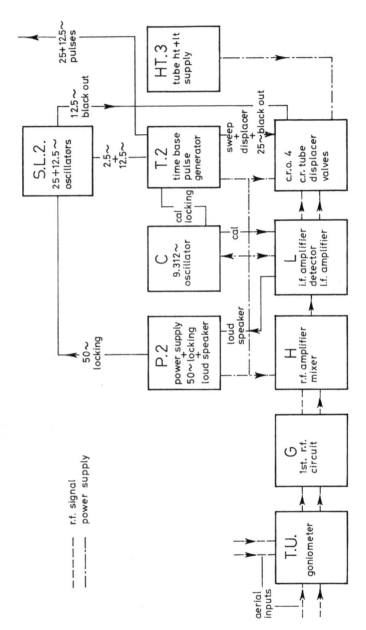

Fig. 5.21   *RF5 receiver schematic*

RF amplification was provided by three push–pull stages, with each stage housed in a screened compartment. Gain control was provided. The first pair of valves were low noise beam tetrodes, with normal pentodes used in the other two stages. Triode-hexode mixing was used with 41 STH compound valves. The intermediate frequency used was 2 MHz, the triode local oscillator (of the triode-hexode circuit) being 2 MHz higher than the station frequency. Two alternative IF bandwidths, 200 kHz and 50 kHz, were available in the four-stage IF amplifier. A full-wave diode detector circuit was used and this was then followed by two power output stages. An audio output was passed on to a monitoring speaker, the principal purpose of which was to help identify interfering signals; it was usual to leave the automatic gain control off when this was being done. Motor ignition, for instance, produced rapid clicks which changed as the car changed speed; a common form of enemy jamming was the use of a frequency-modulated CW signal, and this sounded as a musical note of between 150 Hz and 600 Hz. The 9312 Hz oscillator in the C2 panel supplied a signal to the L panel for the purpose of calibrating the horizontal traces on the cathode-ray display. The circuit was enclosed in an asbestos-lined metal box so as to minimise frequency variations owing to external temperature changes.

A later receiver, the RF7, contained five stages of IF amplification, which were preceded by two sharply tuned circuits known as IFRUs (intermediate frequency rejector units) which were employed to decrease CW type of interference. Bandwidths of 50 kHz, 200 kHz and 500 kHz were employed and stagger tuning was used for the latter two bandwidths. Development of these receivers was a continuous process. A paper by Jenkins [45] describes concisely the development of the receiver between the years 1937 and 1942.

The firing time of the transmitter and the start of the range trace in the receiver display were, of course, synchronised. Several techniques for achieving this were employed: in the final version of the Cossor receiver, the RF8, a pulse generated in the receiver fired the transmitter and then a fraction of the transmitter output was fed back to the receiver to start its time base. The source of synchronisation for all stations was the 50 Hz mains system. paragraph 2.06 of Specification BRS No. 10001/a of June 1937 stated [41]:

> The pulse shall be controlled from a 50 cycle source and shall be capable of adjustment in phase to any angle from 0 to 360 degrees within 1 degree and shall not vary its point of incidence relative to zero phase angle of the 50 cycle supply by more than ±2 microseconds.

Fig. 5.21 shows for the RF5 receiver a 50 Hz supply going from the P2 unit to the SL2 unit where a 25 Hz sinusoidal voltage was generated which went to the T2 unit; there, sawtooth and square-wave signals were then formed. The P2 unit contained a phase shifting transformer to provide a 50 Hz mains supply of continuously variable phase. The 50 Hz supply of variable phase was followed by a 25 Hz master oscillator which oscillated in phase with the mains. This master oscillator, which was a single valve *RC* type, was colloquially referred to as the 'spongy lock', its name being derived from the fact that it could absorb, without affecting its action, small frequency and amplitude variations in the mains. It operated on the principle that one of the resistances of the RC network was replaced by a triode valve: the impedance of the triode was varied by a biasing voltage applied to its grid and this voltage was dependent on the phase relationship between the oscillator output and the mains.

It was customary to stagger the duty cycles of the receivers and the firing times of associated transmitters at the various stations of the chain. This staggering of operating cycles ensured that stations operating on the same frequency did not intefere with each other, but it was able to ensure also that backscatter signals returned via the ionosphere did not interfere with the operation of the chain.

Most of the scattered energy was returned from ranges of approximately 1000 nautical miles and 3200 nautical miles,* equivalent to delays of 12 ms and 40 ms, respectively. Any station firing 12 ms after another station operating on the same frequency would have its trace affected. This method of achieving immunity from scatter can be illustrated as follows. Suppose six stations were operating on the same nominal frequency. If they were all timed by a 25 Hz signal from the master oscillator they underwent cycles of 40 ms duration, receiving, say, for 2 ms and being quiescent for 38 ms. If the stations triggered off at 2 ms intervals, then scattering coming back from 1000 nautical miles would return during the quiescent period of the receivers. The long distance scattering arrived some 40 ms after the firing of a transmitter and, thus, could affect the receivers. To overcome this, a facility was available to halve the pulse repetition frequency to 12·5 Hz and Fig. 5.21 shows a 12·5 Hz sinusoidal oscillation emanating from oscillator unit SL2.

5.9 RDF beam technique – CHL and GCI

Each Chain Home station was a static installation, operating at rather long wavelengths, which floodlit a considerable sector in front of the station. Apart from an inability to pick up low-flying aircraft approaching at angles of elevation below about 2°, the stations served the two requirements of long range detection and subsequent interception of enemy aircraft quite effectively. Until well after the outbreak of war, their construction and development were given priority over all other radar projects. As early as 1935, a beam technique or 'radio lighthouse' was discussed. The obvious advantage of a beam over a floodlighting system is that less transmitter power is necessary to detect any given target. The concept emerged in practical form in the sphere of air-defence as two systems closely similar in many ways, CHL (Chain Station,

---

* Appreciation of this backscatter phenomenon led to the concept of over-the-horizon radar. See also Section 2.12.

Home Type, Low Flying; more briefly Chain Home Low), first available in November 1939, and GCI (Ground Control of Interception), available a little over one year later. Both systems operated on $1\frac{1}{2}$ m wavelength.

The CHL stations were used to supplement the vertical coverage of the Chain Home. Originally, the antenna system consisted of two sets of broadside dipole arrays, rotating in synchronism with each other. This was followed by a transmitting Yagi array mounted over a broadside receiving array. Finally, in the first example of T-R switching in British radar, one common dipole array was employed. This array was mounted either on a 20 ft high wooden gantry for a cliff site or, for flat sites, on 200 ft towers. Photographs of both CHL and GCI stations and equipments are available in the Air Ministry publication 'Photographic record of radar stations (ground)' [46].

Ground control of interception of enemy aircraft directly from the radar station itself was first attempted from CH stations and some success was achieved with very skilled operators. It was attempted for CHL stations with PPI display, but lack of height-finding facilities made it unsuccessful. Hence, the specially designed GCI station, which supplied the interception controller with continuous accurate details of an aircraft's range, bearing and height. In combination with the fighter aircraft's AI (Aircraft Interception) radar, it afforded a successful technique for night-time interception of enemy aircraft. It is instructive to quote briefly from the Minutes of the 23rd Meeting of the CSSAD, which was held on the 31st October 1937 [47].

161(b) Beam Technique
    The suggestion was made that a form of 'radio searchlight' could profitably be used to follow aircraft at relatively close ranges after location at longer range by RDF1 . . . It was recommended that an apparatus to work on the beam technique should be installed at Bawdsey, working on a wavelength of the order of 4 metres, and employing mechanical swinging of the beam.

The stage which the design of a suitable antenna had reached towards the end of 1937 is illustrated by a letter [47] dated 18th November 1937, and headed 'RDF beam technique', from Dr E. T. Paris (for Superintendent Bawdsey Research Station) to the Secretary, Air Ministry (DSR). The letter states:

    On the occasion of a recent visit by Director of Scientific Research, Admiralty, to this Station, mention was made of certain reports that had been prepared for the Admiralty by Mr F. D. Smith. It was understood that these reports dealt with the theory of the production of beams by arrays of vibrating elements and had reference to the design of under-water apparatus for the generation of beams at supersonic frequencies. It is suggested by the Director of Scientific Research, Admiralty, that the theoretical work in these reports might be of service in connexion with the design of radio beams for Air Defence purposes.

Progress was comparatively slow and the situation in mid-1938 may be judged by a short extract from another letter of Paris [47], written on 2nd June 1938 from Bawdsey to the Director of Scientific Research, Air Ministry. It was headed 'Ultra short wave transmitter' and was concerned with the transmitters that could be used in the detection of low-flying aircraft and in the location of ships at ranges of 15 or 20 miles, so as to assist Coast Defence batteries. It recommended that Metropolitan-Vickers be invited to tender for a suitable ultra short wave transmitter. The letter opened as follows:

> The progress of investigations into the use of ultra short wave radio search beams for the detection and location of aircraft and ships is seriously hampered by lack of a suitable high-powered pulsed ultra short wave transmitter.
>
> The research programme of this Establishment is already very heavy and the design and construction of a transmitter of the kind required cannot be undertaken without seriously delaying high priority items . . .

In a letter entitled 'Detection of low flying aircraft', written by Rowe to the Undersecretary of State, Air Ministry, on the 9th March 1939 [48], there is a reference to experiments to be carried out using balloon-supported aerials at heights of up to 2000 ft and starting with an elevated receiving aerial for obtaining range and sense, and then follows the statement:

> We are at present awaiting balloon equipment and personnel before a start can be made on this work.

Later on in the letter Rowe stated:

> The conclusions are that, until large powers can be generated on very short (centimetre) waves, the reflectors necessary would be too large to instal on towers of economical dimensions.

To Butement must go major credit for the development in Britain of beamed radar [26, 49], which branched out into the various $1\frac{1}{2}$ m systems which became known as CD (Coast Defence), CHL, GCI, CDU (Coast Defence U-boat detection) and SLC* (Searchlight Control). The choice of $1\frac{1}{2}$ m wavelength was influenced by Dr Bowen's work on airborne radar (Section 5.10). The first objective was a CD radar, a radar which would provide coastal gun batteries with the necessary range and bearing information to engage enemy ships. A memorandum [47] (on RDF and Coast Defence) from Paris to the President of the Royal Engineer and Signal Board, and dated 23rd November 1937, mentions successful tests done at Bawdsey with 1 metre equipment in which small vessels were detected at three miles. Serious work, however, did not take place until late 1938 when Butement and Eastwood, operating from Bawdsey, took up the threads of previous research and by mid-1939 had a prototype

---

* Colloquially referred to as 'Elsie'.

equipment undergoing trials. Yagi antennae were experimented with initially and then a Sterba* array but finally a 'fir-tree', or broadside, array of dipoles, with wire mesh reflectors, was decided upon for both the transmitting and the receiving arrays. During the experimentation with antennae, sidelobes from the Sterba had proved difficult to suppress. At this time also, a paper published in 1930 by Southworth [50] on directive antennae had proved particularly helpful.

The lack of low-angle coverage available from the chain was felt as war broke out. As an interim measure, on the 12th October 1939 a trawler screen, acting as an observation post, was positioned in the North Sea. The ships reported visual sightings by radio telegraphy back to the Chain stations at Stoke Holy Cross and Stenigot [51]. The trials with CD sets had demonstrated that aircraft flying at low altitude could be detected and tracked successfully [52]. A 'crash' programme for CHL was set in motion under the direction of Ratcliffe and Cockcroft, both of the Cavendish Laboratory, and by the 11th February 1940, in spite of a severe winter, 11 stations were in operation. The urgency with which this programme was carried out arose because of the Germans' use of mine-laying aircraft. A second 'crash' programme began in January 1940 which resulted in a further 7 stations becoming operational by the end of February 1940.

The CD sets were in effect good CHL sets, and indeed the first batch of CD sets produced were diverted into use as CHL sets. One important difference in the operation of a CHL equipment was its use of a PPI display. The CHL sets had no facility for height finding. The antenna system was a four tier array of horizontal full-wave dipoles, with five dipoles in each tier, placed at a distance of $\lambda/8$ in front of a reflecting screen of wire netting. Caledon† turning gear allowed the antenna array to be rotated at speeds of 1·0, 1·5 and 2·3 rev/min. Transmitter peak pulse power was of the order of 150 kW.

The GCI station differed primarily from the CHL station in its height-finding capability. GCI stations were either fixed or mobile. In the latter case the system, which included a ground-to-air VHF radio link and two diesel-electric generating sets (main and standby), comprised 8 vehicles. The usual frequency of operation was 209 MHz. Later systems had three working frequencies, namely 193 MHz, 200 MHz and 209 MHz. Pulse recurrence frequency was variable from 300 Hz to 540 Hz and pulse widths of 3 μs, 5 μs and 8 μs were available. Peak pulse powers of the order of 100 kW were obtainable. Ranges of over 90 nautical miles were readily obtainable.

The antenna was an array of 32 full-wave dipoles arranged in four bays, and it was divided evenly into an upper and a lower part. The antenna beamwidth in azimuth was 15°. The array was rotatable in either a clockwise or a

---

* Developed by E. G. Sterba of A.T.&T. Used in early 1930s on HF transatlantic telephone services between US and Europe.

† One of the manufacturers of antenna turning gear. Information on it is available in A.P. 2888A, Sect. C.

counter-clockwise direction at a rotation rate which could be varied between $\frac{1}{2}$ and 6 revolutions per minute. The turning gear incorporated a Ward-Leonard control system. An A-scope display and a PPI display were used. The bearing of a target was determinable to $\pm 1\frac{1}{2}°$ by noting the centre of its trace on the PPI display.

Siting was a very important factor in determining what coverage would be obtained with a GCI set. A series of coverage diagrams for four specimen sites drawn up for a frequency of 209 MHz and based on a signal/noise ratio of unity was available to the personnel concerned with the siting and operation of the equipment [53]. The four types of sites considered are illustrated in Fig. 5.22: all distances and height were measured to the centre of the antenna. Seven diagrams were, for instance, constructed for the sloping site case for various values of angle of slope, α, for distance along slope and antenna height above level ground. In the situation depicted in Fig. 5.22c the radiation reflected from the ground and striking the target would be incident on the ground at an angle either greater or less than α, and which for a distant target would be the angle of elevation of the target. In selecting appropriate coverage diagrams, the height of the antenna array would be taken as h_1 for angles greater than α and taken as h_2 for angles less than α.

For height finding, a diode switching circuit switched the received signal from the upper half of the antenna and from the lower half, alternately, into the receiver. The diode circuit was operated by two square waves in anti-phase which, at pulse recurrence frequency, rendered the cathode of alternate diodes

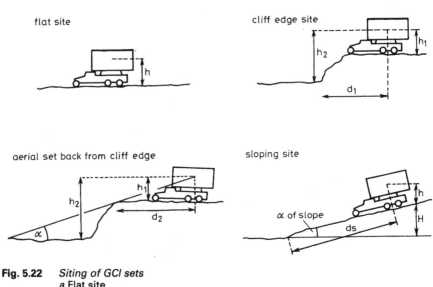

Fig. 5.22   *Siting of GCI sets*
a Flat site
b Cliff-edge site
c Aerial set back from cliff edge
d Sloping site

approximately 35 V negative and hence conducting. Comparison of the strength of the echoes received by the upper and the lower antennae, which were displayed side-by-side on the A-scope enabled, as discussed in Section 5.6.5, the height of the target to be calculated. The order of accuracy of height finding was ±1500 ft. Understandably, the range obtainable using half of the antenna when in the height-finding mode was less than when the antenna receiver switch was at 'normal'. This is illustrated in the performance diagram (referred to as VPD) of Fig. 5.23.

In Fig. 5.23 will also be seen straight lines at 9·2° and at 14°. The line at 9·2° shows, for this particular type of site (flat with antenna array at a height of 12′3″) the point at which the operator should change the transmitter change-over switch from 'phase' to 'anti-phase', while the line at 14° shows where it should be changed back again. The operator's selection of the 'anti-phase' position resulted in the two halves of the array being fed in anti-phase. The gaps in the vertical pattern due to reflection from the ground were modified so that 'gap-filling' occurred, with those angles at which gaps had been produced now being the centre of reflection lobes. In Fig. 5.23 the coverage diagrams assume that the transmitter 'phase' and 'anti-phase' switch was switched to 'anti-phase' at 9·2° and switched back to 'phase' at 14°. It can be seen from the figure that leaving the receiver antenna switch at 'normal' in

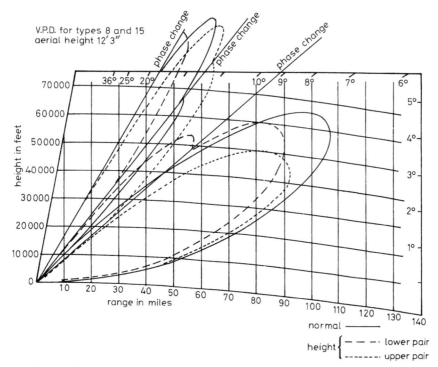

Fig. 5.23   *GCI performance diagram*

the anti-phase region, between 9·2° and 14° resulted in a coverage gap. This was obviated by setting the receiver antenna change-over switch to 'height' when working in this region; indeed, at this particular type of site, it would have been possible to leave the switch permanently at 'height' without undue loss in performance, and without the inconvenience of changing the switch back and forth.

5.9.1 IFF and GCI

Brief mention will now be made of the IFF MkIII system which was colocated with GCI systems.

An outline of the development of IFF was given in Chapter 2. In Britain the MkI set was an experimental set with only 50 units being produced. It operated in the Chain Home band of frequencies. The MkII equipment operated in sequence through the following bands of frequencies: 22 MHz to 30 MHz for Chain Home systems, 39 MHz to 51 MHz for the Mobile or Advance Chain Home (ACH) system, and 54 MHz to 82 MHz for Gun-Laying (GL) equipment. The development of the MkII had begun in the spring of 1939 with the Telecommunications Research Establishment, the Radio Department of RAE and Ferranti all involved. The MkIII system, the prototype of modern secondary surveillance radar, divorced the function of identification from that of detection and operated within the frequency band 157 MHz to 187 MHz [54, 55]. The final development of the MkIII coincided with the entry of the United States into the war and the British authorities put much effort into having it adopted as a standard IFF system by both countries. A large-scale trial of the MkIII, at which American observers attended, took place at Pembroke in December 1941. Eight ground installations (one CH, one Mobile Unit, one GCI, three GL and one SLC) and six types of aircraft (Spitfire, Blenheim, Wellington, Coastal Wellington, Sunderland and Beaufighter) fitted with transponders took part [56]. All parties were impressed with the results.

With the AMES Type 15 GCI equipment, the ground facilities for the MkIII IFF were arranged by placing the interrogator and responsor units in the cabin of the Antenna Vehicle, the interrogator being generally triggered by locking pulses from the main GCI equipment [57]. The frequency of the transmitter (T 3154B) was in the band 165 MHz to 185 MHz with a pulse recurrence frequency of about 150 Hz, which was a submultiple of the main transmitter's recurrence frequency. A peak power output of 0·8 kW to 1·1 kW with a pulse width of about 10 μs was produced. The responsor was a superheterodyne receiver with one RF stage, a diode mixer, five IF stages, which were stagger-tuned* giving a flat response from 9 MHz to 13 MHz, and a diode detector.

---

* A design technique used with cascaded IF amplifier stages whereby individual stages are tuned to different selected frequencies within the overall passband so as to produce a wide and flat-topped response.

The output of the detector was passed through video* amplifiers to the A-scope display in the Operations wagon. The returns from the IFF transmitter were processed by the responsor and were displayed on the A-scope or height–range tube on a second trace above the normal radar trace. This IFF trace was formed by every second sweep of the time-base and displayed returns from the main transmitter in addition to returns from the IFF transmitter. The zeros of the two traces corresponded so that the IFF indication for a particular target appeared above the target's normal radar signal, the former pointing upwards and the latter on both traces pointing downwards. During height finding the IFF trace was removed.

The IFF antenna system was a broadside array of 10 half-wavelength dipoles spaced one half-wavelength apart and placed a distance of $\lambda/8$ in front of a wire mesh reflector. Vertical polarisation was used because of the more satisfactory vertical coverage obtained; as was seen in Chapter 2, the nulls in the vertical pattern produced by ground reflection are less complete than in the case of horizontal polarisation.

5.10 Airborne radar – RDF2

The credit for the design of the first airborne radar goes to Dr Bowen and his small team, who used the facilities of Martlesham Heath airfield for their flight tests. When one considers the relatively large size of the ground radar equipment being developed in Britain in 1936 and 1937, and which was the only radar known to Bowen, then the extent of the challenge in designing a radar set to fit in a small aircraft becomes clear. Apart from obvious constraints in weight, size and electrical power supply, there is for an aircraft installation the overriding factor of 'airworthiness', which, briefly, means that an installation must not jeopardise in any way the flight safety or the operational roles of an aircraft; furthermore, the installation itself must perform correctly in its airborne environment, where such features as vibration, temperature, and air pressure may differ greatly from what pertains at ground level.

Dr Bowen himself [58] gives due credit to Henry Tizard for foreseeing the requirement for airborne radar in British air defence. Tizard, he states, realised that a German air attack made in daylight would be beaten back and that the Germans would then turn to night attack, when ground radar alone would not be sufficient.

The group formed to work on the problem of airborne radar consisted initially of only Bowen himself, but grew to three persons in September 1936 and to six by the spring of 1937. Ground radar was initially known as RDF1 and the airborne radar development programme was then referred to as RDF2; RDF1½ referred to the bistatic system of a ground transmitter illuminating a target aircraft with only a receiver in the interception aircraft.

* Video-frequency, or wide-band.

The first experiments were carried out in June 1937. A Heyford bomber fitted with a receiver operating at 6·8 metres and with an indicator, and with a half-wave dipole wire antenna, connected between the undercarriage spats, circled at a few thousand feet above Bawdsey. Aircraft were picked up at ranges of 8 to 10 miles. Two reasons which prompted the use of the Heyford were the generous space available in the fuselage and the fact that its Kestrel engines were well screened against ignition noise. The wavelength of 6·8 metres was chosen because there was readily available an EMI television receiver of the TRF type tuned to this wavelength which was perfectly suited for the task. A car ignition system was used to provide the HT for the cathode-ray tube. The total weight of the installation was some 50 lbs [59]. This RDF1½ system, which gave excellent ranges even at such an early stage of development, was strongly advocated by Bowen, but Watson-Watt was equally strongly against it, and the latter's viewpoint prevailed.

The antennae used with the first 1½ m equipment which was installed in an Anson were simple dipoles. The transmitting antenna was projected through the Anson's escape hatch, while the receiving antenna was mounted inside the aircraft. The dipoles were rotated and, by observing minimum signal, the bearing of ships could be roughly estimated.

One initial obstacle which Bowen surmounted was that of an adequate electrical power supply in the aircraft to generate worthwhile transmitter power output. Until then, the power supply for aircraft was DC accumulators and DC generator, with maximum available power in a fighter aircraft of some 500 W; the use of AC had not been seriously considered. Bowen succeeded in breaking through convention and had AC generators (80 V, 1200 Hz to 2400 Hz, 800 W), of the same size as existing DC generators, designed and manufactured by Metropolitan-Vickers Ltd, and installed in aircraft. The first installation was carried out on a Fairey Battle fighter-bomber.

The initial objective had been AI. However, because of the success in detecting ships, the thrust of experimentation was switched to developing ASV equipment. The year 1938 was a year of experimentation and progress in ASV radar, with first production models emerging in 1939. Side-looking radar was tried successfully using an array of Yagis mounted along the fuselage of the aircraft; photographs of a sideways-looking radar display showing the Isle of Wight were taken in September 1938.

Towards the end of July 1937, a low power transmitter, operating on 6·8 metres with output less than 100 W and using 'door-knob' triodes (Standard type 316A), was fitted in the Heyford. An A-scope display was used. No aircraft could be detected, but echoes from the coastline and from quays and wharfs at Harwich at ranges of 3 to 4 miles, and echoes from ships at 2 to 3 miles, were received. Bowen wrote [58]:

> This gave a tremendous boost to morale and it can be said that airborne radar was born at that time.

The endeavour was to decrease wavelength and hence antenna and equipment size. Higher frequencies were possible because of the use of new valves such as the Western Electric double-ended doorknob 316A triode valve.* Wavelengths were shortened from 6·8 m to 2·5 m and then to 1·25 m. Because it was easier to tune the 1·25 m sets and obtain better results at a slightly lower frequency, a wavelength of 1·5 m was finally chosen and this became the standard for all ASV (Air to Surface Vessel) and AI (Aircraft Interception) systems before the advent of the cavity magnetron; indeed, the receivers (developed by A. C. Touch) using Acorn 955s in the mixer and local oscillator, and the EMI TRF 45 MHz receiver as the IF amplifier, became the basis also of the 200 MHz CHL, GCI and SLC receivers.

Much effort in 1938 and in the early part of 1939 went into assessing the three types of antenna system which were used for ASV [60]. These comprised:

(*a*) A forward looking system: transmitting antenna radiated in a fan-shaped beam forward of the aircraft, while two receiving antennae gave overlapping lobes which produced an equi-signal when target was directly ahead of the aircraft.

(*b*) A sideways-looking system: this afforded the greatest potential for achieving high gain arrays and long range.

(*c*) An omnidirectional system: a rotating dipole was envisaged. This third system was never developed.

The forward looking system was developed first, while 8 months later, early in 1940, the sideways looking system was developed for long-range ASV.

K. A. Wood, one of the team who carried out much of the airborne tests of the equipment, kept a flight diary for the period August 1937 to September 1938 [61]. This is an excellent source, providing quite detailed accounts of the flights, of the types of equipment used and of the results obtained. For the 3rd September 1937, the following entry is made. The flight took place at 16.50 hours for a period of 2½ hours in Anson 6260, with Sgt Naish as pilot and Bowen and Wood as observers.

Set out to find the Fleet in the Channel. Flew down past Dover just south of Beachy Head. Came upon the *HMS Rodney* and *Courageous* with 5 or 6 smaller craft in attendance. Flew over at varying distances and with varying approaches, the larger craft giving up to 6 kilometres with amplitudes of approximately 1 cm at 3 to 4 kilometres. We also flew along the line of ships receiving each in turn.

This successful demonstration occurred during the autumn exercises of the Fleet and gave airborne radar a considerable boost. The following year, in May 1938, when units of the Home Fleet were passing through the English

* Designed for continuous wave operation, they could deliver 20 W continuous output at 700 MHz.

Channel from Spithead to Portland, an impressive demonstration of ASV, including photographic records of the radar echoes, was carried out.

The ASV MkI system was hurriedly put into service with the Royal Air Force in January 1940. It was used mainly with Hudsons and Sunderlands. The transmitter gave a peak power output of about 7 kW with pulse width and recurrence frequency of approximately 1·5 μs and 1200 Hz, respectively. Its ability to indicate clearly approaching coastlines and the presence of responder beacons made it a valuable aid to navigation and it was perhaps more notable in this than in its ASV role. About 200 sets were produced and were in use until the end of 1940, when the ASV MkII became available. The MkII was basically similar to the MkI, but it was a better engineered set. Its transmitter produced a peak pulse output of some 7 kW, with a pulse width of about 2·5 μs and a pulse repetition rate of some 400 pulses per second. Its receiver comprised two RF stages, a triode local oscillator, pentode mixer and three stages of IF, a diode detector and video amplifier. A typical performance, when at an altitude of 2000 ft, with sideways-looking antennae in the long range ASV role, was [60]:

Surfaced submarine	8 miles
Coastlines	60 miles
10 000 ton ship	40 miles
Destroyers	20 miles

The MkII ASV may well have been the most widely used radar set of the war, with over 10 000 sets being built. It was manufactured in Great Britain, the United States, Canada and Australia [58]. It may be noted that the MkII found employment for a while in uses other than its designated ASV role. It was used as a searchlight director for coast defence purposes, while the Navy employed it for sea and air search on motor torpedo boats, destroyers and other vessels for the first year and a half of operations [58, 62].

It is of interest to consider the types of antenna systems which were used with the ASV MkII equipment. Two aircraft, the Sunderland Flying Boat and the Wellington, have been chosen as examples. The details given, including Figs 5.24 to 5.29, have been taken from S.D. 0249(1) of August 1941 [63].

In the case of the Sunderland, as shown in Fig. 5.24, the transmitting array consisted of two bent antennae positioned on each side of the hull. Each antenna was $\frac{3}{4}\lambda$ in length and had a radiation resistance of approximately 12 Ω. The receiving array comprised two quarter-wave antennae made from $\frac{3}{8}$ inch tubing of 14·96 inches in length and each with a radiation resistance of approximately 20 Ω. The transmitting array provided maximum output straight ahead and at 60° to the line of flight, while the receiving array provided maximum response at 15° and at 60° to the line of flight, as shown in Fig. 5.25. Again, as indicated in Fig. 5.25, the DF (direction finding) ratio increased sharply to 2 : 1 at an azimuthal angle of 10° and rose above 5 : 1 beyond 30°; these patterns were plotted with the aircraft on the ground. Direction finding

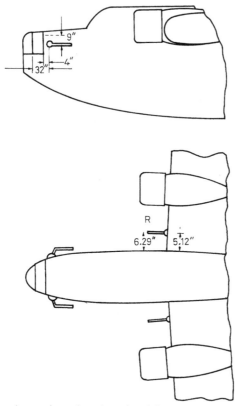

Fig. 5.24   *Sunderland aeroplane: location of aerials*

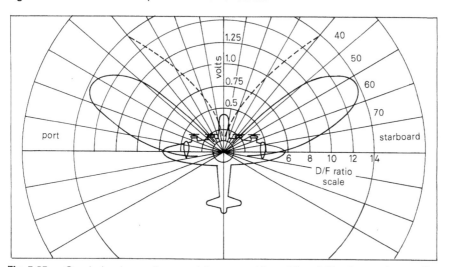

Fig. 5.25   *Sunderland aeroplane: aerial patterns (dotted line, D/F ratio; continuous line, amplitude received on receiver from its own transmitter)*

was achieved rather simply, but adequately, as follows. A receiving antenna was placed on each wing, its axis pointing outwards slightly to either port or starboard, as the case may be. The responses from the antennae were switched in turn to the receiver. When the target was to the right of the line of flight, the response from the starboard antenna exceeded that from the port antenna; when the target was to the left of the line of flight, the response from the port antenna was the greater, and when the two responses on the CR display were equal, the target was assumed to be ahead of the line of flight.

The Wellington had two distinct antenna systems, each of which had separate transmitting and receiving parts. One system was a broadside one for long range ASV (LRASV), while the other was a forward looking one with the receiving antennae positioned so as to allow the same type of simple homing as described above for the case of the Sunderland. Fig. 5.26 gives elevation and

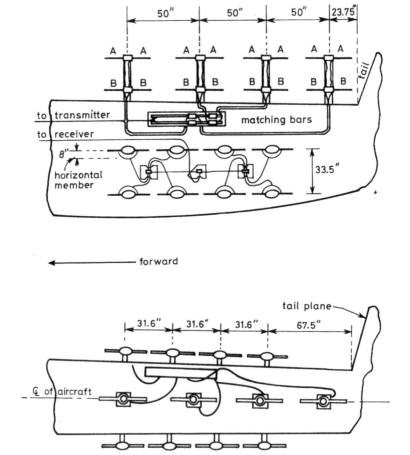

Fig. 5.26   *Wellington aeroplane: broadside aerials*

plan views of the LRASV installation, while Fig. 5.27 shows the layout of the forward-looking system. Fig. 5.28 and Fig. 5.29 show, respectively, the combined polar pattern of the LRASV transmitting and receiving arrays and the DF characteristics of the forward-looking system.

The transmitting array of the LRASV consisted of 8 half-wave dipoles arranged horizontally in two tiers on the top of the fuselage. The corresponding receiving system consisted of two broadside arrays, one on each side of the fuselage and each composed on 8 half-wave dipoles arranged in two tiers. The

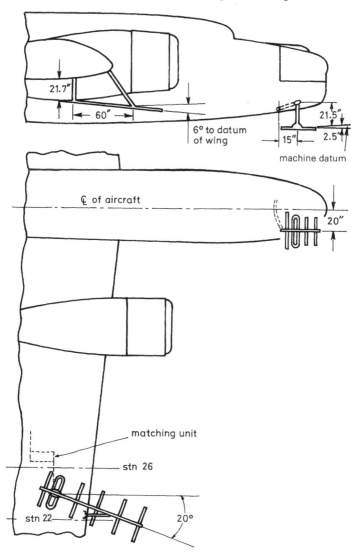

Fig. 5.27   *Wellington aeroplane: homing aerials*

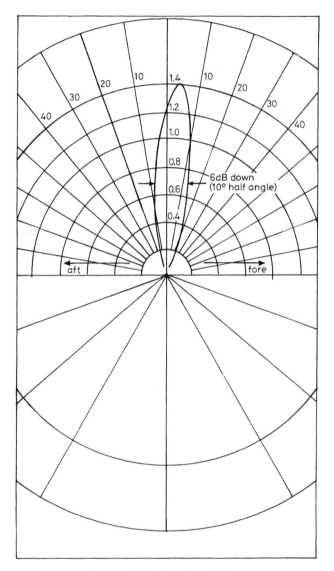

Fig. 5.28   *Wellington aeroplane: aerial pattern broadside arrays*

transmitting array was matched to the transmitter by means of a set of matching bars which was shortened across at one end. The matching of the system was rather critical, so that, once set up, it was important to maintain the transmitter on its correct frequency. The beamwidths of the transmitting and receiving arrays were approximately 26° and 33°, respectively. The homing system used Yagi arrays with a four-element Yagi for transmission and a pair of six-element Yagis for reception. The beamwidth of the transmitting Yagi

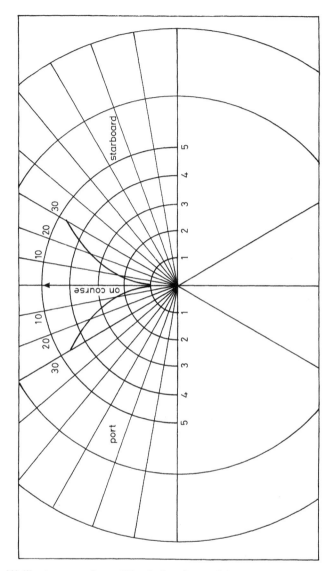

Fig. 5.29   *Wellington aeroplane: DF ratio homing aerials*

was approximately 66°. The receiving antennae were positioned well away from the engines to avoid propeller modulation; and each Yagi was slanted 20° to outboard.

The development of metric AI culminated with the AI MkIV which became available in the summer of 1940 and which continued in service until late 1943. It was essentially similar to the ASV MkII. The initial objective of Bowen's group was the development of an AI radar; then their efforts were directed towards ASV radar. However, towards the end of 1938, and for the first half of 1939, AI development was given priority and in May 1939 the first installation was flown in a Fairey Battle. It could be said, in the case of metric AI and ASV, that the benefits of developments at any particular time in either system became available to the other system.

The broad specification of the AI MkI was:

Transmitter peak power output	2 kW
Receiver sensitivity	between 2 μV and 5 μV
PRF	2000 Hz
Pulse width	1 μs

The antenna system comprised a horizontal single dipole with reflector for transmission and four receiving antennae, two to provide azimuth indication and two to give elevation information.

The various marks of AI appeared with great rapidity. The AI MkII was similar to the AI MkI but was a better produced and a more airworthy installation. Metropolitan-Vickers Ltd, A. C. Cossor and Pye were involved in its production. The AI MkIII had similar circuitry to the MkII, but had an improved antenna system. With the Bristol Blenheim, the first ever radar-equipped night fighter, it was discovered that a squint problem existed when using the horizontally polarised antennae of the azimuth determining array. This was attributed to the position of the engine nacelles and was cured by the use of vertical polarisation, which then became standard for all subsequent AI installations. The performance in range dropped slightly because of the use of vertical polarisation but, with the sensitive receivers then available, this effect proved of no consequence.

All metric AI systems suffered from the disadvantage that, because of the necessarily broad beamwidths of the antennae, strong ground return echoes were obtained; this meant that detection became impossible at ranges less than the height of the aircraft above ground.

On the night of 3rd September 1939, a Blenheim with an AI MkI installed on board, and with an observer from Bawdsey, was flying in the London area. By October 1940, Blenheims fitted with AI MkIII had shot down 6 enemy bombers. In August 1940 Bristol Beaufighter aircraft were being fitted with the AI MkIV. The antenna configuration for a MkIV installation in these twin-engined aircraft comprised a transmitting array in the nose (folded dipole and director), two $\lambda/4$ rods on the starboard wing acting as elevation antennae, and two $\lambda/2$ dipole rods with directors, one set on each wing, acting as an azimuth array.

An important decision, which affected British night fighter policy into the 1950s, was that AI should be operated by a special radar operator and not by the pilot; this was certainly the correct decision for the early designs of AI radar.

Serious difficulties were experienced initially with AI [64, 65]. (Bill Gunston's book 'Night figher' [65] provides an excellent insight into the operational background and problems of AI radar.) There were two principal factors which accounted for an initial lack of success in the use of AI. The first was the Blenheim aircraft itself which, with a maximum speed of about 250 miles/h, was far too slow for the role of interceptor. The second was that a high degree of skill, and therefore special training, was required of both the radar operator and the pilot before successful interception could be accomplished. In the matter of aircrew training, the Fighter Interception Unit (FIU) of the Royal Air Force, formed in 1940, played an important role. It is informative to quote from Bowen [66]:

> Realising the rudimentary state of our knowledge on the practical usage of AI equipment, C-in-C, Fighter Commander, formed the F.I.U., a tactical trials unit under the command of W/Cdr, now G/Cpt, Chamberlain. One cannot emphasise too much the tremendous influence this had on the introduction of A.I. to the R.A.F. It is not too much to say that the transition of A.I. from its experimental form to a useful military weapon was entirely due to this officer's energy, insight and devotion to duty.

5.11 Organisation

The British radar systems which have been alluded to were designed for a predominantly defensive role. As such, they were just one part, one component, of a large defence organisation. In particular, the role that radar played in Air Defence depended for its success on many factors other than availability of dedicated scientists and engineers and the production of reliable equipment. The operators at the Radar Stations, the Filter Rooms, and the Operations Rooms, with their overseers, all played a vital part; as did the Royal Observer Corps watchers whose reports also reached the Operations Rooms. The drills, the standard procedures, and the intuitive insight born of experience were vital to the functioning of the whole system. From the time an aircraft was seen as a 'blip' on a radar screen by an operator until the necessary information was displayed as a plot in the Operations Room in front of those who ordered the fighters into action, the Raid Reporting System performed a succession of functions which were referred to as 'Read', 'Report', 'Filter', 'Identify', 'Tell' and 'Plot'. The organisation and procedures used are outlined in several Air Ministry pamphlets [67–71].

An event of some note occurred when Squadron Leader (later Air Marshal) Raymond George Hart officially opened, in February 1937, a Training Centre at Bawdsey for Royal Air Force personnel who would man the RDF stations. The work of this Training School in preparing crews was of paramount importance in the success of the Chain Home system. Squadron Leader Hart joined Bawdsey on 14th July 1936. His background could hardly have been more suitable for the task in hand. He graduated as a physicist from the Royal College of Science and also as a 'Radio Électricien' from the École Supérieur d'Électricité of Paris. He was an operational pilot during the First World War and qualified as a flying instructor in February 1935. In September 1935 he was posted as Deputy Chief Signals Officer of Fighting Area to Sir Philip Joubert de La Ferté [72–74]. In understanding the requirements of the Royal Air Force, while being in tune with the mind of the scientists, he did much to promote team work in the pre-war and early war years in a rather hybrid and *ad hoc* organisation. Indeed, the whole area of military–scientific organisational co-operation would warrant an in-depth independent study.

With regard to liaison between scientists and the services, it is necessary to mention that from the study of the optimisation of the operation of the Chain Home system there grew the science of Operational Research. The name seems to have been coined by Rowe [75] after the 1938 Air Exercises. The influence of Henry Tizard is again seen in this area; as early as 1936 he had been responsible for the setting up at Biggin Hill of a small team of Royal Air Force officers with Dr B. G. Dickins as scientific officer to study how the proposed radar chain might best be used for the interception of aircraft.

5.12 Radar countermeasures

From the very concept of radar and of the Chain Home system in 1935, the possibility of jamming by the enemy was realised. Reference may be made to Watson-Watt's memorandum of the 9th September (Appendix E) where the danger of deliberate interference, and its counteraction by the provision of several station frequencies, were discussed. Defensive and offensive countermeasures against German communications and radar were developed as the war progressed. A.P. 3407 [76], dealing with radar countermeasures, and Martin Streetly's account [77] of the activities of 100 Group of the Royal Air Force, are important sources for the early history of British electronic warfare.

An important countermeasure, if it were feasible, would be the rendering of one's own aircraft undetectable by enemy radar. The possibility of this was discussed in a report, titled the 'Camouflaging of aircraft at centimetre wavelengths', issued by TRE in August 1941 [78]. The report is reproduced in Appendix H.

References

1   See bibliography, Section 4.1
2   AVIA 10/349: Public Record Office, Kew, London, UK
3   WILKINS, A. F.: 'The early days of radar in Great Britain', Archives of Churchill College, Cambridge, UK
4   Interview with A. F. Wilkins, 7th January 1981

5 GPO Radio Report No. 233, Part V, 'Interference by aeroplanes'. Complete report titled: 'The further development of transmitting and receiving apparatus for use at very high radio frequencies', dated 3rd June 1932. Work carried out by F. E. Nancarrow, A. H. Mumford, P. C. Carter and H. T. Mitchell. Relevant portion of report quoted in Reference 73, pp. 93, 94. Report catalogued under List of Documents, Bowen Papers EGBN 5/46. Comments on report by A. F. Wilkins, Bowen Papers EGBN 1/3, Archives, Churchill College, Cambridge, UK

6 AVIA 10/349: Public Record Office, Kew, London, UK

7 AVIA 10/348: Public Record Office, Kew, London, UK

8 JONES, H. A.: 'The war in the air Vol. III' (The Clarendon Press, Oxford, 1931)

9 Reference 8, p. 79

10 ASHMORE, E. B.: 'Air defence' (Longmans Green and Company, London, 1929), see in particular pp. 93, 94

11 GOLOVINE, N. N.: 'Views on air defence' (Gale and Polden Ltd., Aldershot, 1938)

12 AIR 16/312: Public Record Office, Kew, London, UK

13 AIR 16/319: Public Record Office, Kew, London, UK

14 AVIA 10/349: Public Record Office, Kew, London, UK

15 WATSON-WATT, R. A., HERD, J. F., BAINBRIDGE-BELL, L. H.: 'Application of the cathode ray oscillograph in radio research' (HMSO, London, 1933)

16 Reference 15, p. 207

17 Interview with A. F. Wilkins, 7th January 1981

18 Air Ministry: 'The Second World War 1939–1945, Royal Air Force, Signals Vol. IV, radar in raid reporting' (A.P. 1063, Air Ministry, London, 1950), see p. 6, copy available AIR 41/12, Public Record Office, Kew, London, UK

19 Tape recorded interview by Dr David Allison, Naval Research Laboratory historian, with Dr E. G. Bowen, 16th May 1979. Copy in Historian's Office, Naval Research Laboratory, Washington DC, USA

20 Reference 19, side 2

21 WATSON-WATT, R. A., WILKINS, A. F. and BOWEN, E. G.: 'The return of radio-waves from the middle atmosphere – I', *Proc. R. Soc.*, 1937, **161**, pp. 181–196

22 Air Ministry: 'Introductory survey of radar principles and equipment', A. P. 1093C, Vol. 1, 1948

23 Bowen Papers, EGBN 2/5, Archives, Churchill College, Cambridge, UK

24 Transcript of Dr E. G. Bowen's interview with H. Guerlac on the Origins of British Radar, 27th April 1943. Bowen papers, EGBN 1/9, Archives, Churchill College, Cambridge, UK

25 AIR 10/4132: Public Record Office, Kew, London, UK

26 Record of scientific work undertaken by BUTEMENT, W. A. S., C.B.E., and submitted to Adelaide University for the degree of Doctor of Science, March 1960

27 BELL, J., GAVIN, M. R., JAMES, E. G., and WARREN, G. W.: 'Triodes for very short waves – oscillators', *J. Instn Elect. Engrs*, 1946, **93**, Part IIIA, pp. 833–846

28 An excellent treatment of the principles involved is given in the following: SMITH-ROSE, R. L., and BARFIELD, R. H.: 'A discussion of the practical systems of direction-finding by reception', (Radio Research Board Special Report No. 1, HMSO, London, 1923)

29 SMITH, R. A., and SMITH, C. Holt: 'Elimination of errors from crossed-dipole direction-finding systems', *J. Instn Elect. Engrs*, 1946, **93**, Part IIIA, pp. 575–587

30 FRIIS, H. T.: 'Determining short-wave paths', *Bell Labs Rec.*, 1927, **6**, pp. 359–362

31 Air Ministry: 'Radar Supervisor's manual', A.P.2911R Vol. 1, 1951

32 Air Ministry: 'RDF convertor equipment', A.P.2911M, 1942, available from: AIR 10/3684, Public Record Office, Kew, London, UK

33 AVIA 10/47: Public Record Office, Kew, London, UK

34 For a general survey of blocking oscillators, the following paper is recommended: BENJAMIN, R.: 'Blocking oscillators', *J. Instn Elect. Engrs*, 1946, **93**, Part IIIA, pp. 1159–1175

35 'Report of the Radio Research Board for the Year 1931' (HMSO, London, 1932), see p. 14
36 DODSWORTH, E. J.: 'Radar modulators', AVIA 44/518, Public Record Office, Kew, London, UK. This is a comprehensive monograph on radar modulators
37 HULL, A. W.: 'Characteristics and functions of thyratrons', *Physics*, 1933, **4**, pp. 66–75
38 MELVILLE, W. S.: 'Theory and design of high power pulse transformers', *J. Instn Elect. Engrs*, 1946, **93**, Part IIIA, pp. 1063–1080
39 'Proceedings of Royal Commission on Awards to Inventors', copy available in Science Museum Library, London, UK. See second day of Hearing (13th April 1951), p. 36 of transcript, and fourth day of Hearing (23rd May 1951), p. 45 of transcript
40 WHELPTON, R. V.: 'Mobile metre-wave ground transmitters for warning and location of aircraft', *J. Instn Elect. Engrs*, 1946, **93**, Part IIIA, pp. 1027–1042
41 AIR 2/1969: Public Record Office, Kew, London, UK
42 DODDS, J. M., and LUDLOW, J. H.: 'The C.H. radiolocation transmitters', *J. Instn Elect. Engrs*, 1946, **93**, Part IIIA, pp. 1007–1015
43 'The Cossor story' (A. C. Cossor Ltd, London, 1947)
44 'Description of Air Ministry Receiver R.F.5' (A. C. Cossor Ltd, Highway Grove, London N5, 1939), copy available in Science Museum, London, UK
45 JENKINS, J. W.: 'The development of C.H.-type receivers for fixed and mobile working', *J. Instn Elect. Engrs*, 1946, **93**, Part IIIA, pp. 1123–1129
46 'Photographic record of radar stations (ground)' (S.D.0458, Air Ministry, 1943)
47 AIR 2/2681: Public Record Office, Kew, London, UK
48 AIR 2/3055: Public Record Office, Kew, London, UK
49 BUTEMENT, W. A. S., NEWSON, B., and OXFORD, A. J.: 'Precision radar', *J. Instn Elect. Engrs*, 1946, **93**, Part IIIA, pp. 114–126
50 SOUTHWORTH, G. C.: 'Factors affecting gain of directive aerials', *Proc. Inst. Radio Engrs*, 1930, **18**, pp. 1502–1536
51 Reference 18, p. 84
52 SAYER, A. P.: 'Army radar' (The War Office, London, 1950), p. 118
53 Air Ministry: 'V.P.D's for A.M.E.S. Types 8 and 15', A.P.2912G, 1944
54 Air Ministry: 'Introductory survey of radar principles and equipment, Part 2', A.P.1093D, Vol. 1, 1946, available in AIR 10/2288, Public Record Office, Kew, London, UK. See Chapter 6
55 AIR 2/3049 (IFF Equipment Design Papers: 23rd May 1939 to October 1940), Public Record Office, Kew, London, UK
56 POSTAN, M. M., HAY, D., and SCOTT, J. D.: 'Design and development of weapons' (HMSO, London, 1946)
57 Air Ministry: 'B.S.U. I.F.F. Installation for A.M.E.S. Type 15 Mks. I and II', A.P.2912-C, 1944
58 Bowen Papers, EGBN 2/1. Archives, Churchill College, Cambridge, UK
59 Reference 19, sides 1 and 2
60 SMITH, R. A. (Chief Writer): 'Technical Monograph on Wartime Reserch and Development in M.A.P.: A.S.V.', copy available Bowen Papers, EGBN 2/4, Archives, Churchill College, Cambridge, UK
61 Bowen Papers, EGBN 2/2. Archives, Churchill College, Cambridge, UK
62 Reference 19, side 4
63 Air Ministry: 'Airborne R.D.F. aerial and cabling installations', S.D. 0249(1), 1941, available Science Museum, London, UK
64 Reference 19, side 4
65 GUNSTON, Bill: 'Night Fighters' (Patrick Stephens, Cambridge, 1976)
66 Bowen Papers, EGBN 2/5. Archives, Churchill College, Cambridge, UK
67 Air Ministry: 'An outline of air defence organisation', A.P. No. 3145/1 (Air Defence Pamphlet Number One), 1942, available AIR 10/3757, Public Record Office, Kew, London, UK

68 Air Ministry: 'Radiolocation systems of raid reporting', A.P. No. 3145/2 (Air Defence Pamphlet Number Two), 1942, Available AIR 10/3758, Public Record Office, Kew, London, UK

69 Air Ministry: 'Observer system', A.P. No. 3145/3 (Air Defence Pamphlet Number Three), 1942, available AIR 10/3759, Public Record Office, Kew, London, UK

70 Air Ministry: 'Telling and plotting', A.P. No. 3145/4 (Air Defence Pamphlet Number Four), 1942, available AIR 10/3760, Public Record Office, Kew, London, UK

71 Air Ministry: 'Standard notes for RDF training', CD.0436A, available AIR 10/4131, Public Record Office, Kew, London, UK

72 'Proceedings of Royal Commission on Awards to Inventors', copy available Science Museum Library, London, UK, see seventh day of Hearing

73 WATSON-WATT, Robert: 'Three steps to victory' (Odhams Press Ltd, London, 1957)

74 ROWE, A. P.: 'One story of radar' (The University Press, Cambridge, 1948)

75 Air Ministry: 'The origins and development of operational research in the Royal Air Force' (HMSO, London, 1963)

76 Air Ministry: 'Radio counter-measures', A.P. 3407 (Signals Volume VII), 1950, copy available AIR 41/13, Public Record Office, Kew, London, UK

77 STREETLY, Martin: 'Confound and destroy' (Macdonald and Janes, London, 1978)

78 AVIA 26/22: Public Record Office, Kew, London, UK

Chapter Six

Significance of the magnetron

The cavity magnetron, which was developed from the work of J. T. Randall and H. A. Boot at Birmingham University early in 1940, was a valve oscillator whose resonant tuned circuit was an integral part of the valve and whose physical construction permitted easy and efficient coupling to a radar antenna. The emergence of the high-power cavity magnetron made possible the development of effective microwave radar transmitters. Today the magnetron is widely used in pulse radars, but it is just one of the family of crossed-field valves available, which also includes the amplitron and the dematron [1]. (The term 'crossed-field' derives from the fact that, in these valves, the unvarying component of the electric field, E, and the magnetic field, B, are at right angles to each other.) Microwave valves are divided into crossed-field and linear-beam types, the kylstron and travelling-wave-tube being examples of the latter type.

The discovery of Randall and Boot was helped by their examination of the action of the cavity resonators of the klystron and by their not being unduly prejudiced by the existing literature on the magnetron; they referred to the latter fact as follows [2]:

> Fortunately we did not have the time to survey all the published papers on magnetrons or we would have become completely confused by the multiplicity of theories of operation.

It is of interest to examine the background to the development of the British cavity magnetron and to consider some of the more important work on magnetrons which had taken place in other countries. Firstly, however, it is worth looking briefly at the emergence of the magnetron valve itself some twenty years before the achievement of Randall and Boot in Professor Oliphant's laboratory at the University of Birmingham. Albert W. Hull, writing in 1921[3] remarked that

> 'Magnetron' is a Greeko-Schenectady name, as Mr Lee De Forest calls it, for a vacuum electronic device which is controlled by a magnetic field.

A. W. Hull, of the General Electric Company of the United States, had been involved during World War I in the development of the 'dynatron', a form of triode valve which possessed a negative resistance characteristic. The 'pliodynatron' was a more efficient form of dynatron. John Scott-Taggart wrote[4]:

> In British Patent 130400 (Feb. 15/18)*, the G.E.C. (USA) described a modified form of pliodynatron in which a large number of turns of wire are wound round the tube of a dynatron. Modulating currents, such as those from a microphone, produce a magnetic field which causes the electrons to move towards the anode in a special path which results in many electrons striking the anode. This form of control is described as being used for wireless telephony.

This seems to have been a precursor of the magnetron of 1921, which was a valve with a cylindrical anode and a concentric thermionic cathode whose envelope was enclosed by a solenoid that generated an axial magnetic field. Hull saw the magnetic field as just another means of controlling the current between cathode and anode in a valve, other means being the control of filament temperature or the use of an electrostatic control grid. Hull[3], discussing the applications of the magnetron, wrote:

> The only purpose for which it is actually being used is as a 'synchronous detector' in continuous wave radio telegraphy, in the transoceanic receiving station of the Radio Corporation. In this case, the magnetron is a simple high-frequency valve, opened and closed at approximately signal frequency by a locally generated magnetic field, letting through first the positive peak of the signal and then the negative, giving an audible tone.

One could view the development of the magnetron as a generator of radio-frequency oscillations as occurring in three stages[5][6][7]. Firstly, cyclotron mode oscillations were investigated by August Zácek of Prague in 1924. Then, in 1924 also, Erich Habann of Jena, experimenting with a split cylindrical anode and greatly increased magnetic field, discovered a negative resistance type of oscillation. A third, and what has proved to be the most effective state of oscillation, is a travelling wave mode first studied by K. Posthumus[8] in 1934.

It may be helpful at this point to give some idea of the order of powers attained by various magnetrons. The first British commercially manufactured cavity magnetrons developed a pulse power output of about 10 kW at a wavelength of 10 cm with a pulse duration to cycle duration ratio of 0·001; this was soon increased to 100 kW and by the end of the war powers of over 2 MW at this wavelength had been obtained, while at a wavelength of 1·25 cm powers of over 50 kW were possible. Cyclotron-type magnetrons had given

* 15 February 1918.

continuous-wave powers of 100 W at 500 cm, approximately 1 W at 10 cm and detectable radiation at 0·6 cm. Habann type or negative resistance mode magnetrons had produced continuous-wave power outputs of up to 400 W at 50 cm and 80 W at 20 cm. The various mechanisms of operation of the magnetron oscillator are adequately covered by Collins[7], Harvey[9] and Spangenberg[10], while Pierce writing in 1961[11] provides a useful general overview of microwave valve development.

Professor August Zácek of Charles's University, Prague, made his discoveries on the magnetron when he was carrying out work in his Physics Department on the generation of very short radio waves. On the 31 May 1924, he submitted a patent application on his magnetron and on the 15 February 1926 a Czechoslovak Patent No. 20293 was issued to him[12]. The shortest waves that he generated had a wavelength of 29 cm.

Useful work in investigating the properties of the magnetron was carried out in Japan in the 1920's by Professor Kinyiro Okabe and Dr Hidetsugu Yagi. (Reference is made in section 4.6 to Professor Okabe's later work). Dr Yagi, a student and follower of Dr J. A. Fleming, returned, after his studies in London and Harvard, to the College of Engineering, Tohoku Imperial University at Sendai, Japan. He introduced there a course on the functioning of the diode and this course was directed by Professor Okabe[13]. Experiments with single-anode magnetrons and then with split-anode magnetrons were carried out.

Professor Okabe published his first paper on magnetron oscillations in 1927[14]. In 1928, Dr Yagi visited the United States, where he described to various groups the work of S. Uda on directional antennae and that of Professor Okabe on magnetrons, which had been carried out under his direction[15]. He visited the General Electric Company's laboratory at Schenectady, New York, and as a result of his visit an experimental transmitter and receiver operating at 400 MHz were built and worked successfully. Reference can be made to section 4.6 for later Japanese work on magnetrons.

Mention should be made of theoretical work carried out in the 1930's on the flow of electrons in diodes under high frequency conditions, work which was associated principally with the names of Benham, Llewellyn, Müller, North and Peters[11], [9]. While connected principally with the study of the high frequency limitations of triodes and pentodes, the diode equations provided invaluable background material to the investigation of microwave valves.

Before discussing the development of the British cavity magnetron, a few more examples will be given of work carried out in other countries. The SFR (Société Française Radioélectrique) laboratories in Paris carried out valuable research on magnetrons with multi-segment anodes in the 1930's. They succeeded in obtaining reasonable continuous-wave power (5 W to 10 W) at wavelengths of 10 cm to 20 cm for communication and detection purposes, while employing relatively low anode voltages and magnetic field strengths. H. Gutton and S. Berline were associated with this achievement[16]. The introduction by SFR in 1939 of oxide-coated cathodes was to prove of great benefit to the later British work. Oxide-coated cathodes were also used in magnetrons developed in the late 1930's by Professor Janusz Groszkowski and Dr Stanislau Ryżko of the State Institute of Telecommunications, Warsaw, Poland[17–21]. Groszkowski and Ryżko began, about 1936, to experiment with magnetrons as generators of centimetric waves. They experimented first with split-anode, glass envelope, valves and then with metal enclosed valves with multi-segment anodes. Waveguide matching of the antenna was used and parabolic reflectors were employed. Magnetron transmitters were tested for their suitability in telecommunications links and for distances of the order of 40 km along the Baltic seashore[22].

Mention has already been made in section 4.7 of the Russian demountable water-cooled cavity magnetrons produced between 1936 and 1937 by D. E. Malyarov and N. F. Alekseyev[23][24][25]. The cavity resonators resembled the British design. Continuous-wave powers of about 300 W at a wavelength of 10 cm and at an efficiency of about 20% were obtained with a four-cavity valve. Estimates of power output were made by observing the brightness of an incandescent lamp used as load and also by measuring the heat carried away by the cooling water.

A paper by O. H. Groos[26] is indicative of the work that was carried out in Germany in the 1930's. The magnetrons, including a demountable type, of Ernest Linder of the RCA company of the United States were of some note[27]. A split cylindrical anode one-quarter wavelength long was used. The anode acted as a resonant quarter-wave transmission line and so did not contribute unwanted inter-electrode capacitance, which would have limited the upper level of useable frequencies. The design also promoted efficient heat dissipation. A continuous-wave power output of 20 W at 8 cm wavelength with an efficiency of 22% was achieved.

At the beginning of World War II, the generators of microwaves then available comprised (apart from spark-generators) specially configured triodes, Barkhausen-Kürz valves, magnetrons and velocity modulated valves. The principle of velocity modulation had been first described by Heil[28] in Germany, but what became the cavity resonator 'klystron' was introduced by Russell Varian and the Hansen brothers at Stanford University[29][30]. While the klystron (low-voltage reflex klystron) was used extensively throughout the war as the local oscillator in microwave superheterodyne radar receivers, it is interesting to note that in Britain in 1940, demountable kylstrons existed, which were capable of giving several hundred watts output at wavelengths between 10 cm and 15 cm. Research on kylstrons in Britain was carried out by H.M. Signal School, Portsmouth and by Oliphant's laboratory at the University of Birmingham.

In August 1939 a number of members of the staff of the Physics Department of Birmingham University, led by Professor M. L. Oliphant, visited the Chain Home Station at Ventnor to obtain first-hand experience of radar equip-

Original magnetron of J. Randall and H. Boot of Birmingham, February 1940.

Anode of 6 segment cavity magnetron.

ment[2][31]. They returned at the end of September 1939 and started a programme of research for the Admiralty on centimetric waves. The following arrangement was set up in the Birmingham Laboratory. M. Oliphant, J. Sayers and R. Dawton studied the possibilities of using the klystron as a source of high power; P. Moon and R. Nimmo did theoretical work on the kylstron; E. Titterton worked on the construction of a 50 cm oscillator using the G.E.C. 'micropup' valves, while J. Randall and H. Boot were given two tasks, one of making miniature Barkhausen-Kürz valves as possible receivers and the other of extracting radio-frequency power from plasma oscillations in mercury discharge tubes. No success was attained in the latter work, but some Barkhausen-Kürz valves were made and, in an effort to obtain a suitable radio-frequency source to test them, Randall and Boot were led into their work on magnetrons.

Influenced, no doubt, by the research of their colleagues on the klystron, they concentrated on the high efficiency obtainable from low loss cavity resonators and their possible application to the magnetron. The form of resonator that they choose was a three-dimensional version of Hertz's wire-loop resonator[32], a cylindrical cavity with a slot down one side. They used the formula[33] for the resonance wavelength of a thin wire loop, $\lambda = 7.94 \times$ loop diameter, and so chose 12 mm for the resonator diameter; the slot dimensions were 1 mm wide by 1 mm deep. An anode block of six resonators (see Fig. 6.1(a)) was decided upon and a thick tungsten wire cathode (0.75 mm diameter) was used. Because a modulator was not readily available, and to prevent delay, it was decided to operate the magnetron as a continuous-wave generator. The copper anode block was machined in December 1939 and continuously evacuated high voltage rectifer valves were made for the magnetron's DC supply. The magnetron itself was also continuously evacuated and the first successful operation of the system occurred on the 21 February 1940. It was immediately obvious that the power output was considerable. The best estimate of the output was obtained by noting that low-pressure neon lighting tubes could be lit to a brilliancy equivalent to a power consumption of 400 W. It was feared at first that the oscillations were metric and not centimetric, but measurement, using a 10 ft long lecher wire system, showed that the wavelength was 9.8 cm.

The subsequent sealed off and improved versions of this magnetron were made by the General Electric Company at Wembley and E. C. S. Megaw[34] of GEC Research Laboratories Wembley, who had much experience of conventional magnetrons, played a rather important liaison role between GEC and Birmingham University. The part played by the General Electric Company may be summarised as follows[35][36].

Work of a general nature on magnetrons had begun at Wembley in 1931, but in December 1936 this became part of a specific development programme for the Admiralty. The programme for the Admiralty was concerned with the production, propagation and reception of wavelengths below 60 cm ('U.S.W.',

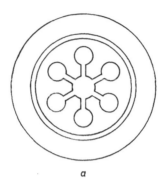

a

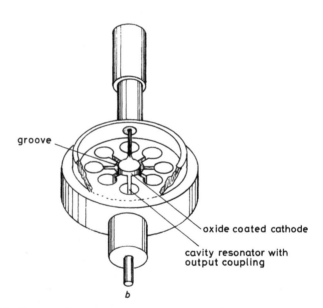

groove

oxide coated cathode

cavity resonator with
output coupling

b

Fig. 6.1   Cavity magnetron
a 6-segment anode block
b Internal construction of the E1189 8 cavity magnetron

or Ultra-Short-Wave, programme).* During a visit to the Signal School on the 30 November, 1938, E. C. Magaw with J. F. Coales of the Signal School carried out tests on the GEC E821 glass magnetron. A pulse output of 1½ kW at a wavelength of 37 cm was obtained, using a pulse width of 0·5 µs, a DC voltage of 8 kV and a magnetic field of 159 kA/m (2000 oersteds), which demonstrated that there was no difficulty in operating magnetrons under pulse conditions.

---

* The valve programme became known as C.V.D., which was derived from 'U.S.W. Communication: Valve Development'.

On the 4 May 1939, at a CVD meeting[35], advice was sought concerning the priority of work on magnetrons which could be used under pulsed conditions and at wavelengths of the order of 20 cm. Telecommunications Research Establishment were interested, but stated that they would not need such valves for some considerable time, whereas the Signal School did not want any work on the shorter wavelengths to hold up progress on their 50 cm programme. Megaw made a private visit to the FR Laboratories in Paris in June 1939. He reported on what he saw there to the CVD Committee and this information was conveyed to the DSR Admiralty. At the time of Megaw's visit, SFR had obtained a pulse output of 50 W at 16 cm with a tungsten filament magnetron which, when used in a radar set, had given a range of 10 km against ships.

In March 1940 GEC designed a four-segment glass magnetron, with internal resonant circuits and with a thoriated tungsten filament, which was intended for 10 cm–20 cm AI radar and which could give up to $\frac{1}{2}$ kW pulse output directly into a waveguide. On the 5 April 1940, GEC heard for the first time of the Birmingham work[35], and on the 10 April they visited Birmingham to discuss the whole question of magnetron production. It is pertinent to briefly divert from the magnetron narrative and to observe that the General Electric Company in late 1939, and independently of its commitment with the Navy, became involved in microwave development for airborne radar[36]. A visit from Henry Tizard on the 6 November 1939 followed by a meeting with Robert Watson Watt on the 29 December 1939 culminated on the 29 December 1939 with an official request from DCD* to take on a development contract for AI radar. It was decided to operate AI radar at a wavelength of 25 cm with a shorter wavelength of 12 cm or therabouts as a second objective. By April 1940 drawings were completed of the first pressurised transmitter using a pair of E1130 ('Millimicropup') valves. On the 18 April 1940, an output of 1 kW at 20 cm from two E1130 valves was reported and on the 26 April 1940 an output of 2 kW at 25 cm was reported. By July 1940, however, Megaw's air-cooled 10 cm resonant cavity magnetron was available for AI working. For a while, development work on AI continued on both 25 cm and 10 cm.

It was agreed between Professor Oliphant and Dr Clifford C. Paterson of GEC that Wembley should produce a sealed-off version of the Randall and Boot valve. A design (E1188) was finished on the 16th May 1940 and the first sample was sent to Birmingham on the 3rd July 1940. It gave the same continuous-wave performance as the original magnetron, but employed a much lighter electromagnet. Wembley also worked on the design of a lightweight magnetron for airborne use at 10 cm. This was achieved in the E1189 magnetron whose design, by Megaw, was completed on the 25th May 1940 and which first operated on the 29th June 1940. It used an existing 6 lb permanent magnet (M4735) which had been developed for GEC by Darwin's Ltd. in 1937 and which had been used in the 50 cm W/T communication set type 74. A

---

* Director Communication Development, Air Ministry.

crucial factor in the magnetron design was the use of a large diameter thoriated tungsten spiral cathode. The E1189 was capable of generating over ten times the pulse output of the original Birmingham design, while requiring a magnet which weighed less than one-tenth of that used with the original valve.

On the 9th May 1940, Dr Maurice Ponte of the SFR Laboratories, with the permission of the French Government, visited Wembley with samples of the SFR M16 resonant-segment magnetron. These magnetrons were an improvement on those seen by Megaw in June 1939; a large oxide-coated cathode was used and power outputs of 0·5 kW at 16 cm were obtainable from the valves.* Measurements were carried out immediately on the M16 and these showed, among other things, that oxide cathodes might be used at much higher anode voltages than previously thought possible and that secondary emission from the oxide cathode was an important factor in obtaining high anode current. Two E1189 samples were then made, one with thoriated tungsten cathode and the other with oxide-coated cathode.† The oxide-coated cathode was so successful that its employment was adopted in the E1190 triode (CV55) and other radar valves. On the 17th July 1940, the output of an E1189 operated with 8.5 kV and 119 kA/m (1500 oersteds) was measured as 12·5 kW into a water load‡ (pulse width was approximately 30 μs and pulsing rate was about 50 pulses/s).

A sample of the E1189 was sent to Birmingham and another to Swanage for testing by AMRE. An urgent demand for further samples developed and several copies were made by GEC using the chamber of a Colt revolver as a drilling jig! [6]. In August 1940, a revised design with 8 instead of 6 segments was made, see Fig. 6.1. This was the type brought to the United States with the Tizard Mission by Bowen.§ By September, 1940, a 10 cm 100 kW design was completed at Wembley. Some interesting information concerned with the early days of magnetron development are available in a CVD report of May 1941 [39].

Instabilities were found to occur in cavity magnetrons in which the valve's output would undergo a sudden change of wavelength [40]. This ability of the valve to operate on more than one frequency was referred to as 'moding' and the phenomenon as 'mode jumping'. J. Sayers of Birmingham found the solution to this in August 1941 by strapping alternate segments of the anode together. Unstrapped magnetrons continued to be manufactured and strapping was not regarded as being essential, but as being highly desirable in mass

---

* After the fall of France the development of the valve was continued in secret and 4 kW of peak pulse power in 1 μs pulses at 16 cm wavelength was obtained [37].
† For optimum performance and for a given size of magnetron, the dimensions of the cathode are rather critical. Luckily, these values had been discovered empirically.
‡ For a schematic of output circuit and water load, together with other details of the E1189, see Megaw [6].
§ The particular sample tested in Bell Laboratories, when X-rayed, caused some consternation; its number of resonant cavities did not quite correspond with the number in the drawings! [38].

produced valves. The real benefit of strapping was that the power level at which moding might occur was much higher for the same size of valve [41].

The German cavity magnetron was copied from the British design. A Stirling bomber carrying an H_2S system, which included a magnetron, crashed near Rotterdam on the 2nd February 1943 and its equipment fell into German hands. By the 22nd June 1943, a prototype of the equipment had been constructed.* The German LMS 10(G) magnetron used was a copy of the CV64. TRE Report T1858 [42] summarises the results of an examination of several samples of the LMS 10 (G).

The visit of the Tizard Mission to the United States with the subsequent setting up of the Radiation Laboratory, as discussed in Section 4.3.3, marked a new phase of growth not only in magnetron development, but also in microwave techniques.

In summary, one could say that the cavity magnetron evolved from the inventiveness of John Randall and Henry Boot, and became a working reality through the resources and experience of the GEC Laboratories at Wembley, assisted by critical information supplied by the SFR Laboratories of Paris. It could also be said that the emergence of the cavity magnetron in 1940 as an efficient generator of microwaves allowed, from that time onwards, an unrestricted growth in microwave radar technology to take place.

References

1 SKOLNIK, Merrill: 'Radar handbook' (McGraw-Hill, New York, 1970), Chapter 7

2 BOOT, Henry A. H., and RANDALL, John T.: 'Historical notes on the cavity magnetron', *IEEE Trans.*, 1976, **ED-23**, pp. 724–729

3 HULL, Albert W.: 'The magnetron', *J. Am. Inst. Elect. Engrs.*, 1921, pp. 715–723

4 SCOTT-TAGGART, John: 'Thermionic tubes in radio telegraphy and telephony' (The Wireless Press Ltd., London, 1921)

5 WATHEN, Robert L.: 'Genesis of a generator – the early history of the magnetron', *J. Franklin Inst.*, 1954, **255**, pp. 271–287

6 MEGAW, E. C. S.: 'The high-power pulsed magnetron: a review of early developments', *J. Instn Elect. Engrs*, 1946, **93**, Part IIIA, pp. 977–984

7 COLLINS, G. B.: 'Microwave magnetrons' (M.I.T. Radiation Laboratory Series, Vol. 6, McGraw-Hill, New York, 1948)

8 POSTHUMUS, K.: 'Oscillations in a split anode magnetron', *Wireless Engr.*, 1935, **12**, pp. 126–132

9 HARVEY, A. E.: 'High frequency thermionic tubes' (Chapman and Hall Ltd., London, 1944), 2nd edn.

10 SPANGENBERG, Karl R.: 'Vacuum tubes' (McGraw-Hill, 1948), Chapter 18

11 PIERCE, J. R.: 'History of the microwave-tube art', *Proc. Inst. Radio Engrs*, 1962, **50**, pp. 978–984

12 Letter from Dr Vladimír Adamec to Professor B. K. P. Scaife of Trinity College, Dublin

13 WHITE, W. C.: 'Some events in the early history of the oscillating magnetron', *J. Franklin Inst.*, 1952, **254**, pp. 197–204

* See seventeenth item of bibliography, Section 4.2.

14   OKABE, Kinjiro: 'A new method for producing undamped extra-short electromagnetic waves', *Proc. Imp. Acad. Japan*, 1927, **3**, p. 204. This paper was followed later in the Proceedings by another which discussed a more effective method of obtaining stronger oscillations, at wavelengths of 12 cm, 8 cm and 42 cm: OKABE, Kinjiro: 'Production of extra short electromagnetic waves by split-anode magnetron', *Proc. Imp. Acad. Japan*, 1926, **3**, pp. 514–515

15   YAGI, Hidetsugu: 'Beam transmissions of ultra short waves', *Proc. Inst. Radio Engrs*, 1928, **16**, pp. 715–741

16   GUTTON, H., and BERLINE, S.: 'Recherches Sur Les Magnétrons: Magnétrons S.F.R. Pour Ondes Ultra-Courtes', *Bulletin De La S.F.R.*, 1938, **12**, pp. 30–46

17   GROSZKOWSKI, Janusz and RYŻKO, Stanislaw: 'A new method of modulating the magnetron oscillator', *Proc. Inst. Radio Engrs*, 1936, **24**, pp. 771–777

18   GROSZKOWSKI, J., and RYŻKO, S.: 'Die Verteilung des elektrostatischen Feldes in Schlitzanodenmagnetronen', *Hochfreq Tech. Elektroakust*, 1936, **47**, pp. 55–58

19   GROSZKOWSKI, Janusz and RYŻKO, Stanislaw: 'Magnetrony z wewnetrznym obwodem oscylacyjnym', *Przegląd Radiotechniczny*, 1937, **15**, pp. 38–41 ('magnetrons with internal circuits of oscillation')

20   GROSZKOWSKI, Janusz and RYŻKO, Stanislaw: 'Magnetron z katoda tenkow ', *Przegląd Radiotechniczny*, 1938, **16**, pp. 17–19 ('Magnetron with oxide-coated cathode')

21   GROSZKOWSKI, Janusz and RYŻKO, Stanislaw; 'Metalowa lampa magnetronowa', *Przegląd Radiotechniczny*, 1939, **17**, pp. 73–75 ('Metal-envelope magnetron')

22   Letter from Dr Boleslow Orlowski to Professor B. K. P. Scaife of Trinity College, Dublin

23   Алексддв, Н. Ф. и Маляров, Д. Е.: 'Получение мощных колебаний магнетроном', *Zh. tekh. Fiz.*, 1940, **10**, pp. 1297–1300

24   ALEKSEEV, N. F., and MALAIROV, D. E.: 'Generation of high-power oscillations with a magnetron in the centimeter band', *Proc. Inst. Radio Engrs*, 1944, **32**, pp. 136–139 (translation of Reference 23)

25   Reference 7, pp. 7, 8

26   GROOS, O. H.: 'Die Erzeugung Von Zwergwellen mit dem Magnetfeldröhrensender', *Hochfreq Tech. Elektroakust*, 1938, **51**, pp. 37–43

27   LINDER, Ernest G.: 'The anode-tank-circuit magnetron', *Proc. Inst. Radio Engrs*, 1939, **27**, pp. 732–738

28   ARSENJEWA-HEIL, A., and HEIL, O.: 'Eine neue Methode Erzeugung kurzer, ungedämpfter, elektromagnetisher Wellen größer Intensität', *Z. Phys.*, 1935, **95**, pp. 752–762, translation available: *Electronics*, 1943, **16**, pp. 164–178

29   VARIAN, R. H., and VARIAN, S. F.: 'A high frequency amplifier and oscillator', *J. Appl. Phys.*, 1939, **10**, pp. 321–327

30   'The very important klystron', *The Lenkhurt Demodulator*, 1959, **8**, pp. 345–352

31   BOOT, Henry, A. H., and RANDALL, John T.: 'Development of the multi-resonator magnetron in the University of Birmingham (1939–1945)', (1945), communicated by Dr H. A. H. Boot to Professor B. K. P. Scaife of Trinity College, Dublin

32   AITKEN, Hugh G. J.: 'Syntony and spark' (John Wiley, New York, 1976), p. 54 *et seq.*

33   MACDONALD, H. M.: 'Electric waves' (University Press, Cambridge, 1902), pp. 7 and 112 (MacDonald derives for the fundamental wavelength of a circular resonator an expression $7.95D$, where D is the diameter of the resonator)

34   MEGAW, Arthur Stanley: 'A backgroom boy' (W. Erskine Mayne Ltd., Belfast, 1960)

35   Research Laboratories of the General Electric Company. Report No. 8717: 'The high power pulsed magnetron: notes on the contribution of G.E.C. Research Laboratories to the initial development', Wembley, 30th August 1945, available Bowen Papers EGBN 3/5, Archives, Churchill College, Cambridge, UK

36 Research Laboratories of the General Electric Company. Report No. 8719: 'G.E.C. development of centimetre wavelength equipment for airborne use', Wembley, 1st September 1945, available Bowen Papers EGBN 3/6, Archives, Churchill College, Cambridge, UK

37 PONTE, M.: 'Sur Des Apports Francais A La Technique De La Détection Électromagnétique', *Annls Radioelect.*, 1945, **1**, pp. 171–180

38 Recounted by Dr E. G. Bowen, 31.1.1979

39 AVIA 7/1275: Public Record Office, Kew, London, UK

40 BOWEN, E. G.: 'A textbook of radar' (University Press, Cambridge, 1954), 2nd edn., p. 50 *et seq.*

41 Private communication, Dr H. A. H. Boot to Professor B. K. P. Scaife of Trinity College, Dublin

42 AVIA 26/86: Public Record Office, Kew, London, UK

Chapter 7

Conclusions

In the previous chapters, an attempt has been made to record the salient events behind the emergence of radar in various countries in the 1930s. This has been done in greater detail for some countries than for others, but the amount of space devoted to any particular country does not in any way reflect a weighted opinion of the relative importance of the achievements within that country; rather, it does underline the linguistic limitations of the writer. An examination of dates will show that, even though the normal channels of scientific communication were purposely closed on the subject, the critical developments in each case occurred almost simultaneously. No particular significance will be attached to this 'simultaneous discovery' of radar, but a definite conclusion will be drawn, namely that, irrespective of the pressures of a threatened war, the emergence of radar by the end of the 1930s was inevitable.

Before setting forth any further opinions or conclusions, one point must be stressed. W. B. Lewis, in making comment on a draft history of British radar, remarked [1] that quite a number of different histories of radar should be written. Likewise, one may view the events of radar history from a number of viewpoints, such as a military historian, as a student of electronic warfare, or as an analyst of the role of centrally organised scientific research and technical development. Thus, one can truly say that there exist many different sets of conclusions. The conclusions drawn will fit the tenor of the previous chapters, which were mainly concerned with the technology of radar, its emergence and its early development.

Some of the key technological achievements of the century, such as the utilisation of atomic energy and the perfection of high-altitude rockets, are usually associated with the Second World War. Included also is radar. While the war enhanced its development, nevertheless, and choosing but one country, Great Britain, as an example, radar technology had advanced considerably before the outbreak of hostilities. Furthermore, one can point both to the Radio Corporation of America and to the Société Française Radioélectrique who, as early as 1935, were experimenting with microwave radars for non-military obstacle detection purposes. Again, the emergence of high-performance and long-range aircraft at the end of the 1930s would, in the absence of war, have promoted the development of sophisticated radio-navigational aids and, since pulse methods were already being experimented with, it is difficult to see that these aids would not have included radar.

Analysis of the growth of radar falls naturally into three subdivisions, namely, the events which gave rise to it, the manner of its development, and, finally, its impact on technology in general. Although this historical study has been concerned primarily with the birth of radar, some brief comments will be made on all three areas.

Just as television, whose technology is akin to that of radar in many respects, became a working reality in the late 1930s, so too did radar. It was assisted on its way in several countries by military policies, but, like television, it would have emerged in any case. The possibility of radar is inherent in the properties of electromagnetic waves, and it is no surprise therefore that in three countries in particular, Great Britain, the United States and Italy, those intimately involved at its inception had been active in studying radio-wave propagation. It is probably true to say, also, regarding those people who experimented with radar type devices or filed patents on them before 1930, that they should be given due credit for their foresight, but that their ideas were too far ahead of the necessary supporting technology; in this latter category one need only consider thermionic valves, cathode-ray tube displays and sensitive wideband receivers.

D. S. L. Cardwell has written [2]:

> (Francis) Bacon* has often been criticized for his suggestion that quite ordinary people should be able to undertake scientific research . . . But he was not wrong, he was clearly right; every large modern research laboratory is a confirmatory instance. In fact, the great discovery of the nineteenth century that ordinary talents can be effectively harnessed for the process of discovery is a vindication of Bacon's judgement in this matter.

This statement has relevance here. While modern precision radar has been made possible by the work of talented theoreticians, the emergence of radar was due to equally talented but practically oriented people, who were competent in practical radio electronics and who understood the methodology of radio propagation measurements.

The scale of development of radar in any particular country was dependent on the resources of that country and then on Government policy and on the level of cooperation between Government, Armed Services, research institutions and industry. Hitler's decision early in the war to prohibit research which could not yield more or less immediate results had an inhibiting effect on

* 1560–1626.

German radar development. Robert Watson-Watt has recounted an incident that occurred in 1940 [3] when Lord Beaverbrook, the Minister for Aircraft Production, informed him that he had decided to discontinue the development of radar because of reports that it was adversely affecting fighter tactics! Political support and direction and willing co-operation between all parties involved were major factors determining the ultimate efficiency of radar development and radar employment in a particular country. These factors appear to have been strong in Britain and the United States and rather weak in Italy and Japan. It is impossible to quantify in any way the effort made in the promotion of any radar programme. One can only surmise that, as in the broader arena of war, strokes of luck and strokes of genius notwithstanding, ascendancy came to the side whose final balance sheet of resources showed the greater credit.

Some of the effects of radar are obvious. It has changed the whole conduct of warfare on land, sea and in the air; it has contributed to safety in the air and on sea and has extended the science of measurement. There have been consequences, however, that are not so readily recognisable. Research on germanium and silicon was boosted by the need for efficient crystal mixers for radar receivers. This research, carried out principally in the United States, was continued through the war years and led to an understanding of the controlled injection of impurities into semiconductors and in 1948 to the emergence of the transistor. Ruggedness and reliability in electronic components and equipment today owe much to the severe demands placed on operational radar equipments in the various theatres of war. On the theoretical side, the study of threshold signals and of noise and random processes in systems progressed as a matter of necessity. Automatic search, ranging and tracking radars promoted advances in control electronics and control theory. The whole modern microwave art developed from radar.

In conclusion, it can be said that radar was a natural, but by no means a foregone, development in any country which had an active radio industry and radio tradition. In the context of the world of modern electronics, radar today might be regarded as commonplace. Its arrival in the 1930s added a whole new dimension to electrical science and technology.

References

1 AVIA 10/348: Public Record Office, Kew, London, UK
2 CROMBIE, A. C.: 'Scientific change' (Heinmann, London, 1963), p. 622 (the book is based on a symposium 'The structure of scientific change', held at Oxford from the 9–15th July 1961)
3 WATSON-WATT, Robert: 'Three steps to victory' (Odhams Press Ltd., London, 1953), p. 231

Appendixes

Appendix A: Reciprocity principle

The reciprocity principle has applications in many areas of physical science including those pertaining to acoustical, mechanical and electromagnetic systems.

Lord Rayleigh in his 'Theory of sound' [1, 2] formulated and discussed various cases of reciprocity. H. A. Lorentz enunciated in 1895 a reciprocal theorem for the electromagnetic field, and John R. Carson of the Bell Laboratories was asked in 1923 [3] to enquire into its validity in radio-communication. In 1924 he published a generalised reciprocal theorem which may be stated as follows [3]:

Let a distribution of impressed periodic electric intensity F' $e^{i\omega t}$ = $F'(x,y,z)$ exp $(i\omega t)$ produce a corresponding distribution of current density u' $e^{i\omega t}$ = $u'(x,y,z)$ exp $(i\omega t)$, and let a second distribution F'' exp $(i\omega t)$ produce a second distribution u'' exp $(i\omega t)$ of current density. Then

$$\int (F'.u'') \, dv = \int (F''.u') \, dv$$

the volume integration being extended over all conducting and dielectric media. F and u are vectors and the expression $(F.u)$ denotes their scalar product.

Much work has been done in analysing the conditions under which reciprocity applies. Some of the limitations in the use and interpretation of the principle have been discussed by Carson [3] and by Millington [4]. A common example in communication practice where reciprocity may not apply is that of a high-frequency link where the radio waves passing through the ionosphere are subject to the action of the earth's magnetic field.

The radio or radar engineer dealing with antennae is concerned with reciprocity because of his ability thereby to deduce the receiving properties of an antenna from its transmitting properties and vice versa. The realisation of this equivalence of properties was, historically, a major step forward in antenna theory and measurements [5, 6]. Before this, receiving antennae were sometimes analysed by questionable methods.

The reciprocity theorem for antennae can be stated as follows [7]:

> Let A and B be two antennae having any orientation and displacement relative to one another. If a voltage V applied to A causes a current I at a given point in B, then the voltage V applied at this point in B will cause a current I in A at the point where V was originally applied.

More concisely, one might say:

> If A and B are, respectively, the transmitting and receiving antennae of a radio link then, if the roles of A and B are interchanged (assuming B to have adequate power handling capability), there will be no alteration in the performance of the link.

In a monostatic radar system, where the same antenna is used for transmission and reception and where one can assume a linear, isotropic and passive medium of propagation, one can say because of the reciprocity principle that the polar diagram, the power gain, and the impedance of the antenna will be the same for both transmission and reception.

References

1 STRUTT, John William: 'Theory of sound, Vol. I' (Macmillan and Co., London, 1894), 2nd edn., pp. 98, 99, 150–157
2 STRUTT, John William: 'Theory of sound, Vol. II' (Macmillan and Co., London, 1896), 2nd edn., p. 145
3 CARSON, John, R.: 'Reciprocal theorems in radio communication', *Proc. Inst. Radio Engrs*, 1929, **17**, pp. 952–956
4 MILLINGTON, G.: 'The reciprocity principle in radio propagation', *Marconi Rev.*, 1971, **34**, pp. 235–252
5 CHU, L. J.: 'Growth of the antennas and propagation field between World War I and World War II, Part I – Antennas', *Proc. Inst. Radio Engrs.*, 1962, **50**, pp. 685–687
6 BURGESS, R. E.: 'Aerial characteristics', *Wireless Engr*, 1944, **21**, pp. 154–160
7 BOWEN, E. G.: 'A textbook of radar' (University Press, Cambridge, 1954), 2nd edn., p. 222

Appendix B: Retarding field generators

The historical importance of retarding field oscillators is that they were used in some of the early radar experiments and in early microwave equipment; indeed, as related in Chapter 6, Randall and Boot at Birmingham University in 1939 were investigating the properties of miniature Barkhausen–Kürz valves.

Retarding field generators, a particular type of electron oscillator, are associated with Barkhausen and Kürz, with Gill and Morrell and also with

Hollmann [1]. In the oscillator, the triode grid is maintained at a high positive voltage and the anode at a slightly negative voltage.

Since the observations of Barkhausen and Kürz in 1920, there was available a means of generating frequencies as high as several gigahertz. Barkhausen and Kürz made their discovery when testing some transmitter valves; they found that oscillations of very high frequency could be maintained in a circuit connected between grid and anode, grid and cathode or between anode and cathode.

The performance of the triode valve deteriorates with increasing frequency owing to valve lead inductances, inter-electrode capacitances and electron transit-time effects. The latter, the most serious factor of the three, is turned to advantage in the positive grid oscillator. No completely satisfactory theory of retarding field oscillators has been given. This point has been discussed by Kelly and Samuel [2]. In many cases, pictorial explanations are resorted to. The basic physical mechanisms in the production of oscillations is as follows. A high positive voltage is applied to the grid and a slightly negative one to the anode. Electrons from the cathode are attracted towards the grid; many of them strike the grid structure and are returned towards the cathode via the grid–cathode circuit; others pass through the grid structure but are driven back by the negative potential of the anode and return towards the grid; they may oscillate back and forth about the grid before striking it. The electrons that suffer retardation in the grid–anode and grid–cathode spaces deliver energy to the electric field and help to set up and to maintain oscillations. The oscillations of the various electrons about the grid structure are not entirely at random, but become synchronised so as to produce useful power.

The wavelength of the oscillations generated in a Barkhausen–Kürz circuit is given approximately by [3]:

$$\lambda \text{ (centimetres)} = \frac{670d}{\sqrt{E_g}}$$

where d is the diameter of the anode in centimetres and E_g is the grid voltage. Hence, the higher the grid potential and the smaller the diameter of the anode, the higher the frequency generated.

λ depends on the geometry of the valve and on the electrode potentials. Harvey [1] gives formulae for arbitrary spacings of cathode–grid–anode, coupled with variations in anode potential.

The Barkhausen–Kürz oscillations are determined solely by the potential field existing within the valve and by the dimensions of the valve, and not by the external circuit, which was generally an open-wire transmission line (Lecher wires) connected across anode and grid. The load was connected by attaching sliding contacts to the parallel wires. What are referred to as Gill–Morrell oscillations [1, 4] occurred when the contacts were moved in closer to the valve. Then the frequency of oscillation increased and was variable, depending only on the effective length of the Lecher wire system.

Triodes were designed specifically for use as retarding-field generators. Barkhausen–Kürz oscillations could be produced in conventional triode valves, but special valves for positive-grid operation were constructed. These differed from the normal triode principally in the construction of the grid and in the arrangement of the leads. A typical performance obtainable from a valve operating at 600 MHz, with the grid set at 500 V, was an output power of 6 W at an efficiency of about 6%.

References

1 HARVEY, A. F.: 'High frequency thermionic tubes' (Chapman and Hall Ltd., London, 1944), 2nd edn., Chapter 3
2 KELLY, M. J. and SAMUEL, A. L.: 'Vacuum tubes as high frequency oscillators', *Electl Engng*, 1934, **53**, pp. 1504–1517
3 TERMAN, Frederick Emmons: 'Radio engineering' (McGraw-Hill, New York, 1937), 2nd edn., p. 388
4 EASTMAN, Austin V.: 'Fundamentals of vacuum tubes' (McGraw-Hill, New York, 1941), 2nd edn., p. 441

Appendix C: Super-regenerative receivers

Super-regenerative receivers are suitable for the reception of RF pulses such as those used in pulse radar. They were employed in some of the early radar experiments and also during the Second World War in IFF (Identification Friend or Foe) and interrogator-responder beacon systems.

The context in which super-regenerative receivers will be briefly discussed here is that of the employment of triode valves in the late 1920s and early 1930s for the reception of relatively weak telephonic or telegraphic signals of rather high frequency. It should be borne in mind that, at that time, useful radio-frequency amplification by conventional triodes was limited to frequencies below about 20 MHz. A super-regenerative circuit, however, could provide voltage amplification of the order of 10^6 with detection in the frequency range of approximately 30 MHz to 600 MHz.

A word about terminology: the terms super-regenerative receiver and super-regenerative detector are synomymous; likewise the reaction or regenerative receiver is synonymous with the regenerative detector. It is fitting to first outline the action of the latter.

The discovery of the principle of the regenerative receiver is usually associated with E. H. Armstrong of the United States. The circuit appeared about 1912 and may be fairly attributed to Armstrong, de Forest and Langmuir of the United States and to Meissner of Germany [1]. In the receiver, positive feedback from the anode circuit to the tuned grid circuit of the triode enhanced amplification and a grid leak configuration was usually used for detection [2].

Feedback was maintained at a level just below the threshold of oscillation. When regenerative receivers were in common use for the reception of broadcast transmissions, it frequently occurred that the circuit, because of too much positive feedback, broke into oscillation and caused interference among neighbouring listeners.

In the super-regenerative receiver, oscillations are allowed to build up, but are quenched periodically at a frequency of the order of 20 kHz or greater (in British 200 MHz radar beacon receivers, the quenching frequency was 300 kHz). Quenching was done either by a separate quenching oscillator or by an internal self-quenching *RC* circuit; mechanical interrupters were on occasion used.

The output of the super-regenerative receiver was obtained by rectification of the pulses of oscillation built up in the *LC* circuit of the valve and usually by means of the nonlinear action of the grid-leak condenser.

Super-regenerative detection is a highly non-linear phenomenon. However, the broad principle of its operation is explained by Appleton as follows [3]:

> For the reception of ultra-short waves, it is possible to employ a very sensitive method known as *super-regeneration*. A valve oscillating nearly at the transmitted frequency has a potential swing of large amplitude applied to its grid (or anode) at a frequency which is just outside the range of audibility. The high-frequency (hf) oscillation is stopped once for each cycle of this auxiliary oscillation because the grid potential becomes too negative to maintain it. It is said that the hf oscillation is 'quenched' with the frequency of the *quenching oscillation*. During the next half of the quenching cycle, the hf oscillation builds up exponentially as given by $E_0 e^{+Kt}$, where E_0 is any E.M.F., however small, which happens to be present due to circuit fluctuations. If there is any rectifying action in the valve the anode current flowing depends on the length of time for which the hf oscillations are allowed to persist. If now a signal of amplitude e_1 is present in the circuit, then the growth of the hf oscillations is given by $E_1 e^{+Kt}$, and it will reach a given (saturation) amplitude in a time which depends on E_1. Since the steady state rectified current depends on the time for which the hf oscillation lasts, it will in turn depend on the signal E.M.F., e_1, and it will fluctuate according to the modulation of the transmitted signal.

The terms 'linear mode' and 'logarithmic mode' are used to describe the valve circuit's action, the former applying when quenching occurs before oscillations are limited by curvature of the valve characteristics, and the latter when the oscillations are allowed to reach a limiting amplitude.

Macfarlane and Whitehead [4] provide an analysis of a super-regenerative receiver in the linear mode, with applications to radar. A quite detailed treatment of super-regeneration and of a typical radar super-regenerative circuit is given in the British 'Inter-Services Radar Manual' [5].

References

1 SWINYARD, William O.: 'The development of the art of radio receiving from the early 1920's to the present', *Proc. Inst. Radio Engrs*, 1962, **50**, pp. 793–798
2 GLASGOW, R. S.: 'Principles of radio engineering' (McGraw-Hill, New York, 1936), pp. 372 *et seq.*
3 APPLETON, E. V.: 'The physical principles of wireless' (Methuen and Company Ltd., London, 1945), 7th edn., p. 95
4 MACFARLANE, G. G. and WHITEHEAD, J. R.: 'The super-regenerative receiver in the linear mode', *J. Instn Elect. Engrs*, 1946, **93**, Part IIIA, pp. 284–286
5 Ministry of Supply: 'Inter-Services radar manual', A.P. 1093E, London, 1950, 2nd edn., pp. 462–477

Appendix D: Watson-Watt's two memoranda

Two memoranda, which were critical to the emergence of radar in Britian and which were placed before the Committee for the Scientific Survey of Air Defence, were composed by Watson-Watt in January/February 1935. The first memorandum, reproduced in (1) below, arose from a request made on 18th January 1935 by Wimperis, the Director of Scientific Research in the Air Ministry, to Watson-Watt for advice on the possibility of a 'death ray'. The conclusions reached in this memorandum led to the compiling of the second one, which was submitted in a first draft form to Rowe, Secretary of the Committee, on 12th February 1935. The final draft, reproduced in (2) below and entitled 'Detection and location of aircraft by radio methods' was submitted on 27th February 1935.

(1) The problems of delivering, by radio beam, energy in amounts adequate for disabling man or craft may be separated into two classes, for quasi-fixed and rapidly-moving targets, respectively. The most favourable case of the simpler class will be examined first, in a roughly quantitative way.

Suppose it is desired to produce physiological disablement in a man remaining for so long as ten minutes in the field of the beam, at a distance of 600 metres. He may be treated as composed simply of 75 kg of water. It is necessary to deliver, over his projected area of 1 sq metre (2 metres high × ½ metre wide) enough energy to raise his temperature by at least 2 °C. Making the very unduly favourable assumptions of black body absorption, of 100% efficiency of conversion, without increased cooling by radiation and convection, the reasonable assumption of negligible absorption en route, and the unfavourable assumption of no aid from resonance in draining an area of front greater than the nett projected area, it is necessary to deliver 1.5×10^4 cal/gm per minute. Using a simple half-wave radiating antenna, it would be necessary to radiate approximately five million kilowatts to give this flux per square metre at 600 metres distance. There is a general concensus of opinion that the economic limit of 'gain' obtainable from an antenna array, constituting a directive reflector system, is about 22 decibels, the use of such a system would enable the radiated power to be reduced to thirty thousand kilowatts. It is not impossible to visualise an 'un-economic' stacked array which would give a further gain of under 20 decibels, reducing the radiated power required to 300 kilowatts.

The maximum power radiated from existing commercial senders, on wavelengths of the order of 15 metres, is about 50 kilowatts, and the inefficiency of conversion to radio-frequency energy involves a supply of roughly twice this amount. Were such a wavelength used, however, the antenna array, which must have an aperture of the order of ten wavelengths, would inevitably become fixed in orientation and would be restricted to a gain much nearer the 22 than the 42 dB specified. Thus on 15 metres wavelength an input power of tens of thousands of kilowatts would be required and the beam would be virtually fixed.

On wavelengths of the order of 5 metres, 200 kW of high frequency energy have been radiated from a single unit. It is not thought impracticable to construct an installation to radiate 100 kW at 3 metres, leaving a threefold deficiency from requirements to be (partially) offset by quasi-resonance which, for a man standing erect, has been measured as occurring at about 3 metres wavelength. Here the full 'uneconomic' gain would be obtainable from a practicable structure of great expense, a beam capable of re-orientation would require approximately ten times the radiated power. In this wavelength range, the input power may be taken as about five times that radiated, so that inputs of 500 to 5000 kW are required.

The technique of radiating substantial power on wavelengths under one metre, enabling good concentration to be obtained from arrays of moderate linear dimensions has not, so far as I know, been developed beyond the radiation of 25 watts only. Such investigation and development lie between this figure and the hundreds of kilowatts required to meet the requirement and to provide a 'factor of safety'.

It must be repeated that these figures depend on the target remaining within the field of a beam, not worse than 5° in semi-angle of divergence, i.e. within a transverse range of 100 metres at 600 metres distance, for ten minutes. The more practical assumption of one minute sends the required power up tenfold and seems to remove the whole scheme outside practicable limits.

It may not be superfluous to point out that any reflector system used for this scheme would have to be much better than any yet used, since in these the minor lobes on the rear of the polar diagram would be sufficient to incapacitate the operator of the projector.

It is of some interest to note that the figures obtained above are in substantial agreement with those computed for a 'searchlight' heat projector to give the same results.

It would be necessary to radiate some 250 kW from an arc projector to meet the requirement, in the absence of absorption. This involves a 1230 kW input – again very much beyond present practical limits.

The limiting factor in the searchlight loading is believed to be the resistance and thermal dissipation of the electrodes. That in the radio wave generator is also the thermal dissipation in the electrodes, which cannot be made large enough for rapid removal of heat without sending the lower limit of the wavelength that can be generated much too high for the purpose now under discussion.

It is possible, though not highly probable, that a spark generator would enable higher powers to be generated at very low wavelengths, but so far as I know the efficiency of conversion to radio frequency is not of widely different order in the valve and spark cases, and I find it extremely difficult to believe that a rotary-gap spark generator for 300 kilowatts at wavelengths under a meter is practically attainable, since the input power would, I believe, have to exceed 8000 kilowatts.

It does not appear useful to deal in the corresponding way with the quickly moving aircraft target until the general conditions of this case have been examined.

The bombing aircraft of the immediate future is assumed to be an all-metal monoplane, with cowled engine and screened ignition system. For any frequency that can be projected in the manner contemplated the pilot, the magneto and the rest of the ignition system must, therefore, be regarded as enclosed in a very effective Faraday cage. Moreover, even were the magneto incompletely protected skin effect would practically confine eddy currents at this frequency to the outer layers of the permanent magnet and demagnetisation would not be at all complete.

It is then provisionally concluded that pilot and ignition system are or can readily be made immune. There remains for attack the radio installation and the few stay and control wires that cannot be enclosed within the main metal sheathings, and those sheathings themselves. This latter, the mainbody and wing metal, may be treated as of such ample section and as so extremely well ventilated as to be immune, the radio installation is, at the worst, susceptible only to destruction of itself without involving the essential control of the craft, and the exposed stays alone remain for examination. They are so well ventilated that, without any elaborate calculation, it may reasonably be concluded that a beam of the short wavelength to which they would resonate cannot be kept trained on them for a period sufficient to heat them to a point much below fusing point – at which their elastic properties would be sufficiently impaired to ensure structural failure. Numerical work will be attempted if, on consideration, these conclusions as to the very effective 'electrical armour-plate' of the modern bomber are shown to be substantially in error.

Meanwhile, attention is being turned to the still difficult, but less unpromising, problem of radio-detection as opposed to radio-destruction, and numerical considerations on the method of detection by reflected radio waves will be submitted when required.

(2) Memorandum by Mr R. A. Watson-Watt, 27th February, 1935

Detection and Location of Aircraft by Radio Methods

It appears unsafe to base any method for the detection or location of enemy aircraft on any of the primary radiations from the craft. Lamps and radio senders will not be used on a scale permitting detection. Sound from engine propeller and structure is steadily being reduced, and is in any case subject to extreme vagaries in propagation which, while still permitting detection, may prevent location. Electro-magnetic radiation from ignition systems is readily screened to very low values. Infra-red radiation from engine is so heavily and variably absorbed in a water-laden atmosphere as to make it an unreliable indicator.

Of the secondary radiations, excited by 'illuminating' the craft by ground installations emitting light, heat, sound or radio-waves, the first two are excluded by atmospheric absorption (especially in cloudy conditions). The use of sound waves above the audible limit has some attractions, but the low-power rating of emitters and the low velocity of propagation – a small multiple of the speed of the craft – are against it. It appears, in sum, that the only moderately promising method of detection and location is that of secondary 'wireless' radiation.

The most attractive scheme is that of setting up zones of short-wave radio 'illumination' through which the approaching craft must fly. The most desirable form of this scheme will be discussed in more detail.

Let it be assumed that the typical night-bomber is a metal-winged craft, well bonded throughout, with a span of the order of 25 metres. The wing structure is, to a first approximation, a linear oscillator with a fundamental resonant wave-length of 50 metres and a low ohmic resistance. Suppose a ground emitting station be set up with a simple horizontal half-wave linear oscillator perpendicular to the line of approach of the craft and 18 metres above ground. Then a craft flying at a height of 6 km and at 6 km horizontal distance would be acted on by a resultant field of about 14 millivolts per metre, which would produce in the wing an oscillatory current of about $1\frac{1}{2}$ milliamperes per ampere in sending aerial. The re-radiated or 'reflected' field returned to the vicinity of the sending aerial would be about 20 microvolts per metre per ampere in sending aerial.

It is at present common practice to put 15 amperes into the sending aerial, giving a received field, from the re-radiating craft, of the order of a tenth of a millivolt per metre after generous allowance for losses. This value can in effect be more than doubled in the pulse technique without overload in the transmitter. If, further, the method proved so reliable that general 'illumination' could be abandoned and a thick sheet of 'illumination' at a convenient inclination could be relied on, this field could be increased at least tenfold by the provision of a suitable beam array, of practicable dimensions and cost, at the ground station. It will be observed that this last improvement is obtained at some sacrifice of easy watch, as an indication is obtained only while the craft is 'illuminated', in the one case the illumination is weak flood lighting of a very large area, in the other it is strong searchlight illumination in an inclined sheet of small thickness.

It is not wholly fantastic to suggest that the span of the machine could be measured to aid identification, by a rapid sweep of the emitted radio-frequency, but without emphasis on this possibility it will be noted that the simpler scheme will lose in efficiency as the emitted frequency fails to fit the resonant frequency of the wing structure.

The resonance curve of the wing and fuselage structures will be very flat; this militates against easy span-measurement but in favour of easy distance-measurement; without change of radio frequency to fit the craft, a variation of two or three to one in span will not much affect sensitivity. On balance, however, it may be concluded that reflected fields of the order of a millivolt per metre are readily attainable at 10 km, rising by the use of an alternative height to the order of 10 millivolts per metre as the craft passes overhead at heights under 20 000 feet. These fields are about ten thousand times the minimum required for commercial radio communication, so that very large factors of safety indeed are in hand for ranges of the order of 10 miles at flying heights of about 20 000 feet.

If now the sender emits its energy in very brief pulses, equally spaced in time, as in the present technique of echo-sounding of the ionosphere, the distance between craft and sender may be measured directly by observation on a cathode ray oscillograph directly calibrated with a linear distance scale, the whole technique already being worked out for ionospheric work at Radio Research Station. In the examples already taken the reflected ray would return after 56 microseconds for 6 km horizontal distance and after 40 microseconds from overhead. I believe these times to be quite manageable within the technique, though they involve a very considerable shortening of the pulse durations now used (about 200 μsec), or an artifice, which we can certainly provide, for reading the time of return even when the reflected pulse is superposed on the primary timing pulse which has arrived at the receiver by a very short ground path. If we are not interested in distances over 300 km, or if other instrumental and propagational limitations prevent us from utilising the method up to such distances, then send a thousand pulses per second and obtain, by superposition of the successive images on a synchronised time base, a very easily visible sustained image permitting close measurement and even showing the advance of the craft. Some compromise pulse-frequency between 50 and 1000 would be selected after experimental trials.

It will be clear that the installation of three such receivers for time-delay measurement alone would enable the equations of position to be solved, by means which could be made partially or wholly automatic, for height and plan projection. The provision of a line of senders over a long front is not prohibitively difficult, since the polar diagrams are such as to permit substantial spacing and the echo-patterns are readily sorted out. Finally, the provision of two parallel lines, roughly perpendicular to expected line of approach, would give still more accurate positional data enabling speed and course to be measured with some precision. There are two main objections to the use of the radio-frequency discussed, to which the whole metal structure of the craft is nearly resonant. The technical one is that echoes from the ionosphere will appear on the received picture and will have to be discounted in observation. This is more than a mere inconvenience, in view of our existing knowledge of what to expect, and even this inconvenience is mitigated by the value of the ionospheric echoes as indicators that the gear is in good order. The time scale can be made very open for the first hundred kilometres – and it is not unreasonable to expect that the technique can be developed to operate on craft up that distance – and the first ionospheric echoes can be crowded into a stand-still period at the end of the time-base. But it is impossible to avoid the ionospheric echoes from, say, $(nk + x)$ km being read as from x km, where k is the distance corresponding to the recurrence frequency of the time-base and $n = 1, 2, 3, \ldots$ except by the exercise of intelligence and experience of ionospheric reflection, or by additional instrumental artifices.

The second objection is one of policy. The ionospheric reflection makes it certain that these special emissions will be audible in foreign countries, and alike on grounds of secrecy and of mitigating interference with communications this is undesirable. The interference problem, can, presumably, be dealt with through the normal machinery, with due regard to the importance of the objective.

The secrecy problem might be best solved by an offer from Air Ministry to Department of Scientific and Industrial Research of facilities for ionospheric investigation and other work for the Radio Research Board at a conveniently flat and isolated site at Orfordness, suitably distant from Slough for special experiments.

It is felt that none of these objections should be allowed to delay the attack which depends on the use of wavelengths, around 50 metres, on which we have adequate experience and adequate radiated power. But as soon as possible the techniques should be developed to cover the wave-lengths under 10 metres, which are not normally reflected from the ionosphere, and which would thus mitigate the interference problem and would help to maintain secrecy after the 'camouflage' already suggested was beginning to wear thin.

The power which can at present be radiated on these shorter wavelengths is about half that attainable in the 50-metre range, and the receivers are probably somewhat less sensitive, so that some sacrifice of sensitivity would at first result. The main reason for preferring the 50-metre wavelength, however, is connected with means for location by reflected pulse signals, other than by the measurement of time-delay as already outlined.

The cathode ray direction-finder, developed at Slough for visual direction-finding on extremely brief signals, has already been used on 50 metres, but not yet on 10 metres. It is almost certain that instruments of this type, working on 50 metres, could be used at the ends of a suitable base-line, the indications being 'piped' to a central control room in which the advance of the craft could be indicated continuously by the movement, on a map, of the point of intersection of two lines of light representing the directly indicated bearings at the two stations.

This technique can doubtless be extended to 10-metre working, but substantial development work is yet required. Closely related experiments down to 15 metres have, however, revealed no acute difficulties.

It may, further, be desirable to supplement or supplant the time-delay measurements by adding to the cathode ray direction-finding measurements another cathode ray technique also worked out (exclusively, as was the direction-finder) at Radio Research Station, Slough. This enables the angle of elevation of descending radio waves to be measured with an accuracy of about half a degree, an accuracy which can almost certainly be improved on demand; the work already in hand has not required higher accuracy. This technique has already been used on wavelengths between 60 metres and 10 metres.

The manner in which the three methods may best be combined for the most rapid deduction of the most convenient positional co-ordinates from these direct and continuous indications of the distance, angular azimuth and angular elevation of the craft, can only be determined by trial and development.

I am, however, convinced that the work can only be brought to a successful issue by the utilisation of the wide range of cathode ray techniques in which Radio Research Station, Slough, has specialised for many years, and in which its experience is unique.

If the foreseen difficulties of the pulse method prove unexpectedly great, or if some major difficulty has not been foreseen, there remain two practicable though less attractive processes. In one the sender would emit continuous-wave signals, and no echo would be detected save from a moving reflector, such as the craft. The rate of approach could be measured from the interference pattern on the cathode ray screen, the plan position could be plotted from cathode ray direction-finders into which were injected suitably phased e.m.fs. to suppress the images due to the direct rays and those reflected from fixed objects.

In the other process the frequency of the sender would be varied over a known range, as in Appleton's frequency-change method of ionospheric sounding. Here the interpretation of the pattern from a moving reflector would appear likely to be slower than is permitted by the practical problem of locating – and intercepting – high-speed enemy craft, and the method is not proposed for consideration until some flaw has been found in the quite unexpectedly favourable indications for the pulse method.

There will also be, for consideration, the problem whether the interval between detection and engagement may not be best reduced to a minimum by having interceptor craft fitted with a keyed resonating array so that they are readily located by the same method as those used on the enemy bombers, but discriminated and identified by the intermissions in their 'reflected' field. The interception operation can then be controlled by radio instructions to the interceptors closing them into the positions indicated for the bombers.

We have already disclosed, in patents and publications, means for making the oscillographs 'follow-up', and these may be relevant to further developments of the present scheme, as for distant repeating, etc.

Appendix E: Watson-Watt's memorandum to the CSSAD on the state of RDF research, 9th September, 1935

The proposals for detection and location of aircraft by radio means, made in detail in February, tested in a single experiment at Daventry in March, and developed, in work at Orfordness since May, have led to the following of metal-framed aircraft to distances of 92 km, to their detection (when expected with a doubt of some twenty minutes) at distances over 60 km, and (when completely unexpected) at distances over 50 km. In all cases where detection is possible the distance of the craft from the observing station can be measured with an accuracy of the order of 1 km. The tests have mainly been made in craft flying above 10 000 feet, with a few trials at 7000 and 5000 feet, and one at 1000 feet. The craft observed include land craft (Bristol R.120, Hart, Valentia, Vildebeeste, Virginia, Wallace) and marine aircraft (Osprey, Scapa, Seal, Singapore, South-ampton), with spans varying over a range exceeding three to one. If failure to follow or to detect at 20 km be taken as the criterion of failure, only three failures have been experienced in the experiments. One, due to excessively heavy atmospheric disturbance and very bad flying conditions, was in observation on a craft which was followed in another test to 55 km and detected at 46 km; one, partly due to engine failure, was on a craft which on another occasion was detected, quite unexpectedly, at 19 km (on leaving Martlesham); the third, due to unfavourable course and flying height, was on a craft with which long ranges have been obtained frequently. Comparable performances have been obtained on wavelengths of 50, 28, 27, 26 and 25 metres. No success has yet been attained in a few trials on 8 metres wavelength.

Experiments on measuring the angle of elevation of the craft detected are now being undertaken, but no tests have yet been made.

For the range of low angles of elevation problem the amplitude of the deflection on the measuring instrument may be taken as proportional to

$$\frac{h^2 H^2}{\lambda^2 d^4}$$

where h = height of (identical) sending and receiving aerials

H = flying height

λ = wavelength used

d = distance of craft from practically coincident sending and receiving stations.

The present state of radio technique does not guarantee the early attainment of ranges on wavelengths under 25 metres, comparable with those obtained on that wavelength, so *that*, for early applications, must be taken as fixed. The limiting range of detection with the present type of installation thus varies as $\sqrt{h.H}$. Since the detection of low flying craft is important, increase of h, the aerial height, is important. It can be obtained by additional mast height or by selecting coastal sites well above sea-level, i.e. on cliffs. The cost of masts varies roughly as h^2 up to 200 feet, and roughly as h^3 between 200 and 500 feet. A 200-foot mast system on a 50-foot cliff would offer valuable improvement in performance, and proposals for direct trials of this system are in hand. The immediate improvements which can be made inside the sending and receiving rooms are believed to justify the acceptance of a 60 km detection range (using 75-foot masts) as a conservative estimate of immediate average performance on craft flying at 13 000 feet, the improvements being taken only as converting the present not infrequent achievement of 60 km into a normal performance.

Table 1* gives a series of estimates of performance based on this experimentally established figure of 60 km for 13 000 feet. Line 1 quotes this basis, line 2 gives the probable performance with 200-foot masts on a 50-foot cliff, for different flying heights. These two lines are doubly conservative, as they do not take account of the certainty of further

* No table was provided in source (reference 18, section 4.1) from which this quotation was taken.

improvements in receivers, and do not take account of the improvements to be derived from simple antenna arrays. Line 3 represents a very conservative estimate of performance based on a qualified optimism on these points. Lines 4 and 5 show, on the same basis as that of line 3, the performance that can virtually be guaranteed now from 200-foot masts on a 200-foot cliff and from an 800-foot system, for example of 400-foot masts on a 400-foot cliff, where such provision is possible and economically justifiable.

While these 'guaranteed' improvements are attainable by high masts or by a high site, interchangeably, great additional advantages from antenna arrays are limited to that part of the total height which is provided by mast height only. The argument outlined may suggest that a high hill inland may be a better site than a low coastal site. While this is true for high-flying craft, closer examination indicates that for low-flying craft it is desirable that the sites of the detecting installations should be within a kilometre or two of the coast. The existing information about the absorption of downcoming waves over imperfectly conducting ground does not permit an exact theoretical prediction beyond this condition in the low-flying case, governing the siting of the main chain of stations. A second line of detection stations on hills over 750 feet high would, however, probably be a very valuable addition to the network, on grounds of reduced vulnerability and of improved watch over land , and it is very desirable that early experiments be initiated with one transmitting and one receiving station on such elevations as are readily available in England. This experiment might be combined with preliminary provision for the defence of the Tyne estuary by using the high country of Durham and Northumberland.

The ranges discussed deal with approach along the axis of maximum sensitivity for such an installation as that at Orfordness, with transmitter and receiver less than a kilometre apart. The most effective and economical distribution of a chain of transmitters and receivers to throw out a 'detection frontier' and a 'location frontier' substantially parallel to and at useful distances from the physical coast line will depend partly on the radio-technical factors, but partly on other factors, especially speed of working and reduction of intercommunciation channels between units. The most expensive single items of equipment are masts, and more masts are required at the receiving than at the transmitting station (because of the need for angle of elevation measurements).

An economical arrangement for giving a good detection frontier would appear to be that in which transmitting and receiving stations are planted alternately at equal intervals of some 30 kms along the coast. For 200-foot masts and 50-foot cliffs the detection frontiers for different flying heights are obtained approximately, as in Fig. 1.* It will be seen that for high-flying

* See footnote previous page.

craft the frontier is good and has no important bays, and that even for low-flying craft the bays are not very deep, while the frontier is sufficiently advanced if the incoming low-flying craft is assumed to be under the necessity of spending time in climbing near or within the coast line before action.

The areas, inside the detection frontier, over which location by rangefinding alone can be effected are shown in Fig. 1. Superficially the situation in respect of high-flying craft would appear sufficiently good, but for low-flying craft there are open corridors to the coast. Moreover, even in the case of high-flying craft there are ambiguities in location where simultaneously detected craft in separate formations have to be dealt with.

It is therefore necessary to improve the location frontier and to remove these ambiguities. Both ends can be met and, in addition, the speed of handling the location data can be much improved by the addition of a transmitting installation, with no substantial addition to the receiving installation, at each receiving site. The detection frontier is then slightly improved, but the location frontier is satisfactorily closed, the areas between them for all save very low-flying craft now being reduced to the quite small butterfly patterns shaded in Fig. 2.

It will be recognised that at this early stage in a new technique there are substantial elements of uncertainty in the estimates given. Only because the circumstances of the moment are emergency circumstances has any estimate at all been attempted now. But if these circumstances should be considered as demanding immediate action, then the following summary is considered a fair and not over-optimistic estimate of what can be done with no methods other than those tested in the four months of experimental work already carried out. The summary is based on plans which allow for the introduction of improvements without substantial scrapping, i.e. the plans call only for elements which are virtually certain to be embodied in the 'standard' installation in its quasi-final state.

A chain of stations with transmitters every 20 miles along the coast to be defended, and with receiving installations at each alternate station, i.e. every 40 miles, is required. The transmitters should have two masts not under 200 feet high, situated on land not less than 50 feet above M.S.L., not more than 2 miles from the coast. The cost of the transmitter from power terminals to aerial, i.e. including transmitter proper, masts and aerials, but excluding land building, power supply and communications, is estimated not to exceed £3000; the crew required need provide only one man for transmitter operation; and the mean power taken is under 5 kilowatts. Replacement of valves, the only important replacement cost, is on a scale not exceeding £1 per 5 hours' running; this may be substantially improved on. Connection to the grid system is desirable but not essential; stand-by equipment for power supply will be required on account of the vulnerability of the grid system. Communciation with the receiving station is not required in operation.

The receiving station required two similar masts similarly sited, and these may be within a kilometre of the local transmitter. Each receiving station utilises its one pair of masts, its one antenna array and its one radio receiver to feed three indicating instruments, one for each of the three transmitters with which it works (i.e. its local transmitter and its immediate neighbour on each side). The receiving station is thus its own local control station, giving location fixes for the craft within its sector, with no need for intercommunication with transmitters or other receivers. The receiving station, with its triple function, costs also approximately £3000, again excluding land, buildings, power supply and communications. In this case, of course, communication to headquarters, probably via one intermediate area control room dealing with four receiving stations, is essential. The actual observing team would probably be three. Replacement costs would be of the order of £1 per 100 hours.

A chain of this character should be able to locate accurately and count roughly any reasonable number – of the order of thirty per sector per five minutes – of metal-framed aircraft between the coast and the location frontiers. These frontiers are not likely to lie within 130 km for craft flying at 13 000 feet, 80 km for 5000 feet, 55 km for 2000 feet and 40 km for 1000 feet, on the costs shown, and can be pushed forward at a substantial additional cost.

The installation outlined can locate in plan position only, and cannot measure flying height. The reasonable certainties discussed above give place to high probabilities only when this problem is discussed thus early. It appears very probable, however, that at an additional cost of £2000 per receiving station, with no additional cost at the transmitters, provision can be made for height estimation to an accuracy which may fairly be expected to reach, after a year or less of research, discrimination to 1000 feet, save in the lowest 2000 feet, at half the maximum location range. A similar accuracy at quarter maximum range can probably be attained, at the same cost, within six months of research, with the team envisaged below. The probabilities of success are so high that if immediate building of any part of an interim defensive chain is undertaken the four masts required for location in plan and elevation, as opposed to two for location plan only, should be provided at each receiving station.

The scheme detailed does not provide for following the craft after they have crossed the coast. While this could be done from the coastal chain, the additional capital cost would not be notably less than that required for separate provision, and while manning costs would be less, the risk of organisational breakdown in a large scale operation is serious. It is therefore considered that the provision for overland following should take the form of a 'second line', the chain of inland hill stations mentioned earlier.

The scheme outlined included provision for minimising the effect of interference, especially of deliberate jamming. Should present expectations in respect of interference elimination be disappointed, the impossibility of keeping secret working wavelength (common to all the transmitters and receivers of one chain) or of providing a sufficiently advanced location frontier with receivers capable of quick change of wavelength, would force a decision as between two possible courses. Two or even three working wavelengths could be provided at transmitters and at plan-locating receivers at no great increase in cost, the increase not reaching 20%. But at receiving stations equipped for height measurement the cost of the receiving equipment might be nearly proportional to the number of wavelengths provided. This may be avoided by further research. Meanwhile the possibility must be noted, but interim planning may proceed on the more optimistic assumption that the anti-jamming design is likely to be so effective that quick wavelength changes will not be required.

It appears essential, especially if the inland chain be added, that our own interceptor and fighter craft should be fitted with means for their identification when located by the radio chain. The fitting of these craft with special tuned aerials, automatically interrupted to give identification, seems possible at a negligible increase in weight and a small increase in drag, but *ad hoc* experiment is required and has not yet been undertaken.

It will be observed that no proposals for directional transmission or reception are embodied in this interim scheme. Were directional reception available on a footing of equal operational ease with the range-finding gear, the cost of the chain could be substantially reduced. But research extending over two years or more may well be required before instantaneous direction-finding of the required high sensitivity can be developed. This work should be undertaken but, since the main aim of the present note is to indicate what may fairly be relied on within the next twelve months, directional methods are excluded from the scheme, which is a flood-lighting scheme involving no 'search' whatever, i.e. no manipulative action at transmitter or receiver, for plan-location, and only the simplest electric manipulation, the turning of one control knob, for height measurement. Directional methods may usefully supplement this static observational system, but they should be supplementary and not integral; research on these lines is proposed.

Such forecasts as have been made above have not touched the problem of the utilisation of ultra-short waves of wavelength 2 to 10 metres, which may offer a solution, applicable to the chain already outlined, of the very difficult problem of the low-flying craft and the, perhaps, still more difficult problem of ship detection. The disappointing results of one or two recent trials, of a rudimentary character, indicate the desirability of basing any urgent application work on the 25-metre results already achieved, and of leaving the ultra-short-wave developments to take the next place in priority.

Still more remote is the ultimate goal of making effective use of waves so short (0·5 m) that they would be useful in detecting a metallic engine supported in an otherwise completely non-metallic craft. Work towards this goal should begin without delay, but the technique, especially on the receiving side, is so backward that several years of work are likely to be spent before useful results are in sight.

In parallel with these investigations there should run the closely related work on the more or less direct control of mechanisms by the cathode-ray or other indicating gear utilised in location. It is doubtful whether, in view of the dependence of this work on that already outlined, any reliance should be placed on the availability of such methods of control within two or three years.

Close consideration has been given to the nature of the provision required for advancing the numerous lines of research indicated herein. The provision of a suitably situated central research and development station, of large size and with ground space for a considerable number of mast and aerial systems, is a first highly urgent necessity. It should provide living accommodation for the resident Director of the radio investigations and for unmarried members of his research team and their assistants. Material provision, even on the considerable scale which is clearly necessary, is not, however, the most difficult provision. The rate of progress will be governed by the availability of a research staff of exceptional quality, and a staff of this quality and with earlier experience of the right kind can never be a very large one.

Detailed discussion of possible recruits, believed to be based on sufficient knowledge of the field of recruitment, has resulted in the finding of only four names to be put forward with a view to appointment to the research staff, in addition to the D.S.I.R., Air Ministry, and Admiralty workers already attached or available. For further part-time assistance special reliance is placed on other members of staff of the Radio Department, N.P.L., who are working on fundamental problems so closely bearing on the work that they should be brought in as consultants without being withdrawn from their present spheres save for very brief periods. The research and development team envisaged in the scheme laid down in this memorandum would thus have a Director and two Scientific Oficers provided by D.S.R., Air Ministry, one or two attached by D.S.R., Admiralty, possibly corresponding officers attached by D.S.I.R. or Air Ministry, together with part-time services from other D.S.I.R. officers, a total of probably twelve full-time and three part-time officers. The ancillary staff on the research and development side should include twelve

Assistants II and III, four laboratory assistants and four to six industrial staff (mechanics, carpenters and labourers).

It would also appear desirable that the training of the operating and observing corps for the chain should begin at a very early date, and that the nucleus of the corps should have its headquarters in the central research and development station so long as the exigencies of the investigational work and the organisation of the corps permitted. It is believed that the individual stations could be best manned by R.A.F. personnel of the Wireless Operator Mechanic type with, perhaps, one officer per 'unit' of one separate transmitting station and one joint transmitting and multiple receiving station. The research and development team would then have the dual role of independent investigation, for which its Director would be responsible to Air Ministry through D.S.I.R., and of consultantship to the Commanding Officer of the new signals unit responsible for the chain. It would, however, be essential that the Director should have authority to resolve minor conflicting claims on his team or to refer major claims to Air Ministry for resolution.

Appendix F: Method of deriving height curves for a Chain Home station

This is based on information provided in 'Radar Supervisor's manual', A.P.2911R, Vol. 1, Amended, Air Ministry, London, 1951.

The following assumptions are made:
(*a*) For any given height-finding pair of antennae (see Fig. 5.10), it was usual for the signal from the top antenna to be fed into the X-coil of the gonimeter and that from the lower antenna into the Y-coil.

The same goniometer and scale as used in direction finding were also used in height finding.

No signal in the upper antenna but a signal in the lower one meant that there was no signal in the X-coil of the goniometer, and this corresponded to the direction-finding case where a signal was being received at right angles to the line of shoot (see Section 5.6.4). In this situation, when the echo was 'D/F'd out', the goniometer would indicate 90° from the line of shoot.

(*b*) The line of shoot varied from station to station and the position of the pointer on the goniometer allowed for this and indicated the actual azimuth, and not the angle relative to the line of shoot. In its height-finding role, the goniometer search coil position (θ_H reading) incorporated the line of shoot angle. In our example here, we have taken the line of shoot angle to be 90° (i.e. due east). The heights above ground of the receiving antennae are here taken to be 215 ft, 90 ft and 45 ft; the A system grouping for height-finding is [215 ft, 90 ft] and the B system is [90 ft, 45 ft].

The operating frequency of the station is taken as 22·69 MHz (13·22 m wavelength), which was one of the scheduled frequencies.

The first gap, above the one at zero degrees, in the vertical radiation pattern of each of these antennae is given by the expression

$$\lambda \text{ (degrees)} = \frac{94\lambda \text{ (metres)}}{H \text{ (ft)}}$$

where λ is the wavelength and H is the height above ground. Thus we obtain the following set of values:

H (ft)	α (°)
45	28·7*
90	13·8
215	5·8

(The above expression for α is based on the approximation $\sin \alpha = \pi\alpha/180$

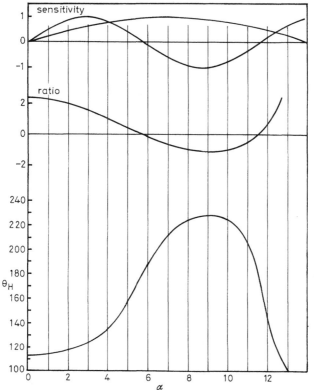

Fig. F1 *Height curves*

---

* The formula would have yielded a value of 27·6°; because α exceeds 15°, this more accurate value has been calculated.

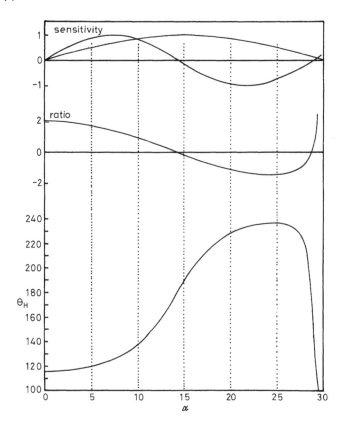

Fig. F2   *Height curves*

which is true for small angles and is certainly usable up to 15°. The minima occur at the following values of θ.

$$\theta = \sin^{-1}\frac{n\lambda}{2H}, \quad n = 0, 1, 2, \ldots$$

where λ and H are expressed in the same units.)

Knowing the angle of the gaps for each antenna, one can draw sensitivity curves for the antennae as shown in Figs. F1 and F2. Fig. F1 shows the curves for the combination [215 ft, 90 ft] and Fig. F2 those for the combination [90 ft, 45 ft]. In each case, the curve of the top antenna goes below the axis after the first gap; this just indicates that the second lobe of the antenna is opposite in phase to the first lobe. These sensitivity curves are but a cartesian representation of the variations in received signal strength with angle of elevation α.

Directly from the sensitivity curves the ratios of the signals in top and bottom antennae for various α are calculated and plotted. From this ratio curve (by means of the arctan function), the angle of the goniometer reading for DF on height, known as θ_H or 'theta height', is plotted against α.

Appendix G: Two maps

The following maps show, respectively:
1. Proposed acoustical mirror system in the Thames Estuary. Work on the construction of the system was suspended in September 1935, see Fig. G1.
2. Chain Home stations operating at the outbreak of war, September 1939, see Fig. G2.

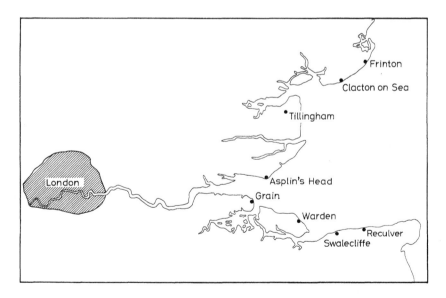

Fig. G1   *Map: sites for sound mirrors*

CHAIN HOME : Stations operating September 1939.

Fig. G2 *Map: Chain Home stations*

Appendix H: Report issued by Telecommunications Research Establishment, 1941

Camouflaging of aircraft at centimetre wavelengths

At Dr Lewis' suggestion (81) further consideration has been given to proposals made (80a) for rendering aircraft almost undetectable by normal RDF at a given frequency. The scheme previously described is not satisfactory, but a modified scheme described below appears to be quite feasible, at any rate as a scientific experiment.

Original Scheme (Quarter Wave Matching Layer)
The intrinsic impedance of aluminium, given by (6) of 80a, is about

$$3.34 \times 10^{-4} \sqrt{f} \, (1 + j) \text{ ohms} \tag{8}$$

where f is the frequency in megacycles/second. Contrary to the impression given in 80a, the reactive component of (8) does not have a large influence on the design of the matching layer. The layer must be substantially a quarter-wavelength of dielectric whose intrinsic resistance is the geometric mean of:

(i) the real part of (8), and
(ii) 377 ohms, which is the intrinsic resistance of free space.

This gives

$$0 \cdot 36 \sqrt[4]{f} \text{ ohms} \tag{9}$$

which, for $f = 3000$ megacycles/second, is

$$2 \cdot 6 \text{ ohms} \tag{10}$$

The matching layer therefore requires a ratio of dielectric constant to permeability of

$$\left[\frac{377}{2 \cdot 6} \right]^2$$

$$= 20\,000 \tag{11}$$

This is quite fantastic, of course. We therefore turn to the use of lumped shunt impedances.

Modified Scheme (Resistive Skin and Stub)
The simplest theoretical way of matching an aircraft to free space is to envelop it in a resistive skin whose surface-resistivity is 377 ohms, and to maintain an air-gap between skin and aircraft of a quarter of a wavelength. In this method we take the view that (8) is so small compared with the intrinsic resistance of free space that we may regard it as a short circuit. The problem then becomes equivalent to that of preventing reflection from the short-circuited far end of a transmission line by connecting shunt impedance across the linear near the short-circuit. The solution suggested is the well-known one of connecting the characteristic resistance across the line a quarter of a wavelength in front of the short-circuit. Other solutions are quite possible, but they do not seem to translate as conveniently into aircraft technique.

The gap between skin and aircraft may be considerably reduced by filling it with a medium in which the wavelength is less than in free space. This may be achieved by employing a medium with a high product of permeability with dielectric constant. For example, if the gap is filled with reasonably pure water (prevented from freezing, of course), its thickness is cut down by a factor 9 to (1/36)th of the wavelength in free space. At a wavelength in free space of 10 centimetres the gap would then be only 2·8 millimetres. A further reduction of perhaps a factor 2 might be obtained by introducing magnetic material into the medium as suggested by Dr Lewis (81). The trouble with the actual medium mentioned by Dr Lewis is that its conductivity is about a thousand times too high for the intrinsic impedance to be mainly real at a frequency of 3000 megacycles/second. The high conductivity would require the skin to have a surface-impedance possessing a capacitative component. This would probably be difficult to arrange in a manner independent of polarisation.

From experiments already made with wave-guides, it is virtually certain that the modified matching system described can prevent reflection at normal incidence from flat metal surfaces. The question is: will it prevent ordinary RDF echoes from aircraft at centimetre wavelengths? The theoretical arguments adduced in 80a in connection with the original matching scheme lose none of their force when applied to the modified scheme. Indeed it would seem that the only insuperable fundamental difficulty that could arise would be diffraction by sharp trailing edges, etc. It would be somewhat surprising, however, if such diffraction is a major cause of echoes from aircraft at centimetre wavelengths.

The great disadvantage of the above scheme, and probably of any scheme of camouflage at centimetric wavelengths, is that it is only effective over a limited range of frequency.

Let

f_0 = frequency at which matching is perfect
R = intrinsic resistance in ohms of medium filling gap between skin and aircraft
ρ = field-strength reflection-coefficient at normal incidence.

Then, at a frequency f near f_0, we have

$$\rho = \frac{30\pi^2}{R} \frac{f - f_0}{f_0} \tag{12}$$

Hence the band-width over which the field-strength reflection-coefficient is less than 5% of what it would be in the absence of matching is

$$\frac{R}{3\pi^2} \tag{13}$$

per cent of the frequency at which matching is perfect. It will be observed that the band-width is proportional to the intrinsic resistance of the medium filling the gap between skin and aircraft. From this point of view pure water would not be a good choice, since its intrinsic impedance (42 ohms) is one of the lowest of known non-conductors. To obtain a large band-width we need to use a medium of low conductivity with a high ratio of permeability to dielectric constant. If this ratio is 4, the intrinsic resistance of the medium is

$$R = \sqrt{4} \times 377$$
$$= 750 \text{ ohms}$$

The band-width (13) then become 25%. For free space ($R = 377$ ohms) the band-with is 13%, and for pure water ($R = 42$ ohms), the band-width is only 1·4%. However, the band-width of an RDF station operating at a frequency of 3000 megacycles/second would hardly be as much as 0·01%. The frequency at which matching is perfect may be varied by varying the thickness of the gap or the medium it contains.

From what has been said above, it appears that the ideal medium with which to fill the gap between skin and aircraft would have:

(i)   a high product of permeability and dielectric constant to secure small thickness;
(ii) a high ratio of permeability to dielectric constant to secure wide band-width;
(iii) a conductivity sufficiently low to keep the intrinsic impedance ((2) of 80a) mainly real and thus avoid the necessity for endowing the skin with capacitative reactance.

If one of these ideals is sacrificed, there is no difficulty in obtaining a suitable medium. Information is obtainable in the Smithsonian Physical Tables.

It is concluded that there is a real scientific possibility of camouflaging an aircraft over a limited frequency-range at centimetre wavelengths. How far large-scale use of such camouflaging may be feasible or useful is for others to decide.

27th August 1941

File 4/18

HGB/MH

S.No. 5273

Appendix I: Table of Japanese naval radars

This table of Japanese Naval radars, as supplied by Shiryo Chosakai (Institute of Historical Research), Tokyo, was prepared at the end of the Second World War.

Warning radar

	Ground-based				Shipborne		Airborne			
Type name	11	12	13	14	21	22	H-6	FM-1	N-6	FK-3
Purpose	Ground shore fixed anti-air	Ground movable anti-air	Ground ship submarine portable anti-air	Ground target type great long range anti-air	Ship anti-air	Ship, submarine anti-surface warning and firing	Large aircraft warning and search	Medium aircraft warning and search	Small aircraft warning and search	Small aircraft warning and search
Completion date	1943 June	1944 April	1943 March	1945 May	1943 August	1944 September	1942 August	1944 September	1944 October	1945 June
Operational status	Used in the war	Used in the war	Used in the war	Used in the war	Used in the war	Used in the war	Used in the war	Under preparation for practical use	Under experiment	Under preparation for practical use
Installed place	Ground key area	Ground key area	Ground ship submarine	Shore line	Ship (surface)	Ship submarine	Large flying boat medium size attack aircraft	Four-engined aircraft	Single-engined 3-seater	Single-engined 3-seater
Wavelength (cm)	300	200 150	200	600	150	10	200	200	120	200
Peak output (kW)	40	5	10	100	5	2	3	42	2	2
Transmitter osc., cct	Parallel two wire	Parallel two wire	Parallel two wire	Parallel two wire	Parallel two wire	Magnetron	Parallel two wire	Parallel two wire	Parallel two wire	Parallel two wire
Receiver Detector	UN-954 RE-3	UN-954 RE-3	UN-954	UN-954	UN-954 RE-3	Crystal	UN-954	UN-954	UN-954	UN-954
Max detection range (km)	Aircraft Group: 250 Single: 130	100 50	100 50	360 250	100 70	35 17	100 70	100 70	70 50	70 50
Weight (kg)	8700	6000	110	30 000	840	Surface ship: 1320 Submarine:	110	70	60	60

Warning radar

	Ground-based				Shipborne		Airborne	
Manufactured quantity	30	50	1000	2	300	2000	20	100
Antenna	Dipole array with mat type reflector S.R. separate use	Dipole array with mat type reflector S.R. separate use	Dipole array with mat type reflector S.R. common use	Dipole array with mat type reflector S.R. common use	Horn type S.R. separate use	Yagi type S.R. common use	Yagi type S.R. common use	Yagi type S.R. common use

Fire control and locating radar

	Ground and shipborne							Airborne	
Type name	41	42	43	23	31	32	33	FD-2	TAMA3
Purpose	Ground anti-air firing	Ground anti-air firing	Ground anti-air searchlight control	Shipborne anti-ship firing	Shipborne anti-ship firing	Shipborne anti-ship firing	Shipborne anti-ship firing	Night-fighter approach	Night-fighter approach
Completion date	1943 August	1944 October	1945 July	1944 March	1945 March	1944 September	1945 January	1944 August	1945 July
Operational status	Used in the war	Used in the war	Used in the war	Not used in the war	Under preparation for practical use	Under preparation for practical use	Not used	Not used	Under preparation for practical use

Fire control and locating radar

	Ground and shipborne							Airborne	
Installed place	Anti-air artillery	Anti-air artillery	Searchlight	Above cruiser class	Ground above medium size ship	Ground large ship	Surface ship	Nightfighter	Nightfighter
Wavelength (cm)	150	150	150	60	10	10	10	60	200
Peak output (kW)	13	13	13	5	2	2	2	25	3
Transmitter osc., cct	Ring parallel two wire	Ring parallel two wire	Ring parallel two wire	Cavity resonance	Magnetron	Magnetron	Magnetron	Cavity resonance	Parallel two wire
Receiver detector	UN-954	UN-954	UN-954	2400	Crystal	Crystal	Crystal	2400	UN-954
Max detection range (km)	Aircraft group: 40 Single: 20	Aircraft group: 40 Single: 20	Aircraft group: 40 Single: 40	Destroyer 13	Battleship 35	Battleship 30	Destroyer 13	Aircraft: 3 Ship: 10	Aircraft: 3
Weight (kg)	5000	5000	500	1000	1000	5000	800	70	70
Manufactured quantity	50	60	121			60		100	10
Antenna	Dipole array with mat type reflector S.R. separate use	Yagi S.R. separate use	Yagi S.R. separate use	Parabola S.R. separate use	Parabola S.R. common use	Square horn type S.R. separate use	Round horn type S.R. separate use	Yagi type S.R. separate use	S.R. common use

	Radio guidance (navigation radar) Ground base (for aircraft)			IFF Small surface ship/boat		Radio noctovision Airborne	Radio altimeter Airborne	ESM warning receiver Shipborne	Shipborne	Airborne
Type name	61	62	63	TH	M-13	51	FH-1	E-27		FT-B FT-C
Purpose	Enemy aircraft height and position measure	Friend aircraft position measure	Long-distance enemy aircraft position measure	Special attack small boat guidance	General aircraft use	Large aircraft use	Low-altitude use	Ground shipborne use warning	Ground shipborne use warning	Aircraft use warning
Completion date	1945 April	1945 June		1945 July	1945 June		1945 February	1944 April		1944 May
Operational status	Under preparation for practical use	Under preparation for practical use	Under preparation for practical use	Not used	Not used	Not used	Used in the war	Used in the war		Under preparation for practical use
Installed place	Ground important area	Ground important area	Ground important area	Shore-side	All kinds of aircraft	Four-engined bomber	Large flying boat	Torpedo attack boat		Bomber reconnaisance aircraft
Wavelength (cm)	60	200	300	150	200	10	88	75-400	3-75	45-370
Peak output (kW)	10	10	40	13	0-05	6	0-0001			
Transmitter osc. cct	Shielded osc	Parallel two wire	Parallel two wire	Ring parallel two wire	Coil and capacitor	Magnetron	Parallel two wire			
Receiver detector	2400	UN-954	RE-3 UN-954	UN-954	UN-955	Crystal	UN-955	UN-955		UN-955 × 2

	Radio guidance (navigation radar)			IFF		Radio noctovision	Radio altimeter		ESM warning receiver	
	Ground base (for aircraft)			Small surface ship/boat		Airborne	Airborne		Shipborne	Airborne
Max detection range (km)	Single: 130	Over 100	160	Under 20	100	20	0·15	300		300
Weight (kg)	15 000	3000	8700	500	10	200	30	40	200	20
Manufactured quantity	1	1	1	100	100	100	100	2500	200	FT-B: 300 / FT-C: 100
Antenna	Parabola S.R. common use	Dipole array S.R. common use	Dipole array S.R. common use	Bend-back 4λ common use		Parabola S.R. common use	Doublet common use			

Appendix J: Military characteristics – detector for use against aircraft (heat or radio) (United States Army)

The following military characteristics were set up by the Chief of Coast Artillery and forwarded to the Chief Signal Officer with approval of The Adjutant General on the 29th February 1935.

1 Military purposes
  (*a*) Detect the presence of aircraft in a sector of approximately 120° in azimuth, elevation limits of 0 to 90°. This sector represents the zone which the device must be able to cover by rapid sweeping in azimuth and elevation; the apparatus to be actuated by the heat radiation or radio waves set up by the airplane.
  (*b*) Register, or indicate, in proper units of measure, the position of the located airplane with respect to the detector or some other designated reference point; this indication of data to be continuous so that the aircraft, while in motion, may be followed by a searchlight or observing instrument laid on such data.

2 Essential qualities of the devices
  (*a*) Under all atmospheric conditions, the accuracy of the indicated data must be such as to permit laying the searchlight, or observing instrument, with a maximum deviation of 1° in azimuth and in elevation from the actual position of the aircraft.
  (*b*) It must begin to furnish data of the above specified accuracy immediately after the device begins to operate.
  (*c*) Under average atmospheric conditions it must function at ranges up to 20 000 yards.
  (*d*) Under conditions of rain, mist, smoke, or fog, during daylight or darkness, it must function at ranges up to 10 000 yards.
  (*e*) The design of the apparatus and the material employed must be such as to permit of operation by the average skilled enlisted personnel.
  (*f*) The material employed must be of such nature as to withstand the conditions of climate and service, including shock from gun fire, which ordinarily obtain in localities where the apparatus would be employed.
  (*g*) The bulk and weight of the apparatus must be such as to permit of its being readily moved from one position to another in a permanent anti-aircraft defense installation. Consideration should also be given to ultimate development of a type suitable for use with a mobile organisation.
  (*h*) The requirements of agencies, such as electric current, to operate the apparatus must be such as to permit of its installation and use in the average anti-aircraft defense system of an area.

(*i*) In case of the radio detector, it must not be subject to interference known as 'jamming' by other radio apparatus nor interference with the operation of friendly communications.

Appendix K: Mathematical analysis

Scattering and the scattering matrix
Refer to Fig. K1 where a set of fixed axes, X_1, X_2, X_3 and the co-ordinate origin are fixed in the target with generally both the transmitter and receiver located at (R, θ, ϕ) [1]. It could be arranged that a common antenna would transmit and receive only a θ component or only a ϕ component, say, of electric field. Other choices of polarisation could be made and for any particular aspect of the target and for arbitrary linear polarisation* a set of scattering data could be composed. In fact, it is possible to formulate a polarisation scattering matrix.

Let E_H^i be the electric field intensity at the target of a horizontal polarised wave incident on it, and let E_H^r be the scattered horizontally polarised component at the receiving antenna which results from E_H^i. The relationship may be expressed by:

$$E_H^r = a_{11} E_H^i \tag{K1}$$

where the coefficient a_{11} relates the two horizontally polarised components of

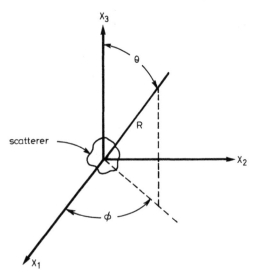

Fig. K1   *Scattering object – co-ordinate axes*

* It may sometimes be convenient to resolve the E field vector into orthogonal circular components instead of into horizontal and vertical components [2].

the field. The incident horizontally polarised wave could, in general, also produce a vertically polarised component of field, after scattering, with a relationship:

$$E_V^r = a_{21} E_H^i \qquad \text{(K2)}$$

Similarly, we can write for the incident vertically polarised wave:

$$E_V^r = a_{22} E_V^i$$

and

$$E_H^r = a_{12} E_V^i \qquad \text{(K3)}$$

An arbitrary plane polarised wave can be decomposed into orthogonal components and the total received field may be written:

$$E_H^r = a_{11} E_H^i + a_{12} E_V^i$$
$$E_V^r = a_{21} E_H^i + a_{22} E_V^i \qquad \text{(K4)}$$

or

$$\begin{bmatrix} E_H^r \\ E_V^r \end{bmatrix} = \begin{bmatrix} a_{11} a_{12} \\ a_{21} a_{22} \end{bmatrix} \begin{bmatrix} E_H^i \\ E_V^i \end{bmatrix} \qquad \text{(K5)}$$

The four elements of the matrix, the reflection coefficients are, in general, complex quantities because of phase differences. One could write the above as:

$$\begin{bmatrix} E_H^r \\ E_V^r \end{bmatrix} = \begin{bmatrix} a e^{j\alpha} & b e^{j\beta} \\ c e^{j\lambda} & d e^{j\xi} \end{bmatrix} \begin{bmatrix} E_H^i \\ E_V^i \end{bmatrix} \qquad \text{(K6)}$$

In the monostatic case when the source and reception points coincide, the matrix is symmetric. This may be demonstrated as follows:

Let $E_H^i = 1$ and $E_V^i = 0$.
Then $E_V^r = c e^{j\lambda}$.
Also, if $E_H^i = 0$ and $E_V^i = 1$, then $E_H^r = b e^{j\beta}$.

In effect, mutually orthogonal transmitting and receiving antennae at the same location have been interchanged and, by the principle of reciprocity (see Appendix A)

$$b e^{j\beta} = c e^{j\lambda}$$

Rearranging the matrix equation and factoring out a quantity $e^{j\alpha}$, which is a phase term, one obtains

$$\begin{bmatrix} E_H^r \\ E_V^r \end{bmatrix} = e^{j\alpha} \begin{bmatrix} a & b e^{j\beta'} \\ b e^{j\beta'} & d e^{j\xi'} \end{bmatrix} \begin{bmatrix} E_H^i \\ E_V^i \end{bmatrix} \qquad \text{(K7)}$$

Thus the general scattering matrix for a co-located transmitter and receiver contains five quantities, three of magnitude and two of phase. One can conclude, therefore, that rigorous experimental determination of cross-section means that, for every aspect angle of the target, seven constants in the bistatic case and five in the monostatic case must be evaluated.

Radar range equation
Assume a common transmitting and receiving antenna of gain, G, and denote the transmitted signal power and the received signal power at the antenna terminals by P_t and P_r, respectively. Then, for a target of cross-section σ at a range R using a wavelength λ, one can see from fundamental definitions and geometrical relationships that:

P_r = (signal power density at target) × (fraction scattered back and intercepted by antenna) × (gain of antenna)

$$P_r = \frac{P_t G}{4\pi R^2} \frac{\sigma}{4\pi R^2} \frac{G\lambda^2}{4\pi}$$

$$= \frac{P_t G^2 \sigma\lambda^2}{(4\pi)^3 R^4} \tag{K8}$$

Free space propagation conditions are assumed and possible atmospheric losses and multipath propagation conditions between target and radar are ignored.

To give an expression for range, the equation may be rewritten in the form

$$R = \left[\frac{P_t G^2 \sigma\lambda^2}{(4\pi)^3 P_r}\right]^{\frac{1}{4}} \tag{K9}$$

If it is assumed that a value of P_r exists, which when processed by the receiver is a minimum detectable value and which may be denoted by $P_{r(min)}$, then the maximum range is

$$R_{max} = \left[\frac{P_t G^2 \sigma\lambda^2}{(4\pi)^3 P_{r(min)}}\right]^{\frac{1}{4}} \tag{K10}$$

The total power amplifying ability of the receiver is an obvious factor in determining detectability and hence maximum range, but noise is the critical decider. Noise power from several sources will become available at the terminals of the receiving antenna and this will be amplified by the receiver circuits which, in turn, add in their own contribution of noise. The signal-to-noise power ratio (S/N) at the output of the receiver pre-detector stage can be seen to be a critical factor in seeking to quantify detectability. (The case envisaged here is for the typical situation where a superheterodyne receiver is used and the detector is the second detector or demodulator.) If the noise generated by the receiver be known this pre-detector S/N can be referred back to an equivalent S/N at the antenna terminals, and at these terminals we may speak of $(S/N)_{min}$ where

$$(S/N)_{min} = \frac{P_{r(min)}}{P_n} \tag{K11}$$

Now P_n, the total pre-detector system noise referred back to the antenna terminals, may be written as kT_sB_n, where k is Boltzmann's constant ($1 \cdot 38 \times 10^{-23}$ J/K), B_n is the receiver noise bandwidth and T_s is the system noise temperature of the receiver (if G is the total pre-detection gain then the noise power at the detector due both to the antenna and to the receiver itself is GkT_sB_n).

$$P_{r(min)} = (S/N)_{min} P_n \tag{K12}$$

Hence one can write

$$R_{max} = \left[\frac{P_t G^2 \sigma\lambda^2}{(4\pi)^3 (S/N)_{min} kT_s B_n} \right]^{\frac{1}{4}} \tag{K13}$$

Some other factors are usually included in the equation. One is the system loss factor, L. L may include many individual and determinable loss factors such as absorption loss and transmission line loss which are multiplicative in their overall effect. (Sometimes the degradations in performance that could occur because of other than ideal operators or because of questionable standards of maintenance are considered as losses. These factors are too inexact for inclusion in the range equation.) If P_t were, in fact,, defined and measured at the transmitter output and not at the antenna terminals then a transmitting loss factor covering losses in the transmission lines and duplexer between transmitter and antenna would have to be included in L. It is essential to be precise in defining all parameters used in a range equation, otherwise a particular factor may unwittingly be introduced into more than one parameter.

The form of the range equation then becomes

$$R_{max} = \left[\frac{P_t G^2 \sigma\lambda^2}{(4\pi)^3 (S/N)_{min} kT_s B_n L} \right]^{\frac{1}{4}} \tag{K14}$$

In the case of a pulse radar of pulse duration τ, there is an optimum receiver bandwidth, which we can denote by $B_{n(opt)}$.

$$B_{n(opt)} = \frac{\alpha}{\tau}, \quad \alpha \sim 1$$

Using the visibility factor V, first introduced by Norton and Omberg [3], one can write

$$V = \frac{E_{min}}{kT_s} \tag{K15}$$

where E_{min} is the minimum available pulse energy from antenna, matched to the receiver, which is required to make the pulse just visible in the presence of noise, and T_s is the system noise temperature of receiver.

One can therefore write

$$V = \frac{P_{r(min)} \tau}{kT_s} \qquad (K16)$$

which, since

$$P_{r(min)} = (S/N)_{min} kT_s B_n$$
$$V = (S/N)_{min} B_n \tau$$

Hence

$$(S/N)_{min} B_n = \frac{V}{\tau} \qquad (K17)$$

Generally, the visibility factor V in the range formula is expressed as $V = V_0 C_b$, where V_0 is a lowest possible value of V corresponding to an optimum value of B_n. C_b, defined in Chapter 2, must be greater than or equal to unity.*

The range equation may now be written

$$R_{max} = \left[\frac{P_t \tau G^2 \sigma \lambda^2}{(4\pi)^3 kT_s V_0 C_b L} \right]^{\frac{1}{4}} \qquad (K18)$$

A pattern propagation factor, F, may be introduced. It is a measure of the departure from free-space conditions of the propagation situation between target and antenna. This departure is caused predominantly by multipath phenomena and by the use of other than the beam maximum of the antenna radiation pattern. In the general situation two factors F_t and F_r, for transmitting and receiving antennae, respectively, would be employed. In the monostatic case with a common transmitting and receiving antenna, $F_t = F_r$ and the range expression becomes:

$$R_{max} = \left[\frac{P_t \tau G^2 \sigma \lambda^2 F^4}{(4\pi)^3 kT_s V_0 C_b L} \right]^{\frac{1}{4}} \qquad (K19)$$

The fourth power of F is used because F is a function of electric field intensity and here we are concerned with a modification to antenna power gain.

In the case of a bistatic radar (see Fig. 2.1b) R^4 is replaced by $R_1^2 R_2^2$ and R_{max} by $(R_1 R_2)^{\frac{1}{2}}_{max}$. G_t and G_r denote transmitting antenna gain and receiving antenna gain, respectively. Thus the range equation for a bistatic pulse radar may be written

$$(R_1 R_2)_{max} = \left[\frac{P_t \tau G_t G_r \sigma_b \lambda^2 F_t^2 F_r^2}{(4\pi)^3 kT_s V_0 C_b L} \right]^{\frac{1}{4}} \qquad (K20)$$

* Skolnik refers to C_b as the *bandwidth correction factor*, see p. 2–6, eqn (9), of Reference 1, Chapter 2.

For a CW radar the form would be

$$(R_1 R_2)_{max} = \left[\frac{P_t G_t G_r \sigma_b \lambda^2 F_t^2 F_r^2}{(4\pi)^3 kT_s (S/N)_{min} B_n L} \right]^{\frac{1}{2}} \tag{K21}$$

3. Electromagnetic Reflection Coefficient of Ground

The electromagnetic reflection coefficient relates the reflected field E_r to the incident field E_i with $E_r = \Gamma E_i$, such that

$$\Gamma = \rho e^{-j\phi} \tag{K22}$$

where

$$\rho = \left| \frac{E_r}{E_i} \right|$$

and $0 \leqslant \rho \leqslant 1$, and $-\pi \leqslant \phi < \pi$. Positive ϕ means that E_r lags E_i.
   The factor

$$\sigma/\omega\varepsilon$$

where $\varepsilon = \varepsilon_0 \varepsilon_r$ and ω is the angular frequency, determines the behaviour of the ground. If $[\sigma/\omega\varepsilon] \gg 1$, the ground is regarded as practically a pure conductor, whereas if $[\sigma/\omega\varepsilon] \ll 1$, the ground may be considered as a pure dielectric.*
   The following values are often adopted for sea and 'good ground', i.e. arable or clay soil.

	σ (S/m)	ε_r (F/m)	Critical frequency $\sigma = \omega\varepsilon_0 \varepsilon_r$
Sea	4	80	900 MHz
Good ground	10^{-2}	10	19 MHz

In a perfect dielectric medium the wave equation for the electric field in the case of a single frequency plane sinusoidal wave is

$$\nabla^2 \vec{E}(\vec{r}, \omega) + \frac{\omega^2}{c^2} \varepsilon_r \vec{E}(\vec{r}, \omega) = 0 \tag{K23}$$

In the case of a lossy dielectric this becomes

$$\nabla^2 \vec{E}(\vec{r}, \omega) + \frac{\omega^2}{c^2} \varepsilon_r - j \frac{\sigma}{\omega\varepsilon_0} \vec{E}(\vec{r}, \omega) = 0 \tag{K24}$$

---

* There is a correspondence here to the 'metallic reflection' and 'vitreous reflection' of physical optics. The formulae for the reflection coefficients are referred to as the Fresnel formulae. Although formulae for radar and for physical optics are similar there is a marked difference between the two situations. In physical optics experiments cannot be carried out, as in the radio or radar case, a few wavelengths away from the reflecting surface.

Thus the propagation factor ω^2/c^2 has become

$$\frac{\omega^2}{c^2}\, \varepsilon_r - \frac{j\sigma}{\omega\varepsilon_0}$$

The relative dielectric constant ε_r has, due to conduction losses, become

$$\varepsilon_r - j\frac{\sigma}{\omega\varepsilon_0},$$

a complex relative dielectric constant which will be denoted by ε_c.

A complex index of refraction, n, is defined by the relation

$$n = (\varepsilon_c)^{\frac{1}{2}} = \left(\varepsilon_r - j\frac{\sigma}{\omega\varepsilon_0}\right)^{\frac{1}{2}} \tag{K25}$$

In SI units this becomes

$$n = (\varepsilon_r - j\, 60\, \sigma\lambda)^{\frac{1}{2}}$$

The Fresnel formulae give expressions for Γ_V and Γ_H and are derived in most standard textbooks on electromagnetic theory [4].

Referring to Fig. 2.4b, one can write

$$\Gamma_V = \rho_V\, e^{-j\phi_V} = \frac{n^2 \sin\psi - (n^2 - \cos^2\psi)^{\frac{1}{2}}}{n^2 \sin\psi + (n^2 - \cos^2\psi)^{\frac{1}{2}}} \tag{K26}$$

$$= \frac{\dfrac{n^2 \sin\psi}{\sqrt{(n^2 - \cos^2\psi)}} - 1}{\dfrac{n^2 \sin\psi}{\sqrt{(n^2 - \cos^2\psi)}} + 1}$$

and

$$\Gamma_H = \rho_H\, e^{-j\phi_H} = \frac{\sin\psi - (n^2 - \cos^2\psi)^{\frac{1}{2}}}{\sin\psi + (n^2 - \cos^2\psi)^{\frac{1}{2}}} \tag{K27}$$

$$= \frac{\dfrac{\sin\psi}{\sqrt{(n^2 - \cos^2\psi)}} - 1}{\dfrac{\sin\psi}{\sqrt{(n^2 - \cos^2\psi)}} + 1}$$

The above equations for Γ_V and Γ_H are generally of value only when expressed in the form of graphs which are computed for various values of ε_r and σ [5, 6]. The latter constants have been determined experimentally at various frequencies for several types of ground.

A cursory examination of the formulae would show that, at grazing incidence. $\psi \simeq 0$ and $\Gamma_H \simeq \Gamma_V \simeq -1$, which means $|\rho| = 1$ and $\phi = \pi$. This allows a common treatment, independent of polarisation or of terrain, for configuration of Fig. 2.4c provided $d \gg h_1, h_2$.

If we assume $|\varepsilon_c| \gg 1$, the expression for Γ may be written

$$\Gamma_V = \frac{\sqrt{\varepsilon_c}\ \sin\psi - 1}{\sqrt{\varepsilon_c}\ \sin\psi + 1} \tag{K28}$$

$$\Gamma_H = \frac{\sin\psi - \sqrt{\varepsilon_c}}{\sin\psi + \sqrt{\varepsilon_c}} \tag{K29}$$

For vertical polarisation, if ψ is large, $\sin\psi \simeq 1$ and $\Gamma_V \simeq 1$. As ψ is decreased, $\Gamma_V \to 0$ when $\psi = \sin^{-1}(1/\sqrt{\varepsilon_c})$. For a perfect dielectric this minimum is well defined and ψ is known as the Brewster angle. For a lossy dielectric the minimum is less well defined and ψ is referred to as the pseudo-Brewster angle. For a perfect conductor ($\varepsilon_c \to \infty$) the pseudo-Brewster angle would become zero.

Hence the pseudo-Brewster angle corresponds to the condition where, with decreasing ψ, the reflection coefficient for vertically polarised waves changes from $+1$ to -1. As ε_c becomes very large (conductor, or lossy dielectric with low frequency) this critical angle approaches zero and leads to the practical result that $\Gamma_V = +1$ for all values of ψ for a perfect conductor. It also means that $\Gamma_V = +1$ for a metric radar working over the sea.

Fig. 2.5 indicates broadly the typical behaviour of Γ for some selected frequency and for ground of a particular dielectric constant and conductivity.

Multipath effects
Consider Fig. 2.4c. The path-length difference, $\Delta d = |AMD - AD|$, can readily be shown to be $2h_1h_2/d$. Because of this path-length difference, the reflected ray is retarded by an angle

$$\frac{2\pi}{\lambda}\left[\frac{2h_1h_2}{d}\right] = \frac{4\pi h_1 h_2}{\lambda d}$$

The phase shift, π, at M brings the total phase difference to

$$\left[\frac{-4\pi h_1 h_2}{\lambda d} - \pi\right]$$

Let E_d be the direct field produced at D. Then, taking acount of the change of phase of AM on reflection, the resultant field at D due to the direct and reflected rays, assuming $|\Gamma| = 1$, will be

$$E = E_\mathrm{d}\left[\left|1 - e^{-\mathrm{j}\left(\frac{4\pi h_1 h_2}{\lambda d}\right)}\right|\right] \tag{K30}$$

$$= E_\mathrm{d}\left[\left|1 - \cos\frac{4\pi h_1 h_2}{\lambda d} + \mathrm{j}\sin\frac{4\pi h_1 h_2}{\lambda d}\right|\right]$$

$$= E_\mathrm{d}\left[\left|\mathrm{j}\sin\frac{4\pi h_1 h_2}{\lambda d}\right|\right], \left(\because \cos\frac{4\pi h_1 h_2}{\lambda d} \simeq 1\right)$$

$$= 2E_\mathrm{d}\cos\frac{2\pi h_1 h_2}{\lambda d}\sin\frac{2\pi h_1 h_2}{\lambda d}$$

$$= 2E_\mathrm{d}\sin\frac{2\pi h_1 h_2}{\lambda d} \tag{K31}$$

Hence the ratio of the power incident at D to that which would be incident if free space conditions prevailed is

$$4\sin^2\frac{2\pi h_1 h_2}{\lambda d} \tag{K32}$$

The signal from the target will return to the radar over the same paths so that the same mechanisms will obtain. Therefore the ratio of power received at the radar to what would be received under free space conditions is

$$16\sin^4\frac{2\pi h_1 h_2}{\lambda d} \tag{K33}$$

P_r in eqn. (K8) will be modified by this factor and hence will vary in value from 0 to 16. Also the range, because of the fourth power relationship, will vary from 0 to 2 times that given by eqn. (K9).

The variation in P_r is as expressed by eqn. (K33) and it will occur concomitantly with the production of a lobed structure in the antenna elevation pattern. The maxima and minima are readily determined from the argument of the sine term and inspection indicates the following two relationships. Maxima occur when

$$\frac{4h_1 h_2}{\lambda d} = 2n + 1, \quad n = 0, 1, 2, \ldots \tag{K34}$$

and minima will occur when

$$\frac{2h_1 h_2}{\lambda d} = n, \quad n = 0, 1, 2, \ldots \tag{K35}$$

When both $h_1 \ll h_2$ and $h_1, h_2 \ll d$, as is usually the case, the angle of elevation of the target $\theta_d \simeq \psi$. (See Fig. 2.4c; AD and MD would be virtually parallel, and one could also say that the angle of elevation of the target, D, would be the same when measured from either A or B.)

The first maximum occurs when

$$\frac{4h_1 h_2}{\lambda d} = 1$$

but

$$\frac{h_2}{d} = \tan \psi \simeq \psi$$

and

$$h_2 \ll d$$

Hence the angle, in radians, of the first lobe maximum occurs at $\lambda/4h_1$ and in general for maxima

$$\theta_d \simeq \frac{(2n+1)\lambda}{4h_1} \tag{K36}$$

and for minima

$$\theta_d \simeq \frac{n\lambda}{2h_1} \tag{K37}$$

A simple illustration of the use of eqns. (K36) and (K37) is helpful. Choose for arithmetic convenience a value of 14λ for h_1 and assume that the transmitting antenna is a dipole or some other simple radiator with a lobe-free pattern. Then for horizontally polarised waves over land, lobes would occur at 1°, 3°, 5°, 7°, etc., with nulls at 0°, 2°, 4°, 6°, etc. For vertically polarised waves over land the lobes would occur at 1°, 3°, 5°, 7°, etc., until the pseudo-Brewster angle at 17°, say, was reached, after which no phase change on reflection would occur; the lobes would then be present at 18°, 20°, 22°, etc., with nulls in between.

Perfect maxima and minima assume $|\Gamma| = 1$. $|\Gamma|$ varies with ψ so that the maxima are not of optimum value and, because the reflected field will be considerably weaker than the direct field, the nulls will not go to zero.

Pulses and their spectra
In considering a succession of rectangular pulses as illustrated in Fig. 2.9a, let $f(t)$ be a periodic function of voltage defined for $-\infty < t < \infty$, with a finite

period $T = 2\pi/p$ and with the time-origin chosen at the centre of one of the pulses, whose duration is τ.

$$f(t) = E, \quad \frac{-\tau}{2} < t < \frac{+\tau}{2}$$

$$f(t) = 0, \quad \frac{\tau}{2} < t < \left(T - \frac{\tau}{2}\right),$$

$$-\left(T - \frac{\tau}{2}\right) < t < \frac{-\tau}{2}$$

$$f(t) = f(t \pm nT), \quad n = 0, 1, 2, \ldots$$

The Fourier series expansion of $f(t)$ is:

$$f(t) = \sum_{n=-\infty}^{\infty} a_n e^{jnpt} \tag{K38}$$

where

$$a_n = \frac{p}{2\pi} \int_{-\frac{\pi}{p}}^{\frac{\pi}{p}} f(t) e^{-jnpt} \, dt \tag{K39}$$

$$= \frac{p}{2\pi} \int_{-\frac{T}{2}}^{\frac{T}{2}} f(t) e^{-jnpt} \, dt$$

$$= \frac{p}{2\pi} \int_{-\frac{\tau}{2}}^{\frac{\tau}{2}} E e^{-jpnt} \, dt$$

where E is the pulse amplitude, and therefore

$$a_n = \frac{E\tau}{T} \frac{\sin(np\tau/2)}{np\tau/2}$$

$$= \frac{E\tau}{T} \, \text{sinc}\,(np\tau/2)$$

Since

$$a_0 = \frac{1}{T} \int_{-\frac{\tau}{2}}^{\frac{\tau}{2}} E \, dt = \frac{E\tau}{T} \tag{K40}$$

$f(t)$ is given by

$$f(t) = \frac{E\tau}{T} \sum_{n=-\infty}^{\infty} \text{sinc}\,(np\tau/2)\,e^{jnpt} \tag{K41}$$

$$= \frac{E\tau}{T} \sum_{n=-\infty}^{\infty} \text{sinc}\,(n\pi\tau/T)\,e^{jnpt} = \frac{E\tau}{T} \sum_{n=-\infty}^{\infty} \text{sinc}\,(n\pi f_r\tau)\,e^{jnpt}$$

Thus one obtains a line spectrum whose envelope is determined by the sinc functions. This equation (K41) for $f(t)$ corresponds to a series of the form

$$f(t) = a_0 + a_1 \cos pt + a_2 \cos 2pt + \dots \tag{K42}$$

(The function $f(t)$ is even, so a cosine series is to be expected.)

a_0 is the steady, or so-called DC component, and represents the average value of the signal $f(t)$. It has a value $E\tau/T = E\tau f_r$. The fundamental component has an amplitude:

$$a_1 = \frac{E\tau}{T}\,\frac{\sin\,(np\tau/2)}{np\tau/2}$$

$$= \frac{E}{\pi}\,\sin \pi f_r\tau$$

and the nth harmonic likewise has an amplitude

$$\frac{E}{n\pi}\,\sin n\pi f_r\tau$$

This assumes a two-sided line spectrum with positive and negative frequencies. If we represent the frequency behaviour by a one-sided or positive-frequency line spectrum, then the magnitudes of the 1st and nth harmonic will be $2E/\pi\sin \pi f_r\tau$ and $2E/n\pi\sin n\pi f_r\tau$, respectively.

Harmonics of zero amplitude occur when $[n\pi f_r\tau]$ is a multiple of π; this occurs at frequencies of $1/\tau,$, $2/\tau$, $3/\tau$, etc.

If T is increased, that is f_r becomes less, then the spacing between the lines of the spectrum decreases. If T becomes arbitrarily large so that one pulse only remains, then the spectrum becomes continuous. A sinc function envelope for the spectrum, with the same zeros but of a much lower amplitude, remains.

One can progress to Fig. 2.9b where a train of pulses modulates the amplitude of a continuous sinusoidal oscillation and does it in such a manner that the phase-coherence of the sinusoidal waveform is maintained from pulse to pulse just as if no disturbance of the waveform were taking place.

$f(t)$, as expressed in eqn. (K42), amplitude modulates a single radio frequency, which may be denoted by $\cos \omega_0 t$, so that

$$f(t) = \cos \omega_0 t\,(a_0 + a_1 \cos pt + a_2 \cos 2pt + \dots) \tag{K43}$$

Eqn. (K43) may be expressed as:

$$f(t) = a_0 \cos \omega_0 t + \frac{a_1}{2} \cos (\omega_0 + p)t + \frac{a_1}{2} \cos (\omega_0 - p)t$$

$$+ \frac{a_2}{2} \cos (\omega_0 + 2p) + \frac{a_2}{2} \cos (\omega_0 - 2p)t + \ldots \qquad \text{(K44)}$$

This result is illustrated in Fig. 2.9d. An RF spectrum is obtained whose carrier has the amplitude of the steady component of eqn. (K42) above and whose sidebands, in a two-sided line spectrum, are spaced exactly as its counterpart in Fig. 2.9c. Nothing has changed except a translation of the complete pattern, with the centre frequency of the pattern moving from zero to the carrier frequency value.

References

1 MENTZER, J. R.: 'Scattering and diffraction of radio waves' (Pergamon Press, London, 1955).
2. BERKOWITZ, R. S.: 'Modern radar' (John Wiley, New York, 1965), p. 565.
3. NORTON, K. A., and OMBERG, A. C.: 'The maximum range of a radar set'. *Proc. Inst. Radio Engrs*, 1947, **35**, pp. 4–24.
4. STRATTON, J. A.: 'Electromagnetic theory', (McGraw-Hill, New York, 1941).
5. BURROWS, C. R.: 'Radio propagation over plane earth—field strength curves', *Bell Syst. Tech. J*, 1937, **16**, pp. 54–61.
6. TERMAN, F. E.: 'Radio engineers' handbook' (McGraw-Hill, New York, 1943).

Name Index

Aitken, H., 268
Alder, L. S. B., 42, 69
Alekseyev, N. F., 142, 171, 261, 268
Allison, D., 80, 159, 161
Andone, M., 154
Appleton, E. V., 53, 54, 57, 80, 278
Arena, N., 167
Armstrong, E. H., 48
Arsenjewa-Heil, A., 268
Ashmore, E. B., 177, 255
Bacon, Francis, 271
Bainbridge-Bell, I. H., 81, 186, 209, 255
Bajpai, R., 80
Baranov, A., 171
Barfield, R. H., 255
Barnett, M. A. F., 47, 53, 54, 80
Barton, D. K., 11, 39
Baxter, J. P., 161
Bay, Z., 145, 146, 173
Bedford, L. H., 86, 147, 230
Beeching, G. H., 147
Bekker, C., 155
Bell, J., 255
Bender, L. C., 114
Benecke, T., 80
Benjamin, R., 255
Berkowitz, R. S., 318
Berline, S., 163, 268
Berne, B., 39
Bion, J., 163, 164
Bird, G., 40
Blackett, P., 84
Blair, W., 112, 113, 114
Blake, L. V., 39
Bley, C., 152
Bomford, G., 80
Bonatz, H., 154

Boot, H. A., 258, 263, 267, 268
Born, M., 39
Bowen, E. G., 18, 29, 37, 40, 90, 148, 186, 187, 196, 209, 243, 244, 255, 256, 269, 274
Brandt, L., 41, 152, 153, 154
Breit, G., 57, 59, 80
Brenot, P., 163
Burch, C. R., 209
Burch, F. P., 209
Burch, L. L., 151
Burgess, R. E., 274
Burkel, H., 154
Burrows, C. R., 39, 318
Bush, V., 160
Butement, W. A. S., 42, 70–74, 255, 256
Cahill, W. J., 111
Calpine, H. C., 148
Carson, J. R., 274
Castioni, L. C., 39, 129, 165, 167
Chemeris, M., 172
Chu, L. J., 274
Clark, R., 147
Coales, J. F., 88, 148, 264
Cockcroft, J. D., 37, 90, 239
Collins, G. B., 267
Cooke, A. H., 40
Crawford, A., 159, 163
Crombie, A. C., 272
Crombie, D. D., 37, 41
Dahl, A., 79
Dahl, O., 80
David, P., 121, 122, 123, 124, 125, 163, 164
Davis, H., 40, 79, 160, 162
Demanche, M., 164
Dewhurst, H., 196
Di Capua, R., 167

Dixon, E. J. C., 207
Dodds, J. M., 256
Dodsworth, E. J., 256
Dominik, H., 46–47
Dowding, H., 34, 180
Dummelow, J., 150
Dziewiro, K., 151
Eastman, A. V., 276
Eccles, W. H., 53
Eckersley, T. L., 14, 37, 39, 41
Eggleston, W., 148
Élie, 163
Englund, C., 159, 163
Erickson, J., 136, 170
Espenschied, L., 42, 64, 66, 79, 80
Everitt, W. L., 63
Fagen, M., 160, 161
Ferrell, E. B., 39
Fertel, G., 40
Fielder, 46
Frayne, T. G., 57
Fried, W. R., 4
Friedman, N., 38, 41, 148, 155, 165, 169
Friis, H. T., 203, 255
Furth, F. R., 1
Garratt, G. R. M., 147
Gavin, M. R., 255
Gebhard, L., 3, 37, 40, 53, 159
Gernsback, H., 45, 47, 79
Getting, I., 161
Giboin, E., 164
Glasgow, R. S., 278
Goldberg, A., 162
Grankin, V., 171
Groos, O. H., 152, 268
Groszkowski, J., 268
Guerlac, H. E., 79, 80, 148, 149, 159, 163
Guillemin, E., 40
Gunston, B., 256
Guthrie, R. C., 107
Gutton, H., 163, 268
Habann, E., 259
Haeff, A. V., 11, 39
Hahn, F., 154
Hanbury-Brown, R., 18
Harris, N. l., 40
Harris, K. E., 41
Harvey, A. E., 267
Harvey, A. F., 168, 276
Hay, D., 150, 256
Heaviside, O., 53, 80
Heil, O., 268
Henderson, J. T., 89

Henney, K., 40, 160
Hepburn, A. J., 109
Herd, J. F., 81, 209, 255
Hertz, H., 2
Hezlet, A., 149
Higgins, T. J., 161
Hightower, J., 160
Hill, A. V., 49, 82, 83, 90
Hinsley, F. H., 81, 150
Hitler, A., 98
Hoffman-Heyden, A. E., 100, 151, 153
Hoffmann, S., 49
Honold, P., 40
Horvat, R., 3
Howeth, L. S., 159, 160
Hugon, J., 163
Hull, A. W., 256, 258, 267
Hulbert, E. O., 80
Hülsmeyer, C., 43–45, 91
Hyde, H., 171
Hyland, L. A., 52, 97, 101, 104
Isted, G., 165
Isumi, S., 168
James, E. G., 255
Jenkins, J. W., 256
Johnson, B., 149
Johnson, J. B., 81
Jones, D. S., 39
Jones, H. A., 79, 255
Jones, R. V., 80, 81, 150
Kayton, M., 4
Keen, R., 40
Kelly, M. J., 276
Kennelly, A. E., 53, 80
Kerr, D., 14, 15, 39
Khoroshilov, P. E., 171, 172
Kinsey, G., 151
Kliukin, I., 171
Knight, R. C., 81
Kobzarev, Y., 171
Künhold, R., 92, 97
Kutzscher, E. W., 50
Larmor, J., 53
Latmiral, G., 166
Lawson, J. L., 11
Lecher, E., 20, 40
Leconte, A., 164
Lewis, W. B., 270
Limann, O., 79
Linder, E. G., 261, 268
Livingston, D., 39
Lobanov, M. M., 171, 172
Lombardini, P., 167

Lothian, Lord, 90
Löwy, H., 59, 80
Ludlow, J. H., 256
MacDonald, H. M., 268
Macfarlane, G., 278
Mackenzie, C. J., 89
Malyarov, D. E., 142, 171, 261, 268
Marconi, G., 9, 51, 79, 80, 126, 127, 128
Matsui, M., 168
Matsuo, S., 81
Megaw, A. S., 268
Megaw, E. C., 14, 39, 265, 267
Melville, W. S., 256
Mentzer, J. R., 39, 318
Mesny, R., 159
Middleton, W. E., 148
Millington, G., 274
Morris, R., 40
Mumford, W., 159, 163
Nakajima, S., 169
Neale, B. T., 151
Newhouse, R., 42, 64, 66, 79, 86
Newson, B., 256
Nicholls, P., 79
Nichols, E. F., 81
Nikolić, N., 3
Norberg, A., 162
North, D. O., 39
Norton, K. A., 9, 11, 39, 318
O'Dea, W., 148
Oger, J., 164
Okabe, K., 130, 132, 260, 268
Oliphant, M. L., 258
Omberg, A. C., 9, 11, 39, 318
Oshchepkov, P. K., 139, 170, 172
Oudat, L., 164
Oxford, A. J., 256
Page, R., 26, 29, 35, 36, 40, 104, 105, 106,
    108, 110, 111, 160, 161
Panlényi, E., 173
Papović, V., 3
Parsons, W. S., 111
Pecora, R., 39
Philips, V. J., 79
Philpott, L. R., 111
Picquenard, A., 39
Pierce, J. R., 267
Plendl, H., 151
Pokrovski, R., 172
Pollard, P. E., 42, 70–74
Ponte, M., 121, 123, 125, 163, 266, 269
Postan, M., 150, 256
Posthumus, K., 259, 267

Praun, A., 154
Price, A., 150
Puckle, O. S., 230, 233
Quick, A. W., 80
Randall, J. T., 258, 263, 267, 268
Ransom, C. F., 81
Ratcliffe, J. A., 239
Ratsey, O. L., 81, 147, 149
Rauch, G. von, 171
Rawlinson, J. D., 148
Reuter, F., 79, 155
Rice, C., 78, 81
Ridenour, L., 3, 4, 24, 39, 160
Rivoire, J., 164
Robinson, D., 151
Rowe, A. P., 82, 83, 84, 148, 175, 176, 179,
    181
Rowlinson, F., 150
Ross, A. W., 147
Runge, W. T., 32, 93, 96, 97, 151
Ryżko, S., 261, 268
Sacco, L., 9, 128
Samuel, A. L., 276
Sandretto, P., 4, 61, 62, 80
Saundby, R. H. M., 34
Sayer, A. P., 81, 147, 148, 256
Scaife, B. K. P., 79, 147, 163, 165, 168,
    169, 173, 267, 268, 269
Schelleng, J. C., 14, 39
Scherl, R., 46, 47
Scholtz, R., 81
Scott, J. D., 150, 256
Scott, O., 148
Scott-Taggart, J., 259, 267
Secor, H., 79
Shearman, E. D. R., 41
Shembel, B. K., 139, 170, 172
Shoshkov, E., 172
Sieche, E., 151, 155
Skolnik, M., 5, 12, 28, 38, 39, 41, 170, 267
Smith, C., 255
Smith, P. H., 160
Smith, R. A., 4, 39, 80, 149, 255, 256
Smith-Rose, R., 255
Soller, T., 40
Southworth, G., 120, 160
Spangenberg, K., 267
Starr, M., 40
Stogov, D. S., 172
Stout, W., 163
Stratton, J. A., 318
Streetly, M., 150, 257
Strutt, J. W., 274

Swinyard, W. O., 278
Taylor, A. H., 42, 80, 101, 102, 103, 108
Taylor, D., 40
Tazzari, O., 166
Tear, J. D., 81
Terman, F. E., 160, 276, 318
Terrett, D., 41, 79, 81, 160, 162
Tesla, N., 2, 3, 43, 47
Theile, U., 152
Thomas, E. E., 81
Thompson, G., 162
Tiberio, U., 9, 39, 128, 130, 165, 166, 167
Timus, W. C., 111
Tizard, H., 83, 89, 175, 243, 254
Trenkle, F., 151, 153, 154
Tuska, C. D., 80, 81
Tuve, M., 57, 59, 80
Udet, General, 97
Vajda, P., 173
Valley, G., 40
Varela, A. A., 108
Varian, R. H., 268
Varian, S. F., 268
Vasseur, A., 163, 164
Vogelsang, C. W., 154
Vollmar, F., 79
Warren, G. W., 255

Wathen, R., 267
Watson, D. S., 148
Watson-Watt, R. A., 29, 31, 34, 81, 148, 174, 176, 181, 186, 187, 209, 233, 254, 255, 272
Weeks, W. L., 40
Weiher, V., 79
Westcott, C. H., 40
Whelpton, R. V., 256
White, J. A., 173
White, W. C., 267
Whitehead, J. R., 278
Wilkins, A. F., 29, 34, 37, 39, 41, 57, 174, 181, 186, 203, 230, 254, 255
Wilkinson, R., 168
Willisen, H. K. v., 94, 151
Wimperis, H. E., 83, 84, 174
Witt, V., 172
Witts, A., 79
Wolf, E., 39
Wood, A. B., 18, 33
Wood, K. A., 40
Woodring, H. A., 115
Yagi, H., 130, 260, 268
Young, L. C., 42, 52, 53, 97, 101, 104, 105
Zácek, A., 260
Zmievski, V., 171

Subject Index

Acoustical mirror system, 179
ADEE (Air Defence Experimental Establishment), 179
ADF (Automatic Direction Finder), 131
ADGB (Air Defence of Great Britain), 179
AI (Aircraft Interception), 120, 237, 244, 251–253
AI MkI, 252
AI MkII, 252
AI MkIII, 252
AI MkIV, 251, 252
Aircraft Altimeters, 60–68
Alexanderson altimeter, 62–63
Amplitron, 258
Antenna duplexing, 26–28
Army radar (United Kingdom), 148, 288
ASV (Aircraft to Surface Vessel), 18, 82, 244–245
ASV MkI, 246
ASV MkII, 246, 251
Barkhausen-Kürz valves, 67, 92, 126, 274–276
Bawdsey, 10, 27, 188, 189, 196, 207, 210
Bay group, 145
Bell Telephone Laboratories, 14, 64, 91
Bellini-Tosi system, 198–205
Bistatic Radar range equation, 11–12
British Munitions Inventions Department, 49
Brewester angle, 17, 18, 313
'BURYA', 140
'BURYA 2', 142
'BURYA 3', 142
Camouflaging of aircraft, 254, Appendix H
Castel Gandolfo, 9
CD (Coastal Defence) radar, 10, 27, 238
CH (Chain Home), 82, 188–236
CH direction finding, 200

CH frequencies, 189
CH height finding, 202
CH receivers, 230–236
CH transmitters, 207–229
CHL (Chain Home Low), 18, 27, 85, 236–239
Coulometers, 147
CSF (Compagnie Générale de TSF), 121
CVD (U.S.W. Communication: Valve Development), 264, 265
CXAM radar, 105, 110–111
CXAS radar, 112
Cyclotron mode oscillations, 259
Daventry experiment, 85, 180–186
David system, 121, 122
Death ray, 75, 175
DECCA, 3
DEM (Détection Électromagnétique), 1
Dematron, 258
Diffraction, 7
Direction finding, 30–32, 200–202
DSIR (Department of Scientific and Industrial Research), 84
Dynatron, 259
EC-1 radar, 129
EC-2 radar, 129, 130
EC-3 radar, 129, 130
EF attachment, 86
Eimac-100 TH valve, 110
Elefant-Russel, 99, 100
Fiat Review of German Science, 152
Flak radar, 98
Flight diary of K. A. Wood, 245
FM radio altimeter, 61, 63–67
Fort Monmouth, 50
Fort Washington, 35
France, 11, 120–126
Frequency bands, 38

Fresnel formulae, 312, 313
Freya, 98
FuMO22, 19
FuMO23, 19
FuMO52, 19
FuMO62, 19
Gap filling, 17, 18
GAU (Glavnoe Artilkerisskoe Upravlenie), 138, 139
GCI (Ground Control of Interception), 85, 210, 236, 243
Gee, 3
GEMA (Gesellschaft für Electro-akustische und Mechanische Apparate), 93, 94, 95, 96, 97, 98, 99
General Electric Research Laboratory, 78
Geodetic surveying, 59, 60
Germany, 11, 50, 91–101
GL (Gun Laying) sets, 85, 86, 87, 89
Great Britain, 11, 50, 82–91, 174–254
'Gufo' radar, 129, 130
Height finding, 202–207
History of British radar, 270
Holland, 142–144
Hollmann valves, 113
Hungary, 144–147
IFF (Identification Friend or Foe), 33–36, 242, 243
IFF MkI, 34, 35
IFF MkII, 34, 35
IFF MkIII, 34, 35, 242
Infra-red detection, 49–51
Institution of Electrical Engineers (London), 14, 38
Ionospheric sounding, 53–59
Italy, 9, 11, 126–130
Jagdschloss, 99
Japan, 11, 130–135
Klystron, 20, 261
LADA (London Air Defence Area), 177, 178
Lecher lines, 20
LEFI (Leningradskii Elektrofizicheskii Institut), 136, 139, 140
LMS 10(G) German magnetron, 267
LMT (Le Materiel Telephonique), 125
Loran, 3, 135
Lorenz landing system, 31, 75
Löwy's U.S. patent, 59
LRASV (Long Range ASV), 248, 249
MADRE (Magnetic Drum Radar Equipment), 36
Magnetron, 132, 142, 258–267

Mammut, 99
Marconi's speech to American Institute of Engineers, 51
Matched filter, 11
Matsuo's altimeter, 67, 68
Metallic reflection, 311
Metropolitan Vickers Ltd., 209, 221, 222, 230
Mie scattering, 7
'Millimicropup' valve, 265
M.I.T. Radiation Laboratory, 11, 120
Monostatic radar, 5, 6, 9
Moon, 146, 147
Multipath propagation, 12–15, 313–315
Munitions Inventions Department, 178
MUSIC (Multiple Storage, Integration, Correlation), 36
Mystery rays, 75
NACA (National Advisory Committee for Aeronautics), 119
Navy radar (United Kingdom), 187, 188
NDRC (National Defence Research Committee), 118, 119
NRL (Naval Research Laboratory, 26, 27, 35, 37, 52, 101, 102, 103, 104, 105, 107, 110, 111
Oboe, 3
Operational research, 254
Orfordness, 18, 37, 186, 188, 189, 190
Oslo report, 75
Pearl Harbour, 35
Performance diagrams, 229, 230
Pliodynatron, 259
Political birth of radar, 82
PPI (Plan Position Indicator), 28, 29, 30
Propeller modulation, 87
Pulse radar, 19–26
Pulse signals, 24–26, 315–318
Pulse tsransformers, 205, 206, 207
PVO (Voiska Protivo-Vozdusknoi Oborony), 138, 139, 140
Radar counter-measures, 254
Radar range equation, 9–12, 308–311
Radiogoniometer, 198–200
'RAPID', 136, 140
Rayleigh scattering, 7
RCA magnetron, 114
RCA valves, 114
RDF, 1
RDF beam technique, 236
RDF 1, 237, 243
RDF 1½, 243, 244
RDF 2, 243
'REDUT', 141

Reflection coefficient, 15–17, 311–313
Resnatron, 120
Resolution, 22–24
Ring-oscillator, 109
Rotterdam, 267
Russia, 11, 135–142
RUS-1, 141
RUS-2, 138, 141
SADIR (Société Anonyme des Industries Radioélectriques), 122, 125
SAFAR (Societa Anonima Fabbricazione Apparechi Radiofonici), 128, 129
Scattering cross-section, 6–9
Scattering matrix, 306–308
SCR-268, 115, 116, 117, 118
SCR-270, 116, 117, 118
SCR-271, 116, 118
SFR (Société Française Radioélectrique), 121, 122
Siemens, 43, 46
Skip distance, 53, 54
SLC (Searchlight Control), 238
Sound locators, 48, 49
'Splashbacks', 37
Split-beam method, 30–32
Stanford University, 261

'STRELETS', 142
Tachi-Tase-Taki designation, 134
Tama Research Institute, 134
Telemobiloscope, 43, 44
TF3 transmitter, 219–221
TNO (Organisatie voor toegepast-natuurwetenschappelijk onderzoek), 143
TR/ATR switches, 26–28
TsRL (Tsentral'naya Radiolaboratoriya), 139
TVA (Torpedo-Versuchsanstalt), 94
Type A radar, 133, 134, 135
Type B radar, 133, 135
United States, 11, 14, 50, 101–120
US Army Signal Corps, 9, 49, 50, 112–118
USW (Ultra-Short Wave) programme, 263, 264
Vatican, 9
Visibility factor, 11, 309
Vitreous reflection, 311
Warning controls, 177
Western Electric altimeter, 63–67
Wurzburg set, 35, 98, 100
XAF radar, 35, 103, 108, 110
Zeppelin air raids, 177

Electronically typeset by Heffers Printers Ltd, Cambridge, England

Printed in the USA
CPSIA information can be obtained
at www.ICGtesting.com
JSHW011518221024
72172JS00008B/60